Healthy Newborn

What Obstetrician and Neonatologist can do

Publication from The FOGSI Perinatology Committee Chairperson
Dr Reena J Wani 2015-2017
Federation of Obstetrics and Gynaecological Societies of India

Healthy Newborn

What Obstetrician and Neonatologist can do

Editors

Reena J Wani MD, MRCOG, FICOG, DNBE, DFP, DGO, FCPS
Obstetrician and Gynecologist
Chairperson, FOGSI Perinatology Committee 2015-2017
Core Committee Member, FOGSI Violence Against Women Cell
Managing Committee Member, MOGS, AMC, Unesco Bioethics
Vice-President, MBPC (Mumbai Breastfeeding Promotion Committee)
Professor (Addl) and Head
Department of Obstetrics and Gynecology
Hindu Hriday Samrat BalaSaheb Thakre Medical College and Dr RN Cooper Municipal Hospital
ex-TN Medical College and BYL Nair Ch Hospital, Mumbai

Alka Kriplani MD, FRCOG, FAMS, FICOG, FIMSA, FICMCH, FCLS
Professor and Head, Department of Obstetrics and Gynecology
Director in-charge WHO-CCR, HRRC and Family Planning
All India Institute of Medical Sciences, New Delhi, India
President, FOGSI (Federation of Obstetrics and Gynaecological Societies of India) 2016

Co-Editors

Madhuri Mehendale DGO, FCPS, DNB
Assistant Professor, Department of Obstetrics and Gynaecology
Lokmanya Tilak Municipal Medical College and General Hospital
Sion, Mumbai, Maharashtra

Pradnya Supe MS
Assistant Professor, Department of Obstetrics and Gynaecology
Lokmanya Tilak Municipal Medical College and General Hospital
Sion, Mumbai, Maharashtra

Rashmi Jalvee MS, DGO, DNB
Assistant Professor, Department of Obstetrics and Gynaecology
Hindu Hridaysamrat Balasaheb Thakre Medical College and Dr RN Cooper Municipal Hospital
Mumbai, Maharashtra

CBS

CBS Publishers & Distributors Pvt Ltd

New Delhi • Bengaluru • Chennai • Kochi • Kolkata • Mumbai
Hyderabad • Jharkhand • Nagpur • Patna • Pune • Uttarakhand

Healthy Newborn
What Obstetrician and Neonatologist can do

ISBN: 978-93-86827-97-5

Copyright © Editors and Publisher

First Edition: 2018

Published by Satish Kumar Jain and Produced by Varun Jain for
CBS Publishers & Distributors Pvt Ltd
4819/XI Prahlad Street, 24 Ansari Road, Daryaganj, New Delhi 110 002, India.
Ph: 23289259, 23266861, 23266867   Fax: 011-23243014   Website: www.cbspd.com
e-mail: delhi@cbspd.com; cbspubs@airtelmail.in.
Corporate Office: 204 FIE, Industrial Area, Patparganj, Delhi 110 092, India
Ph: 4934 4934   Fax: 4934 4935   e-mail: publishing@cbspd.com; publicity@cbspd.com

Branches

- **Bengaluru:** Seema House 2975, 17th Cross, K.R. Road,
  Banasankari 2nd Stage, Bengaluru 560 070, Karnataka, India
  Ph: +91-80-26771678/79   Fax: +91-80-26771680   e-mail: bangalore@cbspd.com
- **Chennai:** 7, Subbaraya Street, Shenoy Nagar, Chennai 600 030, Tamil Nadu, India
  Ph: +91-44-26260666, 26208620   Fax: +91-44-42032115   e-mail: chennai@cbspd.com
- **Kochi:** Ashana House, No. 39/1904, AM Thomas Road, Valanjambalam, Ernakulam 682 016, Kochi, Kerala, India
  Ph: +91-484-4059061-65   Fax: +91-484-4059065   e-mail: kochi@cbspd.com
- **Kolkata:** No. 6/B, Ground Floor, Rameswar Shaw Road, Kolkata-700014 (West Bengal), India
  Ph: +91-33-2289-1126, 2289-1127, 2289-1128   e-mail: kolkata@cbspd.com
- **Mumbai:** 83-C, Dr E Moses Road, Worli, Mumbai-400018, Maharashtra, India
  Ph: +91-22-24902340/41   Fax: +91-22-24902342   e-mail: mumbai@cbspd.com

Representatives

- **Hyderabad** 0-9885175004
- **Jharkhand** 0-9811541605
- **Nagpur** 0-9021734563
- **Patna** 0-9334159340
- **Pune** 0-9623451994
- **Uttarakhand** 0-9716462459

Printed at International Print-O-Pac Limited, Noida, UP, India

Foreword

It gives me a great pleasure to see the release of *Healthy Newborn: What Obstetrician and Neonatologist can do* by Dr Reena J Wani whom I have seen working tirelessly in the Municipal Corporation in the field of Maternal and Child Health since many years. Dr Reena was our student at Seth GS Medical College years ago. It is greatest joy for all of us to see our students gain confidence in themselves, do high grade level work and achieve success. I am extremely proud that she is publishing this book on perinatology. Over last two to three decades, India has been experiencing change in every field. With careers in mind and altered working hours, many working women marry very late and that results into a late precious pregnancy. Many times that is the only pregnancy for these women. Therefore, today perinatology is very important for such precious pregnancies. The Internet, social media, smart phones and tablet computers have been playing a larger role in our daily lives. Hence, a special book like this will definitely help busy practitioners and students of obstetrics to learn more about this special topic.

I would like to congratulate the authors and contributors of this book for editing this excellent book. Chapters of this book are well written and contain experience and wisdom mixed with knowledge. These book chapters will provide a valuable window on perinatology and cover the necessary components for practicing obstetricians. As Director (ME and MH) it is my responsibility and privilege to make many decisions regarding healthcare of women and children in Mumbai amongst other activities to improve health education and awareness. Maternal and childcare is extremely important because it involves good outcome of 2 patients, the mother and the baby. We at MCGM, are striving to reduce the maternal and infant mortality rates and are working towards establishing services to improve perinatal care for needy and underprivileged in the public sector. This book will definitely help to achieve our goal of better MCH services to community.

All best wishes!

Avinash Supe

MS FICS DNBE FCPS DHA, PGDME, MHPE (UIC) FIAGES, FMAS, FAIS
Director (ME and MH), MCGM, Mumbai
Dean, Seth GS Medical College and KEM Hospital
Professor, GI Surgery
Director, GSMC FAIMER Regional Institute
Immediate Past President, IHPBA India

From the Desk of President, FOGSI, 2017

My dear fellow FOGSIANS,

As the 61st President of our esteemed society, I am seeing my vision and efforts shape up in various projects we have undertaken this year. It has been 6 months since I took over as FOGSI president and has been indeed a very busy but satisfying period. My theme for the year 'She Matters' has been well received and worked on across the country. Our social programme *Nari Swasthya Pehal* for free health check-up camps for women is in full swing. We have entered the Limca book of Records for a massive nationwide screening for anemia on 5th June with over 16,000 tested across 11 sites. Amongst these preventive and screening health measures we are continuing with our CMEs and workshops in which maternal and child healthcare issues are being covered.

FOGSI has undertaken many initiatives to help our nation achieve reduction of neonatal mortality rate and stillbirth rate to single digits by 2030. We are working along with government of India on issues such as consensus guidance on retinopathy of prematurity, thyroid screening in pregnancy and influenza vaccination awareness. Through nationwide CMEs on fetal growth restriction, prematurity, gestational diabetes and hypertensive disorders of pregnancy we are reaching out to practicing clinicians and trainees. This we hope will improve training and quality of care to both mother and the neonate. Our various committees, their chairpersons including Dr Reena Wani, Chairperson, Perinatology, and my joint secretary Dr Sarita Bhalerao are working very hard for these activities. We have also released a FOGSI Focus on post-partum care focussing on the issues around childbirth and puerperium and another one on Infections in Pregnancy.

I congratulate our past President Dr Alka Kriplani and our chairperson of FOGSI perinatology committee Dr Reena J Wani who with their team have managed to compile practical pointers to management of peri-partum women and newborns, from both Obstetrician and Pediatrician's perspective. I am sure this book will serve as a ready-reckoner for many clinicians in daily practice and will help budding young postgraduates to gain latest in-depth knowledge of the subject.

Rishma Dhillon Pai
President, 2017
Federation of Obstetrics and Gynaecological
Societies of India (FOGSI)
President (Elect-2018)
Indian Society for Assisted
Reproduction (ISAR)
Vice President
Indian
Association of Gynaecological
Endoscopists (IAGE)

From the Desk of President, FOGSI, 2016

It is a matter of great pleasure to present to you the book *Healthy Newborn: What Obstetrician and Neonatologist can do* from our team.

Perinatology is a subspecialty dealing especially with high-risk pregnancies. It involves various roles in the care of the complicated obstetric patient—from performing imaging and prenatal diagnosis to consultation and interventions concerning vexing medical or obstetric problems. It is vital for all healthcare professionals involved in the management of pregnant women to be abreast with the latest developments in order to give the best possible care.

My theme for Presidential year 2016 "Preventing the Preventable" is truly reflected in this book which was planned by my Chairperson Perinatology Committee Reena J Wani who has been working very hard throughout the year and has done many CMEs and workshops on these topics of maternal and neonatal health.

This book has been designed by us for busy obstetricians as well as postgraduate students for a quick review of maternal and neonatal disorders that we see in everyday practice. We have strived to make the chapters concise, up-to-date, evidence-based approach for management with an emphasis on newer advances in treatment. Written by practising obstetricians and experts in the subjects, the book strives to provide a clear understanding of the subject with the help of tables, flowcharts and figures to simplify and highlight important information.

I invite you to join us on this path to improve maternal and neonatal health through an understanding of the subject. I am sure this book will be immensely useful for both students as well as practising obstetricians.

Alka Kriplani

MD, FRCOG, FAMS, FICOG, FIMSA, FICMCH, FCLS
Professor and Head, Department of Obstetrics and Gynaecology
Director In-charge, WHO-CCR, HRRC and Family Planning
All India Institute of Medical Sciences, New Delhi, India

- "Padma Shri", distinguished service in Medicine by Government of India, 2015
- Awarded Dr BC Roy National Award for "Eminent Medical Teacher", 2007
- Honorary FRCOG: Fellow of Royal College of Obstetricians and Gynaecologists, London, 2007
- President, FOGSI (Federation of Obstetrics and Gynaecological Societies of India) 2016
- President, GESI (Gynaecological Endocrine Society of India) 2011 continuing
- President, elect DGES (Delhi Gynaecological Endoscopists Society) 2015-16
- President, AOGD (Association of Obstetricians and Gynaecologists of Delhi) 2013-14
- Vice-President, FOGSI (2009) Member Governing Council ICOG (Indian College of Obstetricians and Gynaecologists) 2006 continuing
- Honorary Secretary for AOGD (2000-2002)
- Vice-President, AOGD (2002-2003)
- Editor: Asian Journal of Obstetrics and Gynaecology practice 1999 continuing
- National Corresponding Editor: Journal of Obstetrics and Gynecology of India
- Chairman: Endoscopy Training Programme, AIIMS

Chairperson Perinatology (2015-2017) Message and Preface

As obstetricians we shoulder the responsibility of two clients—mother and unborn fetus. We have to ensure that things go as nature planned, with *Watchful Expectancy and Patient Observation* (pun intended) being the hallmark of a good obstetrician. The primary aim of obstetrics is to have a healthy mother and neonate at the end of the 9 months journey. Perinatology is a speciality that focuses on the critical parts of this journey, mainly in the time leading to delivery and extending till the first month post-delivery. In this book we have focussed on many aspects of healthcare which can keep both our valuable patients in optimum health.

It has often been said that the *most dangerous journey one can undertake in life is probably the journey through the maternal birth canal* where many variables can change the outcome. Hence, globally there has been a trend to higher CS rates for what is labelled as a "high-risk" or "precious" pregnancy. However, merely doing a cesarean section is not a guarantee that all will go well and there has been much debate and media coverage regarding spiralling cesarean section rates. There are many issues to be considered when we speak of outcomes in obstetric practice and hence proper maintenance of records is our responsibility to justify our actions.

Labor, birth and the immediate postnatal period is the most dangerous time for mothers and their newborn babies, and there is more that medical professionals need to do to help reduce mortality rates around the world, said FIGO President Professor Sir Sabaratnam Arulkumaran in his plenary lecture in March at the 2014 RCOG World Congress in Hyderabad, India. Our Dr Chittaranjan Purandare is currently FIGO President and has been working on this aspect with over 100 workshops planned to cover related topics. According to UN statistics there has been a decline in maternal mortality rates of 47 per cent since 1990, with Eastern and Southern Asia recording a decline of almost two-thirds. Professor Sir Arul says that while maternal mortality ratios vary greatly between the developed and developing world, the main causes of maternal mortality remain the same, just on different scales.[1]

Moliere had aptly said: *It is not just what you do, but what you do not do, for which you are accountable.* After a woman becomes pregnant, in developing countries it may be a death sentence for her. Past President of FIGO Mahmoud Fathalla had remarked a few decades ago that women are dying because of diseases we cannot treat but because governments and countries are yet to take a decision that their lives are worth saving... this has been changing but still holds true.

Despite the technological advances in various healthcare sectors, when it comes to women's health, our country is still struggling with maternal mortality due to potentially preventable causes like hemorrhage, hypertension in pregnancy and infections. *Maternal Death Review* was started in Maharashtra State as per Government resolution dated 28/5/2010.[2] As per guidelines it was expected that every death whether in institution or at community level, be reported and reviewed. This move has promoted the need for proper record-keeping and has motivated many institutes to streamline the reporting and recording systems.

Our national organization FOGSI (Federation of Obstetrics and Gynecological Societies of India) which now has over 33,000 members is working on these areas in different ways. Every President chooses the theme for their year but pregnancy care is always a part of any chosen agenda. This year's theme is *She Matters—Care. Educate. Transform.* I have had the privilege to work with 3 Presidents as Chairperson: Dr Prakash Trivedi (2015), Dr Alka Kriplani (2016) and Dr Rishma Dhillon Pai (2017). Under the leadership of each President there have been many different programs reaching out to practitioners and patients across the country—through our perinatology committee programs we have laid special emphasis on MCH care.

Everyone wants a good outcome for all cases—yet as medicine is not an exact science, no one can guarantee a 100% result—however the expectations from doctors' services have become sky-high! What do people desire from healthcare services? This is a very complex question and the typical Indian mentality is to expect world-class services at third-world prices!! Hence, we as a speciality dealing with maternal and child healthcare have to walk the fine line between these expectations to do the best we can.

I will request you, dear reader to remember that we can all make a difference in our own way. If each one saves even one mother or baby from a bad outcome, our healthcare indices will improve. Lastly, I wish to share an inspiring message that was given to me: *Never stop doing your best just because someone doesn't understand it ...*

We need to do our best for our satisfaction, not for others appreciation!!

References
1. RCOG Release 2014: FIGO President discusses how to make labour safer for women worldwide
2. Addl Director, State Family Welfare Bureau, Pune, No. SFWB/RCH/CBMDR/report/file 24B/D-10B/41177-220/2012

Reena J Wani
MD, MRCOG, FICOG, DNBE, DFP, DGO, FCPS
Obstetrician and Gynecologist
Chairperson, FOGSI Perinatology Committee 2015-2017
Core Committee Member, FOGSI Violence Against Women Cell
Managing Committee Member, MOGS, AMC, Unesco Bioethics
Vice-President, MBPC (Mumbai Breastfeeding Promotion Committee)
Professor (Addl) and Head
Department of Obstetrics and Gynecology
Hindu Hriday Samrat BalaSaheb Thakre Medical College and Dr RN Cooper Municipal Hospital
ex-TN Medical College and BYL Nair Ch Hospital
Mumbai

Message from Vice-President, FOGSI, 2016

It gives me immense pleasure to pen down the message for the upcoming, book under aegis of FOGSI Perinatology Committee, under chairmanship of dynamic and dedicated Dr ReenaJ Wani.

Perinatology is a field where different specialities share and work for achieving optimum health of mother and newborn. It is joint effort of obstetrician, ultrasonologist, fetal medicine expert, labor room staff and neonatologist. Sharing of knowledge, communication and continuity of care right from conception to postpartum period is main force and direction of FOGSI Perinatology Committee.

The upcoming book will have updated content on diverse field of perinatology and help the readers to improve their clinical protocols as well as obstetric and neonatal work coordination. It will also add onto their skills of many aspects like fetal imaging, essential and advanced neonatal care.

Every step in pursuit of perfection adds to our dimension of thought and action. This book will be an important publication on part of our esteemed Federation of Obstetric and Gynecological Societies of India (FOGSI) and its Perinatology Committee.

Editor of the book and Chairperson of Perinatology Committee Dr Reena J Wani have accomplished many faceted work in our federation. As Vice-President In-charge of Perinatology Committee in 2016 and core committee member she has always impressed us with her continued focussed work and academic distinction. This book will add further to her academic contribution for FOGSI members.

Wishing you happy reading and learning!

Sadhana Gupta
MS, MNAMS, FICOG, FIUMB, FICMCH
Vice-President, FOGSI, 2016
ICOG Governing Council Member (2015-2017)
Corresponding National Editor: JOGI (2017-2020)
Chairperson, Safe Motherhood Committee: FOGSI (2011-2013)
Jeevan Jyoti Hospital and Medical Research Centre
Gorakhpur, India
Email: drguptasadhana@gmail.com

Message from Vice-President, FOGSI, 2015

The spot light today is on the pregnant woman. A healthy pregnancy will result in a healthy baby and healthy generation next. It is up to us, obstetricians, to use this opportunity to manage the pregnancy well—screen and treat disorders of pregnancy optimally.

Care begins ideally in the preconception phase when screening for infections, diabetes, anemia and supplementing folic acid should start. We monitor the pregnancy so as to keep the mother and the baby safe from disorders, which can jeopardize the health of either of them. Pregnancy disorders like GDM/PIH/IUGR/LBW/hypothyroid have an impact on the future life of the neonate.

Post-partum is a important but neglected period—since the woman is lost to follow-up. We should be utilizing this to treat anemia and offer proper contraceptive advice. A long-term follow-up is required for disorders like GDM and hypertension. The most ideal way is to partner with our pediatric colleagues since the mother will take the child for immunization. If we flag the mothers who require follow-up the pediatricians can direct the mother back to us.

Working in sync with the neonatologist and pediatricians will ensure that the non-communicable diseases (which may be detected by screening) are prevented from being transmitted to the next generation. As Vice-President In-charge of Perinatology (2015), with Chairman Dr Reena J Wani, we published a FOGSI Focus on the topic: **The Healthy Generation X** which covered all the above issues in a simple practical manner for the obstetricians to strive for good outcomes for health of the next generation by screening and prevention. We also conducted joint workshops/CMEs in collaboration with our pediatric colleagues across the country to cover these topics.

This book has beautifully captured the essence of the concept of preventing problems by screening, early detection and vigilance, with articles from specialists in obstetrics and neonatology both. I am sure the readers will benefit from the key messages for clinical application.

Uday Thanawala MD, DNB, FCPS
Vice-President, FOGSI (2015)
Consultant Obstetrics & Gynecologist
Thanawala's Maternity Home, Vashi
Navi Mumbai

Acknowledgments

I would like to thank the following people without whom this book would not have been possible.

My patients over the years at Nair, Kasturba and Cooper Hospitals—each case taught me new things!

My family especially my daughter Komal (writer and artist) and son Dr Varun (young graduate) for being my inspiration.

FOGSI Presidents who have encouraged and supported me down the years: Dr Suchitra Pandit (2014), Dr Prakash Trivedi (2015), Dr Alka Kriplani (2016) and Dr Rishma Dhillon Pai (2017).

My three enthusiastic co-editors (Madhuri Mehendale, Pradnya Supe and Rashmi Jalvee) and Ramesh Krishnamachari for their constant help and follow-up.

Each of the experienced contributors for timely submission of their chapters.

Reena J Wani

Contributors

Alka Kriplani MD, FRCOG, FAMS, FICOG, FIMSA, FICMCH, FCLS
Professor and Head
Department of Obstetrics
and Gynecology
Director In-charge
WHO-CCR, HRRC and Family
Planning
All India Institute of Medical Sciences
New Delhi, India

Alok Sharma MD, DHA, MICOG
Kamla Nehru State Hospital
for Mother and Child
Indira Gandhi Medical
College, Shimla, Himachal
Pradesh, India

Anahita Chauhan MD, DGO, DFP
Obstetrician and
Gynecologist
Professor and Unit Head
Seth GS Medical College
and KEM Hospital, Mumbai
Honorary Consultant, Saifee
Hospital, Mumbai
Librarian, Mumbai Obstetric and
Gynaecological Society 2017-18
Secretary and Manager, The Journal of
Obstetrics and Gynecology of India

Ankesh R Sahetya DGO, DNB
Consultant
Medical Speciality
Bandra Bhabha Hospital
Mumbai, Maharashtra

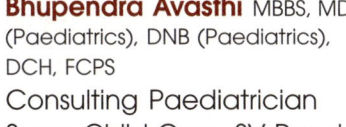

Anuja Rege MD DCH

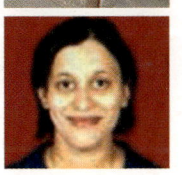

Aparna Sharma MD, DNB, MNAMS
Associate Professor
Department of Obstetrics
and Gynecology
AIIMS, New Delhi

Archana Wani MD, DNB, DGO, DFP
Consulting Obstetrician and
Gynecologist
Archana Maternity Nursing
Home, Panel Consultant
Fortis Hiranandani Hospital

Arpita Thakker Adhikari
MD (Pediatrics), DCH, FCPS,
Fellowship in pediatric Neurology,
PGDDN
Associate Professor
Department of Pediatrics
Lokmanya Tilak Municipal Medical College
and General Hospital, Sion, Mumbai

Baraturam Bhaisara MBBS, MD
Assistant Professor
Depatment of Pediatrics
HBT Medical College and
Dr RN Cooper Hospital
Vile Parle (W), Mumbai

Bhupendra Avasthi MBBS, MD
(Paediatrics), DNB (Paediatrics),
DCH, FCPS
Consulting Paediatrician
Surya Child Care, SV Road
Santacruz West, Mumbai

Chaitra Thunga MD (Obstetrics
and Gynecology)
Senior Registrar
LTM Medical College, Mumbai

Deep Parekh
Pediatrician and Neonatologist
Director, Mom and Me Clinic
Ghatkopar (E), Mumbai

Dheera Samadariya MBBS, MS
Obstetrician and Gynecologist
fellow in Emergency and High Risk
Obstetrics in LTMGH, Sion, Mumbai

Dipti Shende MD, DNB
Assistant Professor, Seth GS Medical
College and Kem Hospital, Parel, Mumbai

Divya Kadam MD, FCPS, DGO
Obstetrics and Gynecology
and Inferility Consultant
Ashirwad Group of Hospitals
Mumbai

Faram Irani MD, FCPS, FICOG,
DGO, DFP
Consultant Obstetrician and
Gynecologist
Southern Cross Nursing Home
United Western Apartments
Ground Floor, Veer Savarkar Marg,
Prabhadevi, Mumbai

Ganesh Shinde
Professor and Head
Department of Obstetrics
and Gynecology
Hindu Hridaysamrat
Balasaheb Thakre Medical
College and Dr RN Cooper
Hospital, Juhu, Mumbai

Gauri Karandikar MD, DGO,
FICOG
Consultant, Obstetrician
Karandikar Hospital, Nasik
Maharshtra, India

Geetha Balsarkar
Consultant, Obstetrics and
Gynecology, Nowrosjee
Wadia Maternity Hospital
Seth GS Medical College
Parel, Mumbai

Hemlata Kuhite
Assistant Professor
Department of Obstetrics
and Gynecology
Hindu Hridaysamrat
Balasaheb Thakre Medical
College and Dr RN Cooper Hospital, Juhu,
Mumbai

Jayashree Mondkar MD, DCH
Professor and Head
Department of Neonatology
In-charge, Human Milk Bank
LTMMC and LTMGH

Juhi Bharti MS, DNB
Assistant Professor
Department of Obstetrics and
Gynecology, AIIMS, New Delhi

Kartikeya Bhagat MD, FICOG,
IBCLC
Consultant, Obstetrician and
Gynaecologist at Grace
Maternity Home, a Baby-
Friendly Hospital
Honorary Assistant

Obstetrician and Gynecologist at the Akurli
Road Municipal Maternity Home, Mumbai
Past President of The Association of Fellow
Gynecologists (AFG), Mumbai
International Board Certified Lactation
Consultant (IBCLC)
Member, Perinatology Committee FOGSI
2006-2009 and 2015-2017
Member, Managing Committee, AMC

Kinjal Shah MBBS, DNB
Consultant, Obstetrician
and Gynaecologist, Bhatia
Hospital, Mumbai

Krishna Ashok Shetye MBBS, DCH, DNB
Senior Resident Medical
College and Hospital
Goa

Madhuri Mehendale DGO, FCPS, DNB
Assistant Professor
Department of Obstetrics
and Gynecology
Lokmanya Tilak Medical
College and General
Hospital, Sion
Managing Council Member Obstetrics
and Gynecology Society, Mumbai

Maimoona Ahmed MS, FNB
(High Risk Pregnancy and Perinatology)
Associate Consultant
Department of Maternal-
Fetal Medicine
Cloudnine, Mumbai

Maitri C Shah
Associate Professor
Department of Obstetrics and
Gynecology
Medical College and SSG
Hospital, Baroda

Mangala Gomare MBBS, DPH, DHA
Deputy Executive Health
Officer, FW and MCH

Mangala Wani

Manjiri Khare MD, FRCOG
Consultant, Maternal-Fetal
Medicine, University Hospitals
of Leicester NHS Trust, UK

Manohar Motwani
Consultant, Obstetrician
and Gynecology
Director, Karbhari Maternity
and Nursing Home, Parle
(East), Mumbai
Honorary Obstetrics and Gynaecologist
and Unit Head, Rajawadi Municipal
Hospital, Ghatkopar, Mumbai
Vice-President, Association of Fellow
Gynecologists, Mumbai (2016-17)

Monika Chhajed DCh, DNB
Consultant, Paediatric
Neurologist, Chaitanya
Hospital, Chandigarh

Nanak Bhagat MBBS
Post Graduate, Obstetrics
and Gynaecology

Neelu Desai MD, DNB
Consultant
Pediatric Neurologist
PD Hinduja National Hospital
and MRC, Mumbai

Neha Singh MBBS, MS (Obstetrics
and Gynecology), DNB
Fellow in Fetal Medicine
Nowrosjee Wadia Maternity
Hospital, Mumbai

Ness F Irani DGO, FCPS, DNB
Consultant
Obstetrics and Gynecology

Nihita Pandey
Senior Resident
Department of Obstetrics
and Gynecology
Hindu Hridaysamrat
Balasaheb Thackre Medical
College and Dr RN Cooper
Hospital, Juhu, Mumbai

Nikit D Kadam DGO, DNB, MNAMS
Clinical Fellow
Doncaster and Bassetlaw Teaching Hospitals
Doncaster, South Yorkshire, UK

Pallavi Untwal MBBS, MS
Obstetrics and Gynaecology
SMO Lokmanya Tilak
Municipal Medical College
Sion, Mumbai, Maharashtra

Pooja Ramchandran MSc, PGDPC, Sc.M (USA)
Genetic Counselor
Personalized Medicine with
Compassionate Care

Pancham Kumar MD
Assistant Professor
Department of Peadiatrics
Indira Gandhi Medical
College Shimla, HP

Prachi Agashe DNB, ICO (UK),
FPOS (Aravind Eye Hospital,
Madurai)
Fellowship in Retinopathy of
Prematurity (LV Prasad Eye
Hospital, Hyderabad)
Pediatric Ophthalmologist
and Squint Specialist
Consultant: Agashe Hospital, Kurla, KB Haji
Bachooali Hospital, Parel
Advanced Eye Hospital, Sanpada, Navi
Mumbai, Kohinoor Hospital Kurla, Surya
Child Care
Santacruz, Mumbai

Pradnya Supe MS
Assistant Professor
Department of Obstetrics
and Gynaecology
Lokmanya Tilak Municipal
Medical College, Sion
Mumbai, Maharashtra

Pragya Tripathi Nichite MBBS,
DGO, DNB
Fellowship in Fetal Medicine
Co-Director and Consultant
Fetal Medicine
TreeTop Fetal Medicine
Centre, Navi Mumbai and
MGM New Bombay Hospital, Navi Mumbai

Prashant Dixit MD, DCH
Fellow Neonatology
Neonatal Intensivist
Scientific Director, Howard
Newborn Centre

Prashant Gangal

Pratima Mittal
Professor and Consultant
Department of Obstetrics
and Gynecology, VMMC
and Safdarjung Hospital
New Delhi

Rahul Wani DNB, DGO, LLB, PGDMLS
Consulting Obstetrician and
Gynecologist
Archana Maternity Nursing
Home
Panel Consultant Fortis
Hiranandani Hospital
Honorary Consultant Mathadi Hospital Trust
Medicolegal Consultant
Vice-President NMOGS
Member State Oversight Committee for
Control of HIV & AIDS

Rashmi Jalvee MS, DGO, DNB
Assistant Professor
Department of Obstetrics
and Gynecology
Dr RN Cooper Hospital
Mumbai, Maharashtra

Reena J Wani MD, MRCOG,
FICOG, DNBE, DFP, DGO, FCPS
Professor and Head
Department of Obstetrics
and Gynecology
HBT Medical College and
RN Cooper Hospital, Juhu
Mumbai

Reeta Bansiwal
Associate Professor
Department of Obstetrics
and Gynaecology
VMMC and Safdarjung
Hospital
New Delhi

Riddhi Shah MS, DNB
Obstetrics and Gynecology
Specialty Medical officer
LTMGH, Sion

Sachin Nichite MBBS, DGO, DNB
Fellowship in Fetal Medicine
Co-Director and Consultant
Fetal Medicine
TreeTop Fetal Medicine
Centre, Navi Mumbai
and MGM New Bombay
Hospital, Navi Mumbai

Sanjay B Prabhu MD, DCH,
IBCLC
Director, Neoplus Criticare
Children Hospital, Borivali
Mumbai

Sarita Bhalerao MD, DNB, FRCOG
Consultant
Obstetrician and
Gynecologist
Bhatia Saifee and Wadia
Maternity Hospitals
Mumbai
Treasure, The Mumbai Obstetric and
Gynecological Society Joint Assistant
Secretary, JOGI

Saurabh Dani
West Zone Co-ordinator
FOGSI Perinatology
Committee
Joint Secretary, ISPAT
Treasure, Association of
Fellow Gynaecologists
Scientific Secretary, Kandivli Medical
Association
Co-Founder, Health N Wellness
Co-Founder and Director, Prenatal Care
and Support LLP
Founder Member, Patient Safety Alliance

Shailesh Kore MD, DGO, DNB,
FCPS, MNAMS, DFP, DICOG, MICOG
Professor and Unit Chief, LTM
Medical College, Mumbai
Chairperson, Genetic and
Fetal Medicine Committee
FOGSI
Secretary, FOGSI Website Committee

Shantala Vadeyar MBBS, MD,
Advanced Obstetric Ultrasound
Diploma, FRCOG, DM (UK), CCST
Consultant
Obstetrics, Fetal and
Maternal Medicine

Shreya Goenka
Senior Resident
Department of Obstetrics
and Gynecology, All India
Institute of Medical
Sciences, Raipur (CG)

Shweta Chawla MD, DCH
Consultant, Paediatrician
and Neonatologist
Scientific Director, Howard
Newborn Centre

Siddesh Mahadev Iyer MBBS,
DGO, DNB (Obstetrics and gynecology)
Assistant Professor
HBT Medical College and
Dr RN Cooper Hospital
Juhu, Mumbai

Sushma Malik MD (Pediatrics)
Professor, In-charge NICU
Neonatology Division
Department of Pediatrics
Medical College and BYL
Nair Hospital, Mumbai

Swati Manerkar MD, DCH
Fellowship in Neonatology
Associate Professor
Department of Neonatology
LTMMC & LTMGH

Uday Thanawala MD, DGO,
FCPS, DNB
Consultant
Thanawala Maternity Home
Navi Mumbai
Vice-President, FOGSI (2015)

Vaidehi Dande MD FIAP
(Neonatology)
Assistant Professor
Department of Pediatrics
Division of Neonatology
BJ Wadia Hospital for Children
Parel, Mumbai

Vandana Bansal MD, DGO,
DNB, MRCOG
FNB (High Risk Pregnancy and
Perinatology)
Associate Professor
Department of Obstetrics
and Gynecology
Nowrosjee Wadia Maternity
Hospital and amp; Seth GS Medical College
Director and Head
Department of Fetal Medicine
Surya Mother and Child Hospitals
Mumbai, India

Vidya A Thobbi
Professor and Head
Department of Obstetrics
and Gynecology
AL-Ameen Medical College
Vijayapur

Vikas Bhatia
Founder MERD India
Parent Support Group

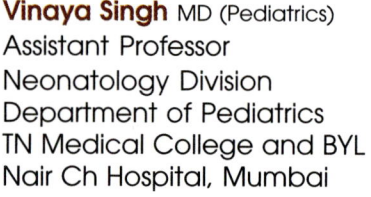

Vinaya Singh MD (Pediatrics)
Assistant Professor
Neonatology Division
Department of Pediatrics
TN Medical College and BYL
Nair Ch Hospital, Mumbai

Contents

Section I: Back to Basics

Section II: What an Obstetrician can do?

Section III: What a Neonatologist can do?

Back to Basics

1. Role of Genetics in Perinatology
2. The Concept of Gestational Age—Viability, Preterm, Postdatism
3. Basic Embryological Development
4. Biochemical Markers for Detection of Fetal Aneuploidies
5. Role of Invasive Tests (Amniocentesis, CVS and Cordocentesis) in Perinatology
6. Ultrasound and Aneuploidy
7. Parental Concerns with Previous Affected Child

Role of Genetics in Perinatology

Pooja Ramchandran

INTRODUCTION

During preconception and prenatal checkups, obstetricians and gynecologists have several screening and testing options available for women and their families. India has been going through a transition from infant mortality and morbidity due to infections to the emergence of genetic conditions. The World Health Organization (WHO) recommends the introduction of genetic services in countries with infant mortality rates (IMR) less than 50. India, with an IMR of 40, is ripe for such an introduction to and implementation of genetic services. This chapter introduces the ways that obstetricians and neonatologists, together, can incorporate genetics in their practice in order to address the growing need of risk assessment, screening, diagnosis, and management of genetic conditions in the perinatal period. For the purpose of structure and practicality, this chapter focuses on the genetic aspects in infertility, pregnancy, neonatology during the preconception period. By the end of this chapter, readers will be able to:

i. Improve identification of women, pregnancies and neonates at increased risk for genetic disease

ii. Provide timely referrals, recommenda-tions, and support for ordering genetic tests and managing genetic risk factors

iii. Stay informed and adhere to standards of care based on professional society guidelines

iv. Provide relevant information about tests and risks to patients

v. Improve patient outcomes based on standards of care.

1. INFERTILITY

Worldwide, the prevalence of couples with infertility over a period of a year ranges from 3.5% to 16.7%. The reasons for patients seeking assisted reproductive technology can be attributed to male infertility in 1/3rd couples, female infertility in 1/3rd couples and, in 1/3rd of couples, the reason for infertility remains unidentified. Genetic testing in cases of infertility is important for the purpose of identifying appropriate treatment options and preventative measures for the patient as well as future offspring. Genetic factors surrounding infertility include chromosomal abnormalities, single gene mutations and multifactorial conditions, and warrant a genetic referral.[1, 2] There are about 300 genetic mutations (70 of which are considered syndromic) known to cause reproductive disorders, including early pregnancy loss. For practical reasons, this section focuses only on the most common genetic reasons for infertility. Flowchart 1.1 provides a guide to the genetic evaluation in the case of infertility, recurrent early pregnancy loss and miscarriage.

3

Flowchart 1.1: Genetic evaluation of the infertile couple

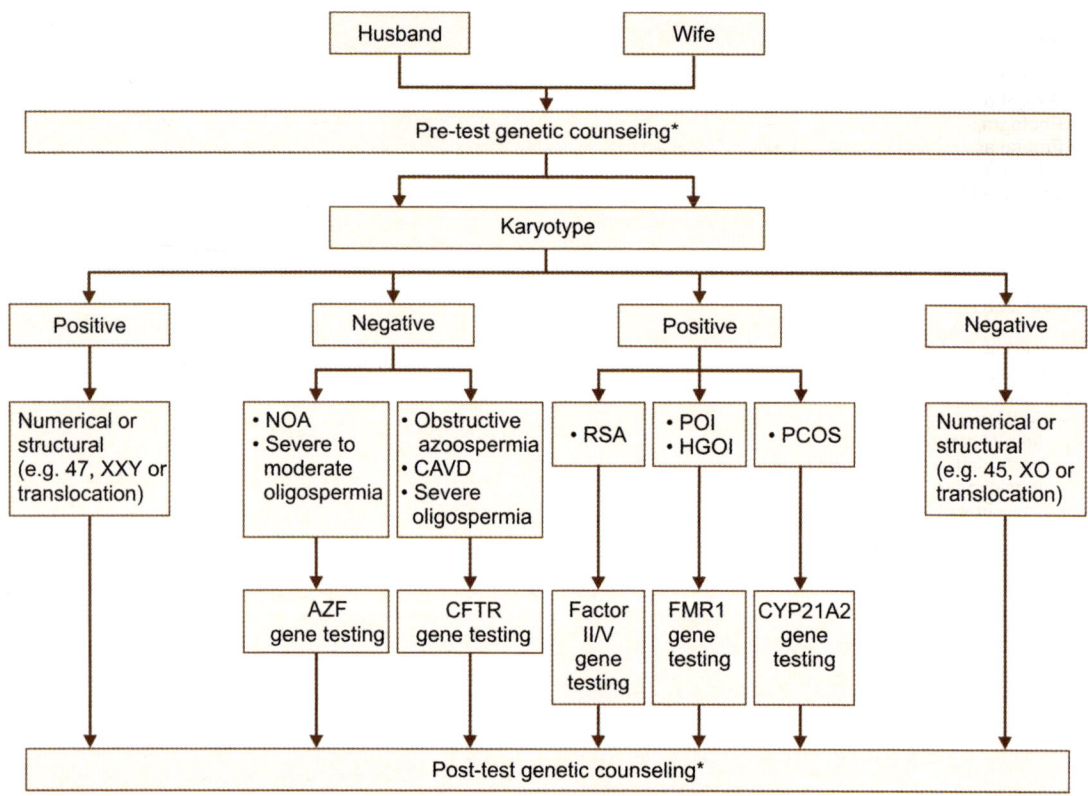

NOA:Non-obstructive azoospermia, CAVD:Congenital absence of the vas deferens,
RSA:Recurrent spontaneous abortions, POF:Premature ovarian insufficiency
(formerly referred to as premature ovarian failure),
HGOI:Hypergonadotropic ovarian insufficiency, PCOS:Polycystic ovarian syndrome

*Most genetic counseling services are provided in-person or via telemedicine. By contacting your local genetic counselor, they can get your patients scheduled for a genetic counseling consultation following which, the genetic counselor will communicate back to you. The patient remains under the care of the referring physician.

A comprehensive discussion of the genetic factors associated with infertility and pregnancy loss is beyond the scope of this chapter. Further, some genetic conditions are known to cause stillbirth, and many of these are associated with nonimmune hydrops. The ACOG Practice Bulletin for the Management of Stillbirth[3] delineates the following important tests in the evaluation of a stillbirth:

• Chromosome analysis (via fetal tissue sampling, or preferably, through amniocentesis before delivery)
• Fetal autopsy
• Placental pathology

The algorithm is provided in Flowchart 1.2.

The evaluation of the genetic cause of infertility, pregnancy loss, and stillbirth is important not only to provide patients a cause and closure, but also to inform future risks and reproductive decision making. Examples of a few genetic conditions associated with pregnancy loss include hemoglobinopathies, thrombophilias, aminoacid disorders, peroxisomal disorders, storage diseases. Alpha thalassemia (HAA), glutaric aciduria, type II (ETFA, ETFB, and ETFDH), Smith-Lemli-Opitz (DHCR7), Zellweger syndrome (PEX), Gaucher disease (GBA), activated

Flowchart 1.2: Algorithm for the management of stillbirth

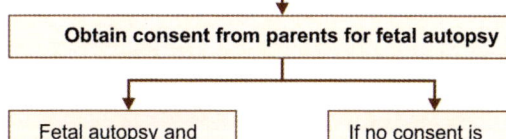

Fetal and placental inspection

• Weight, head circumference and length of fetus
• Weight of placenta
• Photographs of fetus and placenta
• Frontal and profile photographs of whole body, face, extremities, palms, and any abnormalities
• Document findings and abnormalities

↓

Obtain consent from parents for cytologic specimens

• Obtain cytologic specimens with sterile techniques and instruments
• Acceptable cytologic specimens (at least one)
  – Amniotic fluid obtained by amniocentesis at time of prenatal diagnosis of demise: particularly valuable if delivery is not expected imminently
  – Placental block (1 × 1 cm) taken from below the cord insertion side on the unfixed placenta
  – Umbilical cord segment (1.5 cm)
  – Internal fetal tissue specimen, such as costochondral junction or patella: Skin is not recommended
• Place specimens in a sterile tissue culture medium of lactated Ringer's solution and keep at room

↓

Obtain consent from parents for fetal autopsy

↓   ↓

| Fetal autopsy and placental pathology (may include fetal whole body X-ray) | If no consent is given for autopsy, send placenta alone for pathology |

protein C resistance (factor V), factor II (PT G20210A), Rett syndrome (MECP2, CDKL5), incontinentia pigmenti (NEMO) are a few examples. Referral to genetics/genetic counseling may achieve suitable genetic evaluation of infertility, pregnancy loss and stillbirths.

2. PRENATAL SCREENING AND DIAGNOSIS

In 2007, ACOG issued Practice Bulletins that changed the recommendations for prenatal testing. These guidelines recommended offering diagnostic and screening testing to all women, rather than limiting the offer of prenatal testing to mothers considered "AMA". It has been 9 years since those guidelines were published and, since then, clinical practice has seen the advent of cell free DNA screening (no-invasive prenatal screening, NIPS). Informed by the NIPS and the years of clinical practice since 2007, ACOG issued new guidelines regarding diagnostic and screening testing in 2016 as follows.

Prenatal Screening

ACOG defines prenatal genetic screening as a test "designed to assess whether a patient is at increased risk of having a fetus affected by a genetic disorder".[4] The purpose of such screening is to provide risk assessment for the more common aneuploidies. Prenatal screening has become more nuanced with the advent of newer tests; Practice Bulletin No. 163, released May 2016, provides the latest guidelines for screening for fetal aneuploidies.

A detailed discussion of these guidelines is beyond the scope of this chapter; a summary of the recommendations is outlined.

Pre-test Counseling

ACOG recommends that patients be counseled regarding:

• Available prenatal screening options that are available
• The advantages, disadvantages and limitations of each test (along with the screening test's detection rate, positive and false positive rates)
• Patients should also be counseled about diagnostic testing
• After appropriate counseling, patients may decline for any reason

Note: Documentation of genetic counseling referral is important for medico-legal purposes. Genetic counseling must also

The American College of Obstetricians and Gynecologists
WOMEN'S HEALTH CARE PHYSICIANS

Society for Maternal·Fetal Medicine

(Published Electronically Ahead of Print on March 1, 2016)

PRACTICE BULLETIN

CLINICAL MANAGEMENT GUIDELINES FOR OBSTETRICIAN–GYNECOLOGISTS

NUMBER 163, MAY 2016

(Replaces Practice Bulletin Number 77, January 2007)
(See also Practice Bulletin Number 162, Prenatal Diagnostic Testing for Genetic Disorders)

Screening for Fetal Aneuploidy

The wide variety of screening test options, each offering varying levels of information and accuracy, has resulted in the need for complex counseling by the health care provider and complex decision making by the patient. No one screening test is superior to other screening tests in all test characteristics. Each test has relative advantages and disadvantages. It is important that obstetrician–gynecologists and other obstetric care providers be prepared to discuss not only the risk of aneuploidy but also the benefits, risks, and limitations of available screening tests. Screening for aneuploidy should be an informed patient choice, with an underlying foundation of shared decision making that fits the patient's clinical circumstances, values, interests, and goals.

Fig. 1.1: Excerpt from ACOG Practice Bulletin Number 163

include documentation and assessment of a 3-generation pedigree with a focus on identifying other genetic risks to the pregnancy in order to identify and offer the most appropriate prenatal test. Genetic counselor's responsibilities include documentation of consent/dissent of testing as well as reasons for dissent if testing was declined.

First Trimester Screening

- Should be offered to all women regardless of age.
- Performed between 10 and 13 6/7 weeks.
- Nuchal translucency (NT) combined with maternal serum screening is recognized and recommended as the first trimester screen.
- "Meticulous technique" is key in addressing the serious potential of inaccuracies in results caused by even the slightest errors in measurements.
- If positive, patients must be provided "detailed counseling" and should be offered cfDNA screening and/or CVS or amniocentesis.

Second Trimester Screening

- Performed between 15 and 22 6/7 weeks (ideally between 16 and 18 weeks).
- Should be offered if first trimester screen was missed.
- AFP screening should be offered for open neural defects regardless of whether first trimester screening was conducted.
- Multiple aneuploidy screening should not be conducted independently of each other to avoid the chance of reporting a false positive and creating confusion.
- If positive, patients should receive "detailed counseling" and be offered cfDNA screening and/or amniocentesis.

Ultrasound Screening

- Ultrasound should not be used in isolation to diagnose or exclude Down syndrome.
- Isolated increased nuchal thickness confers the highest risk of genetic syndromes and aneuploidy; patients should be offered targeted ultrasound examination and fetal echocardiography in the second trimester.

- Isolated renal pelvis dilation, echogenic bowel, or shortened humerus or femur length may benefit from referral for detailed ultrasonography and follow-up.
- If a soft marker is identified, serum screening and prenatal testing should be offered if it had not been already.

Practice Bulletin Number 163 provides recommendations for the management of ultrasonographic markers for aneuploidy (Table 1.1).

Cell-free DNA Screening

- May be offered anytime from 10 weeks
- cfDNA screening remains a "screening" test with false positives and false negatives
- "Unreportable" results must be treated as a report of increased risk as such results have

Table 1.1	Management of ultrasonographic markers for aneuploidy, ACOG Practice Bulletin Number 163, May 2016		
Soft marker	*Imaging criteria*	*Aneuploidy association*	*Management*
First trimester: Enlarged nuchal translucency	Certified ultrasonography measurement ≥ 3.0 mm or above the 99th percentile for the CRL	Aneuploidy risk increases with size of NT. Also associated with Noonan syndrome, multiple pterygium syndrome, skeletal dysplasias, congenital heart disease, and other anomalies	1. Genetic counseling 2. Offer cfDNA or CVS 3. Second trimester detailed anatomic survey and fetal cardiac ultrasonography
First trimester: Cystic hygroma	Large single or multilocular fluid-filled cavities, in the nuchal region and can extend the length of the fetus	If septate, approximately 50% are aneuploid	1. Genetic counseling 2. Offer CVS 3. Second trimester detailed fetal anatomic survey and fetal cardiac ultrasonography
Second trimester: Echogenic intra-cardiac foci	Echogenic tissue in one or both ventricles of the heart seen on standard four-chamber view	LR 1.4–1.8 for Down syndrome. Seen in 15–30% of Down syndrome and 4–7% euploid fetuses	1. If isolated finding, aneuploidy screening should be offered if not done previously 2. If aneuploidy screen result is negative, no further evaluation is required
Second trimester: Pyelectasis	Renal pelvis measuring ≥ 4 mm in anteroposterior diameter up to 20 weeks of gestation	LR 1.5–1.6 for Down syndrome	1. If isolated finding, aneuploidy screening should be offered if not performed previously 2. Repeat ultrasonography in third trimester for potential urinary tract obstruction
Second trimester: Echogenic bowel	Fetal small bowel as echogenic as bone	LR 5.5–6.7 for Down syndrome. Associated with aneuploidy, intra-amniotic bleeding, cystic fibrosis, CMV	1. Further counseling 2. Offer CMV, CF, and aneuploidy screening or diagnostic testing

Contd...

Table 1.1	Management of ultrasonographic markers for aneuploidy, ACOG Practice Bulletin Number 163, May 2016 (Contd...)		
Soft marker	*Imaging criteria*	*Aneuploidy association*	*Management*
Second trimester: thickened nuchal fold	≥ 6 mm from outer edge of the occipital bone to outer skin in the midline	LR 11–18.6 with 40–50% sensitivity and >99% specificity for Down syndrome. Most powerful second trimester marker	1. Detailed anatomic survey 2. Further detailed genetic counseling and aneuploidy screening or diagnostic testing
Second trimester: Mild ventriculo-megaly	Lateral ventricular atrial measurement between 10 and 15 mm	Associated with aneuploidy LR 25 for Down syndrome	1. Genetic counseling 2. Second trimester detailed anatomic ultra-sound evaluation 3. Consider diagnostic testing for aneuploidy and CMV 4. Repeat ultrasound in third trimester
Second trimester: Choroid plexus cysts	Discrete cysts(s) in one or both choroid plexus(es)	In isolation, no aneuploidy association	1. Second trimester detailed anatomic survey and fetal cardiac ultrasound 2. No further follow-up if isolated 3. Consider aneuploidy screening or diagnostic testing if other markers are present
Second trimester: Short femur length	Measurement <2.5 percentile for gestational age	LR 1.2–2.2 for Down syndrome, can be associated with aneuploidy, IUGR, short limb dysplasia	1. Second trimester detailed fetal anatomic evaluation for short limb dysplasia 2. Further detailed counseling 3. Consider repeat ultrasonography in third trimester for fetal growth

Abbreviations: CF, cystic fibrosis; cfDNA, cell-free DNA; CMV, cytomegalovirus; CRL, crown-rump length; CVS, chorionic villus sampling; IUGR, intrauterine growth restriction; LR, likelihood ratio; NT, nuchal translucency

shown to have a higher association with being true positives

- cfDNA screening is not recognized for microdeletions due to lack of clinical validation
- cfDNA screening is not recommended for multiple gestations

Preimplantation Genetic Screening (PGS)

- PGS may be performed in an IVF cycle; however, being a "screen", false positives and false negatives cannot be ruled out
- All pregnancies conceived through IVF (even after PGS) must be offered routine prenatal screening and diagnostic testing

Post-test Counseling

- Screening test results "should be reported as either positive or negative, and the adjusted numerical risk should be provided."
- Patients with a positive screening result should be offered "detailed counseling"
- Patients with a positive conventional screen result but a negative cfDNA screen should be informed that there is still a 2% residual risk of a chromosomal abnormality.
- "All patients with a positive cfDNA result should have a diagnostic procedure before any irreversible action, such as pregnancy termination, is taken."

Prenatal Diagnosis

ACOG describes prenatal genetic testing as a test aimed at identifying "health problems that could affect the woman, fetus, or newborn"[5] aimed at providing information to facilitate a fully informed decision about pregnancy management. Prenatal testing should focus on the patient's individualized risks, reproductive goals, and preferences. It is crucial for patients to understand the benefits, risks and limitations of all prenatal diagnostic testing, including conditions that can and cannot be diagnosed through testing. In the latest guideline (Practice Bulletin Number 162, 2016), ACOG retains the previous recommendation that all women should be offered diagnostic testing regardless of maternal age or other risk factors, ideally at their first visit.

Pre-test Counseling

- Information about possible detectable conditions should be provided to patients before making a decision to proceed with diagnostic testing
- A healthcare professional with genetics expertise "can help with counseling,

The American College of Obstetricians and Gynecologists
WOMEN'S HEALTH CARE PHYSICIANS

Society for Maternal·Fetal Medicine

(Published Electronically Ahead of Print on March 1, 2016)

PRACTICE BULLETIN

CLINICAL MANAGEMENT GUIDELINES FOR OBSTETRICIAN−GYNECOLOGISTS

NUMBER 162, MAY 2016

(Replaces Practice Bulletin Number 88, December 2007)
(See also Practice Bulletin Number 163, Screening for Fetal Aneuploidy)

Prenatal Diagnostic Testing for Genetic Disorders

Prenatal genetic diagnostic testing is intended to determine, with as much certainty as possible, whether a specific genetic disorder or condition is present in the fetus. In contrast, prenatal genetic screening is designed to assess whether a patient is at increased risk of having a fetus affected by a genetic disorder. Originally, prenatal genetic testing focused primarily on Down syndrome (trisomy 21), but now it is able to detect a broad range of genetic disorders. Although it is necessary to perform amniocentesis or chorionic villus sampling (CVS) to definitively diagnose most genetic disorders, in some circumstances, fetal imaging with ultrasonography, echocardiography, or magnetic resonance imaging may be diagnostic of a particular structural fetal abnormality that is suggestive of an underlying genetic condition.

Fig. 1.2: A summary of the salient takeaways from Practice Bulletin No. 162 is described

choosing the right test, and interpreting the test results" if a fetal genetic abnormality is suspected

Note: Documentation of genetic counseling referral is important for medico-legal purposes. Genetic counseling must also include documentation and assessment of a 3-generation pedigree with a focus on identifying other genetic risks to the pregnancy in order to identify and offer the most appropriate prenatal test. Genetic counselor's responsibilities include documentation of consent/dissent of testing as well as reasons for dissent if testing was declined.

Chorionic Villus Sampling (CVS)

- Performed between 10 and 13 weeks
- Calculated procedure-related loss rate is 0.22%* (~ 1 in 455)
- Risk of limb-reduction defects is not significantly greater than in the general population (as long as CVS is performed at/after 10 weeks)

Amniocentesis

- Performed between 15 and 20 weeks (not recommended before 15 weeks)
- Estimated procedure-related loss rate is approximately 0.11%* (~ 1 in 909)

* Specified loss rates for amniocentesis and CVS are found in high-volume, experienced centers. The low loss rates "may not apply to other situations," i.e. "among health care providers with less cumulative experience".

Fluorescence in situ Hybridization (FISH)

- Considered a screening test due to false-positive and false-negative results
- Clinical decision should not based after confirmatory diagnostic results or consistent clinical information and not solely on FISH results, e.g. abnormal ultrasound findings or a positive

Table 1.2	Ultrasound findings in genetic syndromes

Current evidence supports the application of first trimester ultrasound as a screening tool for select genetic syndromes. Second trimester ultrasound provides the most information on fetal abnormalities and should be carried out in a structured and systematic manner. It is advised that all pregnant women undergo a fetal anatomic survey (level 2 ultrasound) in the second trimester after about 18 weeks. A genetics professional will be able to recognize patterns seen with individual syndromes. Identification of these abnormalities can lead to informed recommendations for definitive diagnostic testing, reproductive decision-making and/or preparation for the post-natal period. Because of the large number of genetic disorders with assosiated ultrasound findings, it is not possible to discuss them all within the scope of this sub-section. For example, a few genetic syndromes and their associated ultrasound findings are listed below. ACMG recommends a genetics referral in the case of abnormal ultrasound findings.

Syndrome	Ultrasound findings
Turner syndrome	Cystic hygroma, hydrops fetalis, CHD (e.g. coarctation of aorta, VSD, TOF, dilated right ventricle), renal abnormalities, short femur, echogenic intracardiac focus, echogenic bowel
22q11.2 deletion syndrome	Cardiac anomalies (e.g. tetralogy of Fallot, VSD, truncus arteriosus, pulmonary atresia, interrupted aortic arch), polyhydramnios, thymic hypoplasia, renal anomalies, microcephaly, cleft palate
Cri-du-chat (5p deletion) syndrome	Growth restriction, microcephaly, micrognathia, and hypertelorism, nasal bone hypoplasia, ventriculomegaly, cardiac defects
Beckwith-Wiedemann syndrome	Omphalocele, macroglossia, macrosomia, nephromegaly, and polyhydramnios
Noonan syndrome	Cystic hygroma, increased nuchal translucency, cardiac defects, polyhydramnios

screening result for Down syndrome or trisomy.

Chromosomal Microarray Analysis (CMA)

- Should be made available to any patient choosing CVS/amniocentesis
- In the case of an ultrasound finding of fetal structural abnormality, CMA is recommended as a primary test (Exception: If the abnormality is "strongly suggestive" of a particular aneuploidy, karyotype may be offered before CMA)

Preimplantation Genetic Diagnosis (PGD)

- Errors are possible as only few cells are used
- Therefore, "confirmation of results with CVS or amniocentesis is usually recommended."

Post-diagnosis Counseling

- "The patient should receive detailed information, to the extent that information is available, about the natural history of the specific condition"
- The option of pregnancy termination is to be discussed on prenatal detection of a genetic disorder or major structural abnormality detected prenatally
- "Referral to parent support groups, counselors, social workers, or clergy may provide additional information and support for some patients."

The guideline further emphasizes that counseling must include:

- Family education and preparation
- Obstetric management recommendations (including fetal surveillance, intrapartum monitoring, and mode of delivery)
- Referral to pediatric specialists and a tertiary care center for delivery, if appropriate
- Availability of adoption or pregnancy termination

- Perinatal palliative care services and comfort care for delivery if diagnosis or fetal presentation is incompatible with survival.

Newborn Screening

Newborn screening (NBS) is aimed at achieving the earliest possible identification of genetic conditions in order to prevent the serious consequences through timely and appropriate intervention. It is important to bear in mind that NBS is not a confirmatory diagnosis and requires further investigations. In the United States, NBS is a mandatory state-based public health program focused on providing all neonates with presymptomatic testing and necessary follow-up testing and care for a range of medical conditions. The goal of this essential public health program, according to ACOG's Committee on Genetics, is "to decrease morbidity and mortality by screening for disorders in which early intervention will improve neonatal and long-term health outcomes".[6] ACOG emphasizes the importance of integrating NBS education during prenatal visits as it prepares prospective parents for having their child undergo NBS and receiving NBS results. ACOG's Committee Opinion refers to the ACMG-published Executive Summary[7] that established guidelines for NBS programs in the United States. This summary included a uniform panel with "core conditions" for NBS (Table 1.3). Each condition on the panel:

- Has a screening test that can be performed within 24–48 hours after birth
- Can be treated, and
- Has a known natural history

This panel of core conditions consists of five main categories of disorders:

1. Hemoglobinopathies
2. Organic acid disorders
3. Amino acid disorders
4. Fatty acid oxidation disorders, and
5. Miscellaneous disorders, such as cystic fibrosis, hypothyroidism, and hearing loss.

Table 1.3	ACMG recommended uniform newborn screening panel of core conditions
Disease categories	**Diseases**
Inborn errors of organic acid metabolism	Isovaleric acidemia
	Glutaric acidemia type I
	3-Hydroxy-3-methylglutaric aciduria
	Holocarboxylase synthase deficiency
	Methylmalonic acidemia (methylmalonyl-CoA mutase)
	3-Methylcrotonyl-CoA carboxylase deficiency
	Methylmalonic acidemia (cobalamin disorders)
	Propionic acidemia
	β-ketothiolase deficiency
Inborn errors of fatty acid metabolism	Medium-chain acyl-CoA dehydrogenase deficiency
	Very long-chain acyl-CoA dehydrogenase deficiency
	Long-chain L-3 hydroxyacyl-CoA dehydrogenase deficiency
	Trifunctional protein deficiency
	Carnitine uptake defect/transport defect
Inborn errors of amino acid metabolism	Classic phenylketonuria
	Maple syrup urine disease
	Homocystinuria
	Citrullinemia, type I
	Argininosuccinic aciduria
	Tyrosinemia, type I
Hemoglobinopathies	S, S disease (sickle cell anemia)
	S, β-thalassemia
	S, C disease
Miscellaneous multisystem diseases	Primary congenital hypothyroidism
	Biotinidase deficiency
	Congenital adrenal hyperplasia
	Classic galactosemia
	Cystic fibrosis
	Severe combined immunodeficiency
Newborn screening by methods other than by heel stick	Hearing loss
	Critical congenital heart disease

*Recommendations as of April 2013. For updated information, please see Health Resources and Services Administration. Recommended uniform screening panel

A list of the American College of Medical Genetics core conditions for NBS is provided below.

However, in India, NBS has taken a second seat on account of policies that target mortality and infectious diseases. While this has succeeded in reducing the rates of infant mortality, it has, unfortunately, been met with a marked increase in disability that could have been prevented. The scenario is now beginning to slowly change with the plan for NBS screening in India showcased in May 2015 President's Page of the Indian Academy of Pediatrics[8] as detailed in Table 1.4.

Following implementation of the above mentioned ACMG Uniform Core Conditions for NBS in the United States, the Centers for Disease Control and Prevention, in 2012, identified the five most commonly diagnosed conditions in the United States as:

1. Hearing loss
2. Primary congenital hypothyroidism

Table 1.4		National Plan for NBS Public Health Program in India	
Category	*Criteria*		*Disease*
A	All newborns		• Congenital hypothyroidism • Congenital adrenal hyperplasia • G6PD deficiency • Sickle cell disease • Thalassemias • Hearing loss
B	High risk screening • Consanguinity • Previous child(ren) with intellectual disability/seizure/unexplained death • Critically ill neonates • Neonates/children with symptoms or investigations suggestive of inborn errors of metabolism (e.g. poor feeding, vomiting, diarrhea, respiratory distress, ketosis, hyperammonemia, hypoglycemia, seizures ataxia, lethargy, coma, etc.)		• Phenylketonuria • Homocystinuria • Alkaptonuria • Galactosemia • Sickle cell anemia and other hemo globinopathies • Cystic fibrosis • Biotinidase deficiency • Maple syrup urine disease • Medium-chain acyl-CoA dehydrogenase deficiency (MCADD) • Tyrosinemia • Fatty acid oxidation defects
C	Expanded screening or in "resource-rich" setting		Screening for r30–40 inherited metabolic disorders may be offered to 'well-to-do' families, especially in urban settings where facilities for sending sample to laboratory are available

3. Cystic fibrosis

4. Sickle cell disease, and

5. Medium-chain acyl-CoA dehydrogenase (MCAD) deficiency

While these findings clearly demonstrate the significance of screening for these disorders in newborns, the fact that the national implementation of NBS in India covers these conditions as part of phase one and two should serve as impetus to implement NBS as part of routine neonatal care. Further, the WHO recommends genetic services to be introduced in countries having an infant mortality rate (IMR) less than 50. India, with an IMR of 40, needs to introduce NBS and its requisite genetic services. The ACOG's Committee on Genetics' opinion paper titled 'Newborn Screening and the Role of the Obstetrician-Gynecologist' (Ob-Gyn) emphasizes the Ob-Gyn's role in providing newborn screening information at various times during prenatal visits, and these can be implemented in India as well:

i. During the first trimester new obstetric visit along with other patient education materials

ii. Later in pregnancy, e.g. at the time of glucose tolerance test or group B streptococcal screening in the third trimester

iii. "During a discussion of past adverse pregnancy outcomes related to a positive newborn screening test result or birth defect, at the same time that options for prenatal or preimplantation genetic screening or diagnostic testing are considered."

Carrier Screening

A joint statement[9] of the American College of Medical Genetics and Genomics, American College of Obstetricians and Gynecologists, National Society of Genetic Counselors, Perinatal Quality Foundation, and Society for Maternal-Fetal Medicine describes carrier screening for inherited genetic conditions as "an important component of preconception and prenatal care", the purpose of which is "to identify couples at risk for passing on genetic conditions to their offspring". The conditions recommended across all ethnicities may be a good place to begin carrier screening. In the affording preconception/early prenatal populations, pan-ethnic expanded carrier testing may be offered. The three conditions applicable to India for carrier screening are:

   i. *Hemoglobinopathies*
     a. If anemia + MCV less than 80 fL, evaluate iron deficiency
     b. If iron studies are normal, perform hemoglobin electrophoresis
     c. If hemoglobin electrophoresis is abnormal, refer to genetic counseling for appropriate molecular testing
     d. If hemoglobin electrophoresis is normal, a referral to genetic counseling for molecular testing for alpha-thalassemia is indicated as silent carriers have normal haemoglobin and MCV
   ii. *Cystic fibrosis*
     a. Screening panel of 23 pathogenic variants in the CFTR gene
     b. Sequencing of entire CFTR gene is not appropriate for carrier screening
     c. Genetic counseling referral if screen positive
   iii. *Spinal muscular atrophy*
     a. ACMG emphasizes the need for genetic counseling

If carrier screening reveals only one partner to be a carrier of an autosomal recessive condition, "the chance that the couple will have an affected pregnancy is significantly reduced and no further testing of the partner should be offered; prenatal diagnosis is not indicated". Counseling needs to include that while a residual risk persists, the partner is unlikely to have a mutation for the same disorder.

If both partners are identified as carriers of the same autosomal recessive condition, there is a 25% risk (1 in 4 chance), with each pregnancy, to have an affected child. Genetic counseling by a genetics professional is indicated for preconception patients, and should review prenatal diagnosis and management as well as preimplantation genetic diagnosis and use of noncarrier donor gametes as additional options. For pregnant patients, genetic counseling and prenatal diagnosis should be offered.

If an affected fetus is identified, available reproductive options must be discussed including:
- Prenatal management
- Delivery planning and coordination of care for the child
- Pregnancy termination

This joint statement emphasizes that carrier screening results should be made available to the patient, and counseling should include an explanation of the condition and its inheritance, the implication of this information to first degree relatives (e.g. parents, siblings) who may also be carriers. It is also important to provide patient with written information regarding the availability of carrier screening that they can share with relatives.

CONCLUSION

In summary, the time is ripe for the incorporation of genetic in obstetric and neonatal care, and the standards of care based on professional society guidelines are updated and clear. Table 1.5 lists common reasons for a genetic referral suggested by the Professional Practice and Guidelines Committee of the American College of Medical Genetics. While this table is certainly not intended to be exhaustive or comprehensive, it may serve as a guide for obstetricians. A genetics consultation may be helpful under the following circumstances for preconceptional or prenatal patients. This is followed by a practical sample of a genetic intake sheet (Form 1.1) that may be incorporated into clinical practice. Any 'yes' on page 2 may be considered an indication for a genetic referral.

Table 1.5	ACMG practice guideline—indications for genetic referral
Prenatal or preconceptional patient who is or will be	
Finding	*Reason to consider consultation*
Age 35 years or older at the time of delivery (for a singleton pregnancy)	Discuss testing options for identifying an age-related chromosome anomaly
Age 33 years or older at the time of delivery (for a twin pregnancy)	Discuss testing options for identifying an age-related chromosome anomaly
A close blood relative of her partner (consanguineous union)	Review pedigree and assess degree of relatedness; discuss potential additional fetal risks and testing options before and/or after delivery
Prenatal or preconceptional patient who has:	
Finding	*Reason to consider consultation*
An abnormal first or second trimester maternal serum ± nuchal translucency screening test	Discuss risks to pregnancy and testing options
Exposure to a teratogen or potentially teratogenic agent during gestation such as radiation, high-risk infections (cytomegalovirus, toxoplasmosis, rubella), drugs, medications, alcohol, etc.	Discuss risks to pregnancy and testing options and rule out significant fetal ± maternal risks
A fetal anomaly or multiple anomalies identified on ultrasound and/or through echocardiography	Discuss risks to pregnancy and testing options
A personal or family history of pregnancy complications known to be associated with genetic factors such as acute fatty liver of pregnancy	Rule out significant fetal risks ± maternal risks, including a metabolic disorder
Either member of the couple with:	
Finding	*Reason to consider consultation*
A positive carrier screening test for a genetic condition such as cystic fibrosis, thalassemia, sickle cell anemia, Tay-Sachs, etc.	Discuss additional testing strategies and inheritance
A personal history of stillbirths, previous child with hydrops, recurrent pregnancy losses (more than two), or a child with sudden infant death syndrome (SIDS)	Rule out a chromosomal, metabolic, or syndromic diagnosis that may be associated with an unexplained neonatal death or SIDS
A progressive neurologic condition known to be genetically determined such as a peripheral neuropathy, unexplained myopathy, progressive ataxia, early onset dementia, or a familial movement disorder	Discuss a potential diagnosis, the differential diagnosis, inheritance, and testing options
A statin-induced myopathy	Discuss a potential mitochondrial disorder, inheritance and testing options

Contd...

Table 1.5	ACMG practice guideline—indications for genetic referral (*Contd...*)
Either member of the couple with a family or personal history of:	
Finding	*Reason to consider consultation*
A birth defect such as a cleft lip ± palate, spina bifida, or a congenital heart defect	Discuss recurrence risks and testing options; discuss folate supplementation, if appropriate, for subsequent pregnancies
A chromosomal abnormality such as a translocation, marker chromosome, or chromosomal mosaicism	Discuss risks to the fetus and testing options
Significant hearing or vision loss thought to be genetically determined	Discuss risks to the fetus and testing options
Mental retardation or autism	Discuss risks to the fetus and testing options

Prenatal Intake and History Form

Name: _____
Your age at delivery: _____

Partner's name: _____
Partner's date of birth:_____

Current Pregnancy

First day of last menstrual period: _____ Due date: _____ Twins or multiples? _____

When was your most recent ultrasound? _____ Your current weight? _____ Height? _____

Was your current pregnancy conceived using fertility treatments, such as IVF? ☐ Yes ☐ No
    If yes, was a donor used?   ☐ Egg   ☐ Sperm   Donor egg age: _____
    If a frozen embryo was used, what was your/donor's age at the time of retrieval? _____

Any pregnancy complications, such as cramping, spotting or bleeding? _____

Have you had any illnesses, infections, or fevers or any exposures to alcohol, cigarettes, medications, drugs, radiation, chemicals, or gases during the pregnancy? If yes, please describe. _____
_____

Pregnancy History

How many pregnancies have you had (*including the current pregnancy*)? _____

Name of living child(ren)	Date of birth	Any health problems, birth defects, or genetic disorders?

	How many?	Weeks gestation/age of child	Causes (if known)
Miscarriages			
Terminations			
Infant or child deaths			

Personal and Family History

Do you or your partner have a personal or family history (including aunts, uncles, cousins, grandparents, etc) of the following? *If yes, please describe; if no, leave blank.*

	You	Your partner
Genetic conditions (Down syndrome, cystic fibrosis, muscular dystrophy, etc)		
Birth defects (heart defects, cleft lip, spina bifida, etc)		
Infant/child death		
Muscle weakness		
Developmental delay, intellectual disability or learning disabilities		
Autism or behavior disorder (ADHD, ADD, etc)		
Premature ovarian insufficiency (early menopause) **or** adult tremors		
Other medical conditions		

Carrier Screening

Have you or your partner ever had carrier screening for genetic conditions (such as cystic fibrosis, Tay Sachs disease, sickle cell anemia, etc)?

☐ You   ☐ Your partner   Where? _____

If yes, were you or your partner found to be a carrier of a genetic condition?

☐ You   ☐ Your partner   Which condition(s)?_____

Questions?

Do you have any specific concerns you would like to discuss today? _____

Form 1.1: Genetic intake sheet

REFERENCES

1. ACMG Practice Guideline, Vol. 9 No. 6, Indications for genetic referral: a guide for healthcare providers, June 2007.
2. ACOG practice bulletin, Management of recurrent early pregnancy loss, Number 24, February 2001.
3. ACOG practice bulletin, Number 102, March 2009, Management of stillbirth.
4. ACOG practice bulletin, Number 163, May 2016, Screening for Fetal Aneuploidy.
5. ACOG practice bulletin, Number 162, May 2016, Prenatal Diagnostic Testing for Genetic Disorders.
6. ACOG Committee Opinion (Committee on Genetics), Number 616, January 2015, Newborn Screening and the Role of the Obstetrician-Gynecologist.
7. Newborn screening: toward a uniform screening panel and system—executive summary. American College of Medical Genetics Newborn Screening Expert Group. Pediatrics 2006; 117:S296–307.
8. S Kamath, President's Page, Newborn Screening in India, Indian Pediatrics, Vol. 52, May 2015.
9. Janice G. Edwards, Gerald Feldman, James Goldberg, Anthony R. Gregg, Mary E. Norton, Nancy C. Rose, Adele Schneider, Katie Stoll, Ronald Wapner, Michael S. Watson. Expanded Carrier Screening in Reproductive Medicine-Points to Consider, Obstetrics & Gynecology 2015; 125:653–62).

The Concept of Gestational Age—Viability, Preterm, Postdatism

Sarita Bhalerao, Kinjal Shah

INTRODUCTION—GESTATIONAL AGE

Gestational age (GA) refers to the length of pregnancy after the first day of the last menstrual period (LMP) and is usually expressed in weeks and days. This is also known as menstrual age. Conceptional age (CA) is the true fetal age and refers to the length of pregnancy from the time of conception.

The 3 basic methods used to help estimate gestational age (GA) are menstrual history, clinical examination, and ultrasonography. The first 2 are subject to considerable error and should only be used when ultrasonography facilities are not available.

In women who conceived following assisted reproduction techniques, the date of embryo transfer is known and may help to date the pregnancy accurately.

The median length of human pregnancy is 280 days of amenorrhea (from the first day of the LMP).

In women with regular cycles and a certain LMP, the EDD is calculated by adding 7 days to the first day of the LMP and adding 9 months. This is known as the Naegle rule.[1]

ULTRASONOGRAPHIC ASSESSMENT OF GESTATIONAL AGE

The introduction of obstetric ultrasonography in the early 1970s led to a marked improvement in the evaluation of fetal and placental anatomy, as well as fetal growth. Now, it is by far the most accurate technique for estimating gestational age (GA).

In the early first trimester, when no structures are visible within the gestational sac, GA may be estimated from the sac diameter. Several formulas can accomplish this. A common method is to measure the mean sac diameter (MSD), by calculating the mean of the 3 sac diameters. GA is then determined by consulting a table. An alternative simpler method is to add 30 to the sac size in millimeters, to give GA in days.

GA in the first trimester is usually calculated from the fetal crown-rump length (CRL). This is the longest demonstrable length of the embryo or fetus, excluding the limbs and the yolk sac.

The correlation between CRL and GA is excellent until approximately 12 weeks' amenorrhea.

The GA estimate has a 95% confidence interval of plus or minus 6 days, and it is most accurate between 7 and 10 weeks' amenorrhea.

In twin pregnancies, the CRL of the smaller fetus is more accurate in determining gestational age.

The biparietal diameter measured between 9 and 13 weeks' gestation has recently been shown to be at least as accurate as the CRL.

Fetal biometry in the second trimester can yield acceptably accurate estimates of GA from 12 to approximately 22 weeks of amenorrhea. Recent work has shown that the accuracy of ultrasonographic biometry at 12–14 weeks' gestation is at least as good as biometry performed after 14 weeks. The best parameters are the biparietal diameter (BPD) and the head circumference (HC), which are virtually linearly related to GA.

The femur length (FL) can also be used and is nearly as accurate as head measurements.

Fetal biometry in the third trimester is subject to much greater individual size variations than in the second trimester. Its accuracy for GA assignment is reduced considerably, and estimates may have confidence intervals of plus or minus 3 weeks.

PRETERM

According to the WHO[2], preterm babies are those born before 37 completed weeks of pregnancy.

There are sub-categories of preterm birth, based on gestational age:
- Extremely preterm (<28 weeks)
- Very preterm (28 to <32 weeks)
- Moderate to late preterm (32 to <37 weeks).

POSTDATISM

A postdate pregnancy usually is defined as a pregnancy lasting more than 294 days, or 42 completed weeks after the first day of the last menstrual period.

Postdate pregnancy, prolonged pregnancy, and postmaturity syndrome should not be used as interchangeable terms.

The postmaturity syndrome was described in detail by Clifford[3] and advocates the use of a staging system to quantify increasingly severe clinical manifestations of placental dysfunction. Stage I is typified by a long, lean infant with wrinkled, peeling skin. Stage II includes the clinical findings of stage I and adds greenish meconium staining of amniotic fluid, fetal skin, and placental membranes. Stage III is characterized by a high incidence of fetal distress and yellow-brown meconium staining, indicative of the presence of meconium for several days. The incidence of the postmaturity syndrome increases with the length of pregnancy; at 42 weeks, about 20% of fetuses have stigmata of postmaturity.

VIABILITY

According to studies between 2003 and 2005, 20 to 35 percent of babies born at 23 weeks of gestation survive, while 50 to 70 percent of babies born at 24 to 25 weeks, and more than 90 percent born at 26 to 27 weeks, survive. It is rare for a baby weighing less than 500 g (17.6 ounces) to survive.[4,5]

According to Websters Encyclopedic Unabridged Dictionary of the English Language, a viable fetus is a fetus which has reached a stage of development so as to be capable of living, under normal conditions, outside the uterus. Viability exists as a function of biomedical and technological capacities, which are different in different parts of the world. As a consequence, there is, at the present time, no worldwide, uniform gestational age that defines viability.

Today, the prospect of survival is only about 1 in 10 at 23 weeks, and if the child lives it is more likely to be handicapped. At 24 weeks the chance of a normal survivor is about 50%, and after this the odds are in favor of a normal survivor. Considering this data, intensive care should be an optional choice for fetuses at 23 and 24 weeks of gestation and should be offered to every fetus at 25 weeks or more.[6]

REFERENCES

1. Nägele FC. Lehrbuch der Beburtshilfe fur Hebammen. 3rd ed. 1836.
2. WHO fact sheet http://www.who.int/mediacentre/factsheets/fs363/en/.
3. Clifford Postmaturity, with placental dysfunction; clinical syndrome and pathologic findings. J Pediatr. 1954 Jan;44(1):1–13.

4. Tyson JE, Parikh NA, Langer J, Green C, Higgins RD (April 2008). "Intensive care for extreme prematurity--moving beyond gestational age". N. Engl. J. Med. 358 (16): 1672–81. doi:10.1056/NEJMoa073059. PMC 2597069. PMID 18420500.

5. Luke B, Brown MB (December 2006). "The changing risk of infant mortality by gestation, plurality, and race: 1989–1991 versus 1999–2001". Pediatrics. 118 (6): 2488–97. doi:10.1542/peds.2006-1824. PMID 17142535.

6. The American College of Obstetricians and Gynecologists (September 2002). "ACOG Practice Bulletin: Clinical Management Guidelines for Obstetrician-Gynecologists: Number 38, September 2002. Perinatal care at the threshold of viability". Obstet Gynecol. 100 (3): 617–24. PMID 12220792.

Basic Embryological Development

Ankesh R Sahetya, Nikit D Kadam

INTRODUCTION [1-3]

Embryonic development can be divided into different developmental phases. Like every biological developmental process, embryological development is a process in time and as such the visible stages appear as a continuous process in time

Embryology explains how structures in the body come to be what they are and explains their interrelationships. It holds the keys to where development may go wrong, resulting in anatomical defects, malfunction and biochemical abnormalities; all under the umbrella term congenital errors. This is an area of medical speciality, but even to non-medical persons, it can explain many common phenomena that we encounter in this field.

GAMETOGENESIS [3,4]

Process of formation and development of specialized generative cells—gametes. The male gamete is a spermatozoon (the short form 'sperm' is equally acceptable). It is a small cell, with very little cytoplasm and has a tail which makes it motile. The female gamete is an ovum. It is one of the largest cells in the body, with a diameter of over 100 μm. It is often called the 'egg'.

Gametes have 23 single chromosomes, as opposed to 23 pairs in other cells. The reduction in the number is due to meiotic cell division (meiosis).

In the male, some cells are 'set aside' in the testis for this purpose. These are called spermatogonia. After puberty, these cells divide by mitosis, some of the daughter cells undergo meiosis while others stay as spermatogonia to maintain the 'pool'. This process continues throughout life. Thus, a male can produce sperms indefinitely.

SPERMATOGENESIS (Fig. 3.1)

In contrast, the cells set aside in the female form a fixed pool. They begin meiotic division even before the female is born. But this division is not completed—a female is born with a fixed number of such cells, in a state of suspended meiosis. After puberty, approximately every month a few of these cells (oocytes) mature and proceed with meiosis. Normally only one of these is released for fertilisation by a sperm. This cell completes the second stage of meiosis after the entry of the sperm. Therefore we say that it is an oocyte that is released for fertilisation.

An oocyte has a non-cellular covering in addition to the cell membrane. This covering is the zona pellucida (the 'clear zone', so called because of its translucent pink appearance under the microscope).

Outside the zona pellucida is a layer of 'supporting cells'. One layer of these supporting cells surrounds the oocyte even as it is released. This layer appears like a radiating

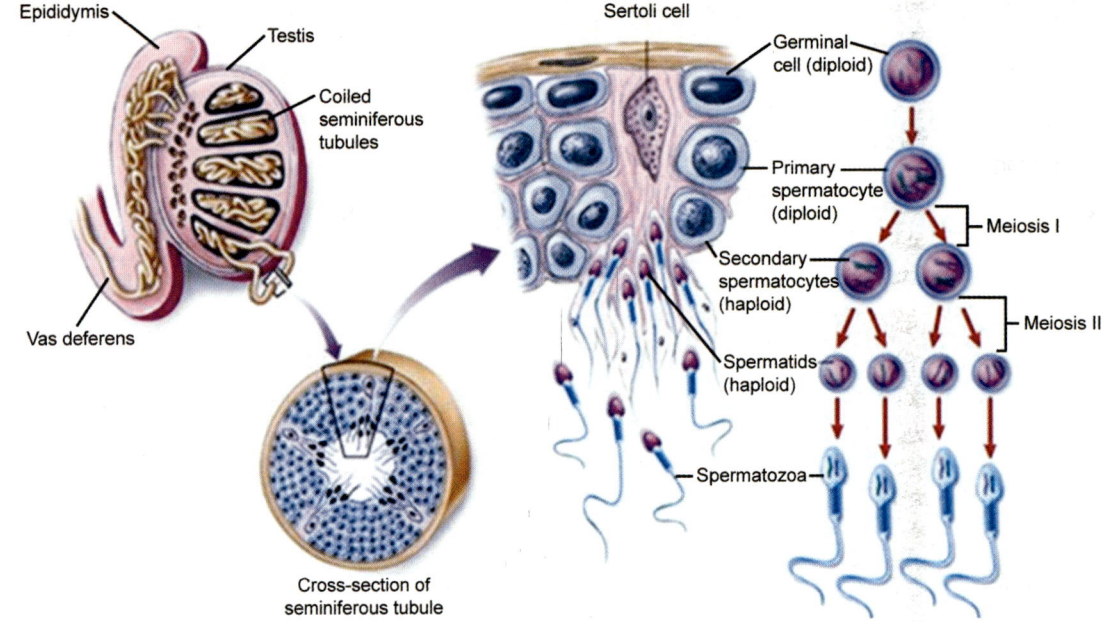

Fig. 3.1: Spermatogenesis

crown in a section and is indeed called the corona radiata.

FERTILIZATION[3-5]

Fertilization is a necessary condition for the development of a new human being. Thus, gametogenesis can be considered as a basic exemplary phenomenon which points to the creation of new possibility: The development of the zygote. At fertilization both oocyte and sperm cell each can live only one or two more days and then will die if fertilization does not take place.

Sperms are deposited deep into the vagina just below the opening of the uterus. Though sperms are motile, their speed is far from adequate to carry them to the oocyte in time. Remember that the oocyte has a 'long' journey along the length of the tube! Other factors operate in transporting sperms to the oocyte, probably movements in the smooth muscle wall of the uterus.

The nucleus and the scanty cytoplasm of the sperm is all in the 'head' of the sperm. At the very tip is a tiny structure called acrosome.

(*akron* = highest point, *soma* = body, here taken to mean a structure or a feature). The acrosome (shown as a tiny green line in Fig. 3.2) has enzymes which allow it to penetrate the zona pellucida. In the female reproductive tract, the acrosome undergoes changes which allow the acrosomal enzymes to be exposed. This is called capacitation. Fertilisation takes place most commonly in the dilated part of the uterine tube, the ampulla. When a sperm comes in contact with the zona pellucida, the acrosomal enzymes act on it and facilitate entry of the sperm. The cell membranes of the sperm and the oocyte fuse and the sperm enters the oocyte. The zona pellucida undergoes molecular changes which prevent the entry of any other sperm (Figs 3.3 and 3.4).

The fertilised oocyte immediately completes meiosis II and the second polar body is released. The fertilised cell (also called the zygote) has two haploid nuclei. These are called the male and female pronuclei. The pronuclei enlarge as the DNA within them is replicated. The nuclear envelopes disappear and the mixed lot of 23

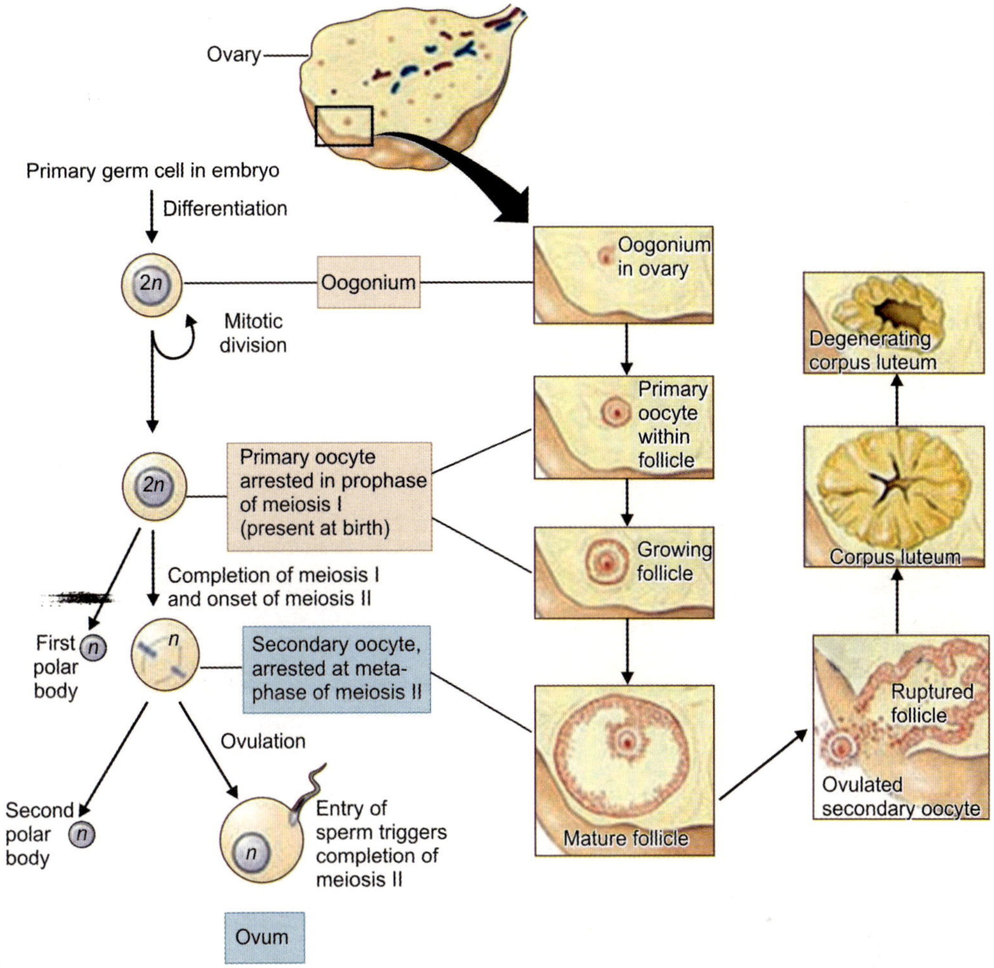

Ovary

Primary germ cell in embryo

Differentiation

2n — Oogonium — Oogonium in ovary

Mitotic division

2n — Primary oocyte arrested in prophase of meiosis I (present at birth) — Primary oocyte within follicle

Completion of meiosis I and onset of meiosis II

First polar body — n — Secondary oocyte, arrested at meta-phase of meiosis II — Growing follicle

Ovulation

Second polar body — n — Entry of sperm triggers completion of meiosis II — Mature follicle

Ovum

Degenerating corpus luteum

Corpus luteum

Ruptured follicle

Ovulated secondary oocyte

Fig. 3.2: Oogenesis

pairs of replicated chromosomes are lined up for separation.

Depending upon the sex chromosome present in the sperm, the sex of the individual is determined at fertilisation. The oocyte can only have an X chromosome, the sperm can have either X or Y.

The process of fertilization has several consequences. One is the fact that the number of chromosomes becomes 'normal' (diploid) after the fusion of the haploid gametes. Also the sexual identity of the organism is determined.

After fertilization, the zona pellucida undergoes a striking change. From the

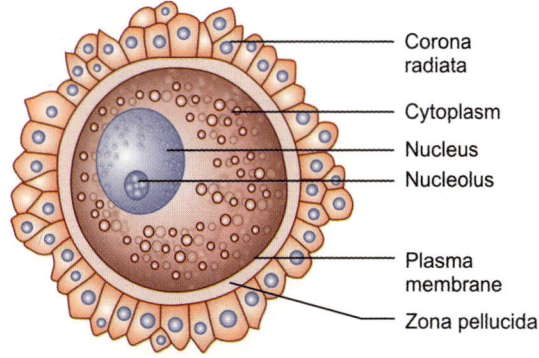

Corona radiata

Cytoplasm

Nucleus

Nucleolus

Plasma membrane

Zona pellucida

Fig. 3.3: Ovum

moment of fusion of the gametes onwards, the zona pellucida changes to become an impermeable membrane, separating the zygote physically and physiologically from the environment.

Cell Cleavage[3, 4]

As a result of fertilization, cleavage is initiated (Fig. 3.5). From this moment on, the zygote will develop a number of daughter cells called blastomeres. Characteristic for this process is the fact that during cleavage each new cell (blastomere) contains half the volume of cytoplasm of the mother cell. This process will go on until a specific ratio is reached between the cell volume and the volume of the nucleus: The volumetric ratio between nucleus and the cytoplasm has gained a value characteristic of the human organism in question. The amount of cytoplasm of the zygote is so large that this does not take place till the zygote reaches the 16-cell stage. This specific ratio is a necessity for the cell to have the bio-activity to continue, including protein synthesis. During the process of cleavage the overall size of the zygote does not change.

Compaction

About four days after fertilization, the formed cluster of blastomeres, now called the morula, undergoes the process of compaction. In compaction the peripheral cells begin to stick together in much closer contact than before, forming a more dense structure. It is a process comparable to epithelialisation. These peripheral cells will give rise to the trophoblast. The inner cell mass will give rise to the embryo proper, and is therefore called the embryoblast.

Growth

From the moment the blastula comes into contact with the uterine mucosa, the blastocyst grows rapidly. Growth, which now happens in the sense of increase in volume and cell mass, is obvious. This growth is not the same in different parts of the blastocyst. In this phase of development, growth primarily takes place at the periphery of the blastocyst.

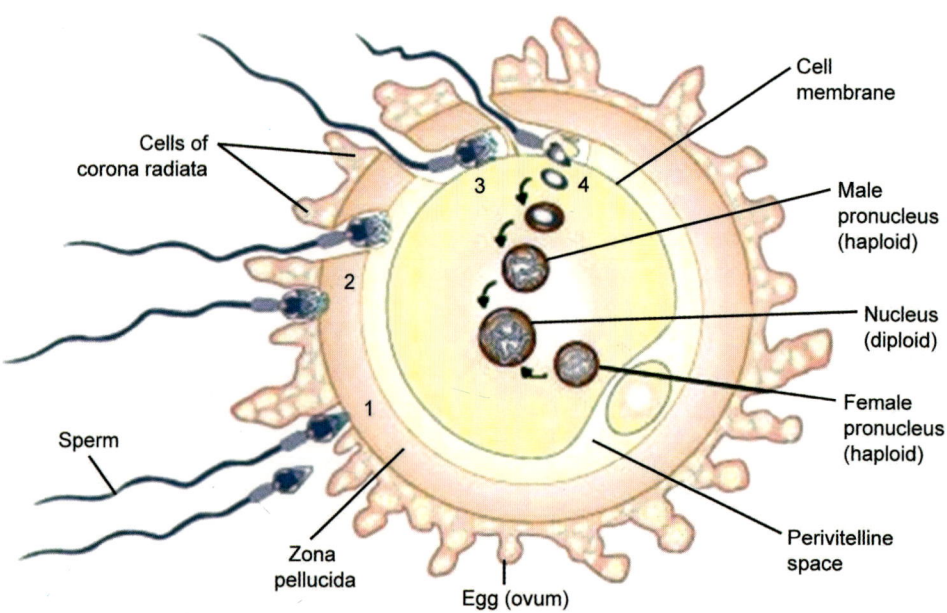

Fig. 3.4: Fertilization

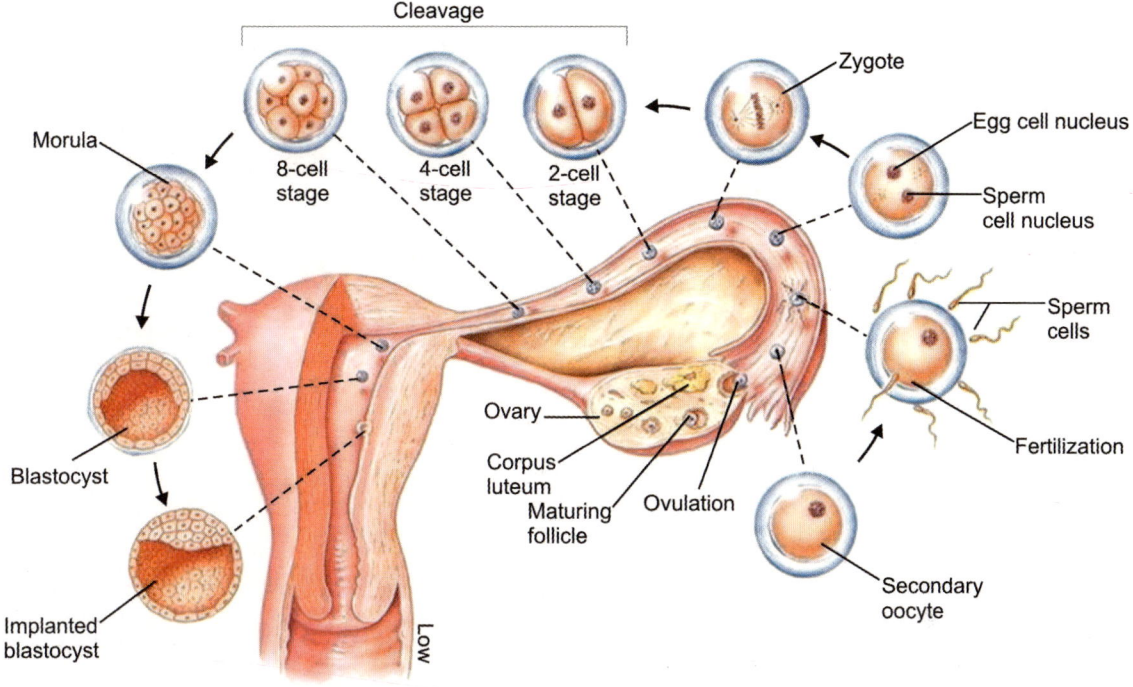

Fig. 3.5: Cell cleavage

Differentiation

The trophoblast develops rapidly in a centrifugal direction. At the same time it will undergo histological differentiation into two layers: The cytotrophoblast, consisting of well-differentiated cells, and the syncytiotrophoblast, in which individual cell structure is lost. The embryoblast also differentiates into two layers: Epiblast and hypoblast.

Implantation

The syncytiotrophoblast has an invasive property. When it comes in contact with the endometrium, it burrows into the endometrium by loosening the epithelium and destroying the connective tissue. Gradually the embryo sinks deeper into the endometrium until finally it is covered by the uterine epithelium, completing this process called implantation. Implantation begins at the end of the first week and is complete in the second week (Figs 3.6 and 3.7).

Metabolism[6-8]

In the syncytiotrophoblast, vacuoles fuse to form lacunae. Cells of the syncytiotrophoblast cause erosion to the maternal blood vessels. Cells in the lacunae will, from about the twelfth day, come into contact with maternal blood. Blood circulation can be seen as a phenomenon belonging to the form of increased metabolism that the blastocyst needs.

Production of HCG by trophoblast cells prevents degeneration of the corpus luteum. This means that the embryo is not only active on a morphological level but also on a physiological level. The production of HCG is the secretion process that enables the blastocyst to interact physiologically with the maternal organism, thereby tremendously enlarging its 'biological environment'. This process has its morphological counterpart in the peripheral expansion of the trophoblast. Both processes show an invasive tendency. Giving up its own boundaries, morphologically

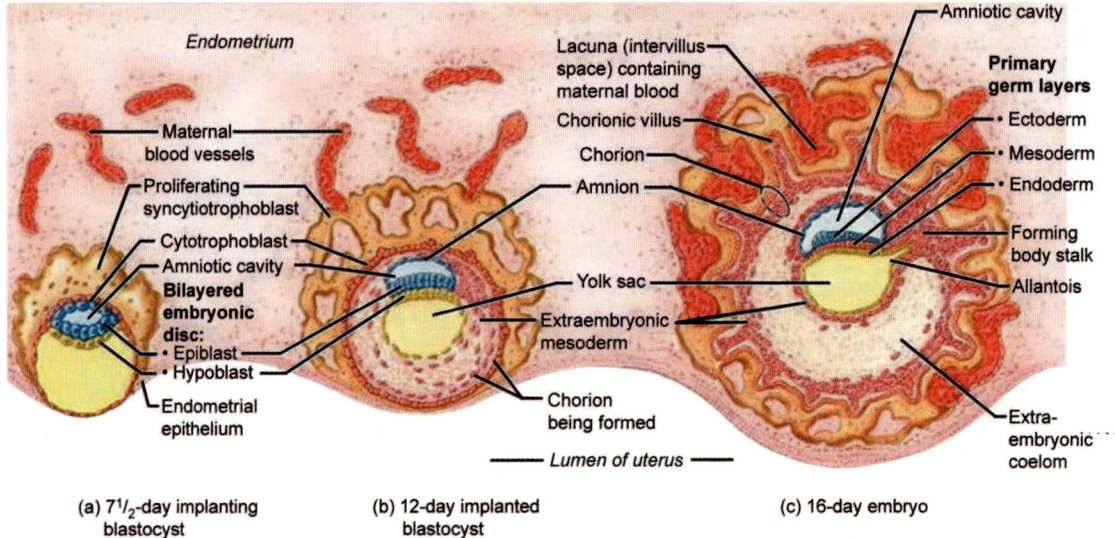

(a) 7½-day implanting blastocyst   (b) 12-day implanted blastocyst   (c) 16-day embryo

Fig. 3.6: Implantation

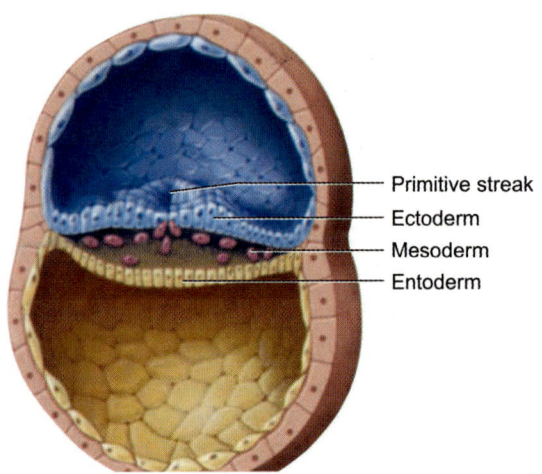

Fig. 3.7: Implantation

and physiologically, the embryo comes into contact with a wider periphery.

Bilaminar Germ Disc

The differentiation of the embryoblast in the second week results in the formation of the bilaminar germ disc, the next developmental stage of the primitive embryonic body.

Because of the round shape of amnion and primitive yolk sac, their contact surface constitutes a circular-shaped bilaminar disc.

This means that there is radial symmetry in the embryonic disc.

When we look at a 12-day-old blastocyst we can also find this radial symmetry in the total 'body form'.

The primitive streak, a temporary structure, plays a vital role in establishing the 'form' of the embryo.

The streak is seen as a faint ridge of the epiblast side of the embryo. At one end the streak has a knot-like swelling, the primitive node. This establishes the axis of the embryo. The part of the embryo beyond the node will the head end of the embryo. The other end, near the narrow tip of the streak is the tail end of the embryo.

Formation of the Germ Layers

Cells of the epiblast tend to 'heap up' at the primitive streak. From the streak, they sink deeper inside. The 'first wave' of these cells displace the hypoblast and forms the endoderm (also spelt entoderm). More cells proliferate and pour between the epiblast and the endoderm, forming the mesoderm. The remaining cells of the epiblast form the ectoderm. By the middle of the third week the embryo is thus trilaminar.

The head end of the embryo begins to differentiate first. The sequence of development is often described a cephalocaudal. (From head to tail, *kephale* = head, *cauda* = tail). Even as the head end begins to differentiate, movement of mesoderm in the caudal part of the embryo continues well into the 4th week.

Thus, in the final reckoning, the three layers are all formed by the epiblast, through the medium of the primitive streak.

The trilaminar embryo has two cavities around it. The cavity on the side of the ectoderm is the amniotic cavity, as mentioned earlier. The cavity, on the other side, undergoes modifications (details not essential).

It now faces the endoderm and is named the yolk sac.

Notochord

While these changes are taking place, cells from the tip of the primitive streak grow and form a strip in the midline, in the plane of the endoderm. This midline strip detaches itself and forms a solid column, the notochord. In vertebrates, it exists as a continuous column only during embryonic life. After the primitive streak disappears, it is the definitive 'axis' of the body. Moreover, it 'directs' a strip of ectoderm to develop into the nervous system. This process, where one structure influences the development of another, is called embryonic induction.

The Spread of the Mesoderm[7-10]

Mesodermal cells spread out laterally and towards the head end in a definite pattern. They follow arc-like courses between the epiblast (later ectoderm) and the endoderm. When the spread is complete, the mesoderm is seen mainly as three columns along the length of the embryo (Fig. 3.8).

Cells from the head end of the streak keep close to the midline, spreading in 'narrow' arcs. This column of mesoderm is close to the midline or the axis. This column is the paraxial

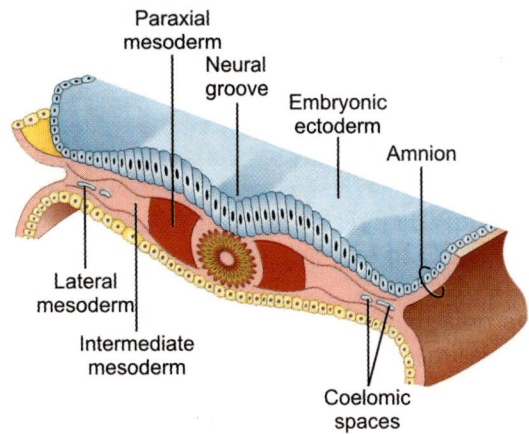

Fig. 3.8: Spread of mesoderm

("para + axial") mesoderm. Paraxial mesoderm organizes into segments in craniocaudal sequence forming somites, each of which has its own sclerotome (cartilage and bone), myotome (muscle) and dermatome (skin) and nerve component.

Cells from the middle part of the streak occupy a position just lateral to the paraxial mesoderm. This form a column called the intermediate mesoderm. Intermediate mesoderm forms segmental and unsegmented masses that ultimately generate components of the urogenital system.

Cells from the tail end (caudal part) of the streak spread out in broad arcs and finally form a plate close to the lateral boundaries of the embryo. This part of the mesoderm is called the lateral plate mesoderm. Lateral plate mesoderm splits into parietal and visceral layers. The parietal mesoderm along with overlying ectoderm will form the lateral and ventral abdominal wall, while that surrounding the intraembryonic cavity form the mesothelial membranes lining the peritoneal, pleural and pericardial cavities. The visceral layer with the underlying endoderm will form the wall of the gut, or form the serous membranes around the organs.

Cells migrating through the extreme tail end of the streak move outside the boundaries of the embryonic disc and form a part of the mesoderm outside the embryo proper. This is called extra-embryonic mesoderm. Some cells from the region near the head end are special—they reside at the cranial tip of the lateral plate and indeed form a horseshoe-like band at the head end of the embryo. These cells form the mesodermal precursor of the heart (cardiogenic area).

The spread of the mesoderm to the head end, however, leaves a small area where ectoderm and endoderm are in contact. This is called the oropharyngeal membrane. At the tail end there is a similar small area where ectoderm and endoderm are in contact with no mesoderm in between. This is called the cloacal membrane.

Fate of the Germ Layers[11-13]

The ectoderm gives rise to the epidermis—the epithelium of the skin. Note that the skin comprises epidermis and dermis, the latter is a connective tissue layer which develops from the dermatome part of the somites. Ectoderm also gives rise to hair and nails. Besides these, a specialised area of the ectoderm close to the midline forms the nervous system. The endoderm gives rise to the epithelial lining of the digestive tube, plus specialisations of this lining. Almost everything else in the body comes from the mesoderm. This includes almost all muscle tissue and all connective tissue including cartilage and bone, blood cells and cells of the defence system. The mesoderm also gives rise to parts of the urinary and reproductive systems and a part of the adrenal gland (Fig. 3.9).

Formation of the Neural Tube (Neurulation)[2,5,12]

A band of ectoderm across the midline becomes thickened to form a plate called the neural plate. Since it forms nervous tissue and it is a part of ectoderm, it is called neuro-ectoderm. As we have mentioned above, the notochord 'induces' the formation and subsequent modification of the plate. The plate becomes grooved along the midline. Its junction with the rest of the ectoderm is like a lip on either side. Soon, the groove closes to form a tube as if the edges are zipped together. The fusion begins in the middle of the embryo and proceeds towards both ends. The tube 'sinks' in the surrounding mesoderm, losing its contact with the ectoderm on the surface. The 'surface ectoderm' becomes continuous across the midline when the tube closes. The lips of the groove separate out as a long column on either side of the tube. These form the neural crest, which also sinks deeper along with the tube (Fig. 3.10).

The neural tube forms the brain and the spinal cord. The main derivatives of the crest are the ganglia of peripheral nervous system and sensory nerve fibres.

Ectoderm	Mesoderm	Endoderm
• Epidermis of skin and its derivatives (including sweat glands, hair follicles) • Epithelial lining of mouth and rectum • Sensory receptors in epidermis • Cornea and lens of eye • Nervous system • Adrenal medulla • Tooth enamel • Epithelium of pineal and pituitary glands	• Notochord • Skeletal system • Muscular system • Muscular layer of stomach, intestine, etc. • Excretory system • Reproductive system (except germ cells) • Dermis of skin • Lining of body cavity • Adrenal cortex	• Epithelial lining of digestive tract • Epithelial lining of respiratory system • Lining of urethra, urinary bladder and reproductive system • Liver • Pancreas • Thymus • Thyroid and parathyroid gland

Fig. 3.9: Germ layers

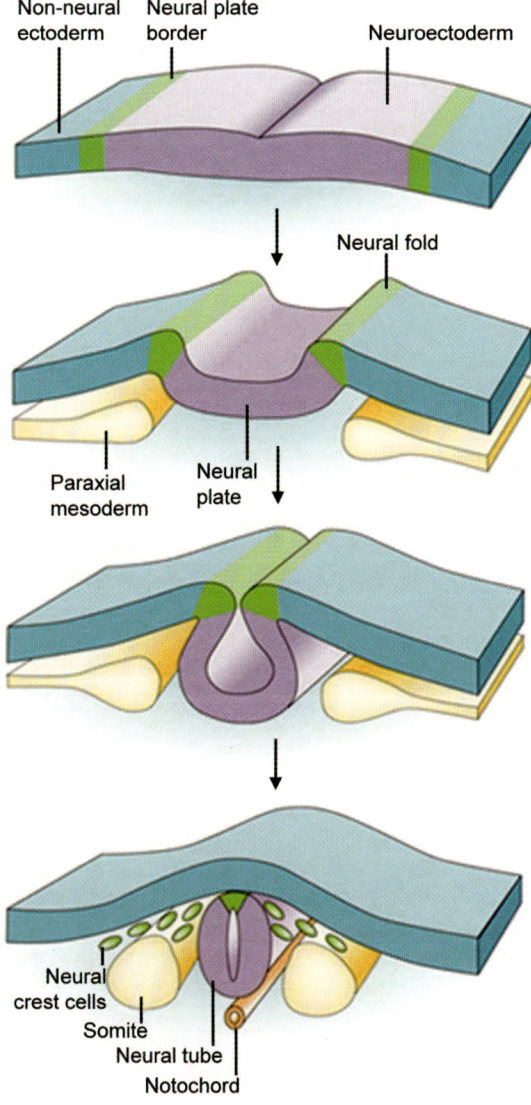

Fig. 3.10: Neurulation

Folding of the Embryo

The flat trilaminar embryo undergoes complex changes in form, with the head and tail ends and the sides folding up making it a three-dimensional tubular structure. Actually, it folds more like an envelope with four flaps—one flap at the head and tail ends each, and two flaps on the sides.

REFERENCES

1. Barsoum I, Yao HH-C: The road to maleness: from testis to wolffian duct. Trends Endocrinol Metab 2006; 17: 223–228.
2. Baskin LS: Urethral seam formation and hypospadias and others. Cell Tissue Res 2001;305: 379–387.
3. Basson MA: Sprouty1 is a critical regulator of GDNF/RET-mediated kidney induction and others. Dev Cell 2005; 8: 229–239.

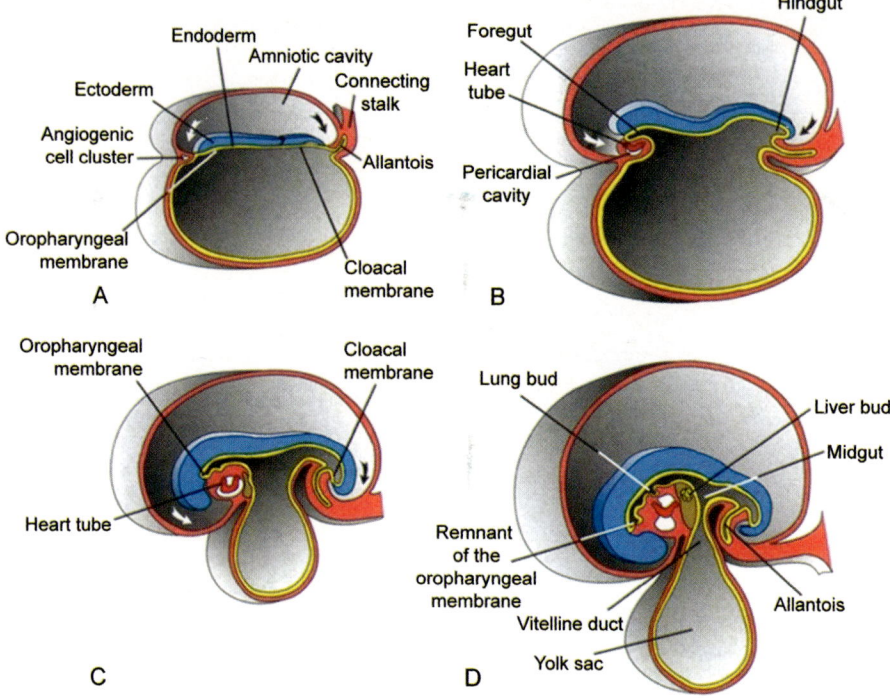

Fig. 3.11: Cephalo-caudal folding

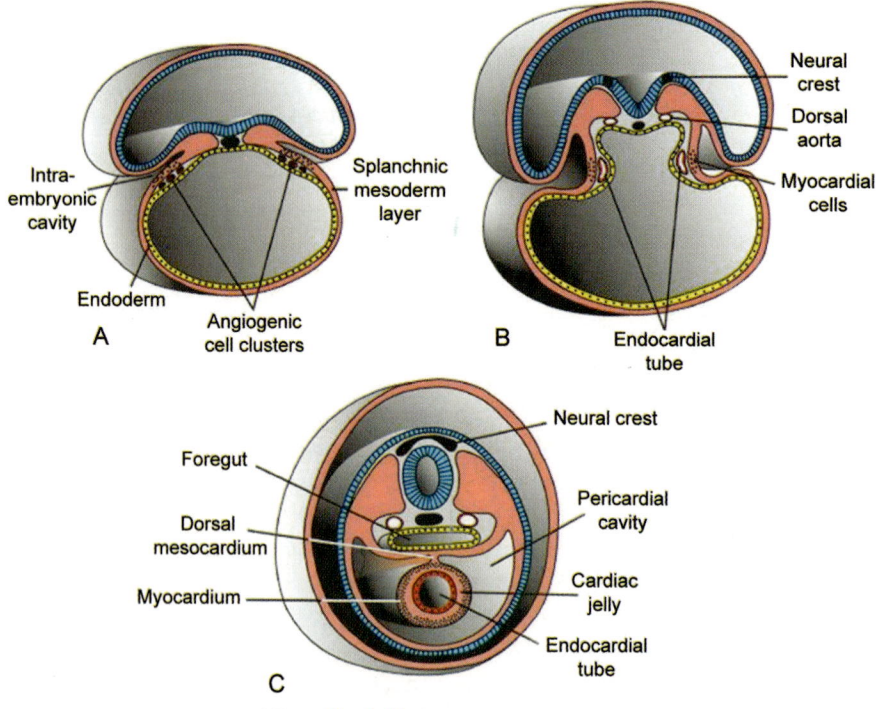

Fig. 3.12: Lateral folding

4. Bowles J, Koopman P. Retinoic acid, meiosis and germ cell fate in mammals. Development 2007; 134: 3401–3411.

5. Boyle S, de Caestecker M. Role of transcriptional networks in coordinating early events during kidney development. Am J Physiol Renal Physiol 2006; 291: F1-F8.

6. Brennan J, Capel B. One tissue, two fates: molecular genetic events that underlie testis versus ovary development. Nat Rev Genet 2004; 5: 509–521.

7. Brenner-Anantharam A. Tailbud-derived mesenchyme promotes urinary tract segmentation via BMP-4 signaling, and others. Development 2007; 134: 1967–1975.

8. Cartry J. Retinoic acid signalling is required for specification of pronephric cell fate. and others Dev Biol 2006; 299: 35–51.

9. Cohn MJ. Development of the external genitalia: conserved and divergent mechanisms of appendage patterning. Dev Dyn 2011; 240:1108–1115.

10. Combes AN. Endothelial cell migration directs testis cord formation. and others. Dev Biol 2009; 326: 112–120.

11. Costantini F. Renal branching morphogenesis: concepts, questions, and recent advances. Differentiation 2006; 74: 402–421.

12. Costantini F, Kopan R. Patterning a complex organ: branching morphogenesis and nephron segmentation in kidney development. Dev Cell 2010; 18: 698–712.

Biochemical Markers for Detection of Fetal Aneuploidies

Alka Kriplani, Aparna Sharma

Aneuploidy is having one or more extra or missing chromosomes. The loss or gain of large chromosomal segments disrupts significant amounts of genetic material and often results in a nonviable pregnancy . In the case of a surviving newborn, it results in congenital birth defects, failure to thrive and functional abnormalities like mild-to-severe intellectual disability.

Down's syndrome is the most frequently encountered autosomal trisomy with severe intellectual disability, heart defects and a spectrum of metabolic disorders. Although the propensity for occurrence is more in the elderly women, all pregnant women are at a risk of having a baby with trisomy 21.[1]

Because of its profound social and economic impact on the family, there is a great emphasis on early detection during pregnancy giving a choice to the parents as regards continuation of pregnancy. The availability of high definition ultrasound and serum markers for the detection of aneuploidies have revolutionized the concepts of prenatal care with respect to aneuploidy screening.

1. Prenatal Screening versus Diagnosis

The screening tests are done to assess whether a pregnant woman is at increased risk of having a fetus affected by aneuploidy. In contrast, prenatal diagnosis is intended to determine, with as much certainty as possible, whether a specific condition is present in the fetus. It is imperative to follow-up the various screening protocols with appropriate diagnostic tests.

2. Who should be Screened?

It is a common misconception that only elderly mothers should be offered screening for aneuploidies. However, it is now generally accepted that all women should be offered screening for aneuploidies and based on the results further diagnostic testing should be made available. A diagnostic test rather than screening can be offered for women of any age at high risk of Down syndrome or other fetal aneuploidies, such as women with:

- A previous pregnancy complicated by fetal trisomy
- At least one major or two minor fetal structural anomalies in the current pregnancy
- Chromosomal translocation, inversion, or aneuploidy in the pregnant woman or her partner.

3. Screening Modalities Available

- Ultrasound (first and second trimester)
- Biochemical markers (first and second trimester)

4. Biochemical Markers

There are five analytes (commonly referred to as markers) measured by the laboratory that are used to calculate the likelihood of a pregnancy being affected by Down's, Edwards' or Patau's syndromes—six if human chorionic gonadotropin (hCG) and its free beta subunit are considered as two separate analytes.[2]

Pregnancy Associated Plasma Protein A (PAPP-A)

PAPP-A is a large zinc glycoprotein produced by the placenta where it is thought to regulate the activity of factors responsible for the growth of the placenta. First trimester PAPP-A levels increase by 30 to 50 percent per week between 10 and 13 weeks of gestation.

Human Chorionic Gonadotropin (hCG)

hCG is a glycoprotein of 244 amino acids produced by the developing embryo and later by the placenta. It is a dimeric molecule composed of an alpha and a beta subunit. The alpha subunit is common to several other hormones [luteinising hormone (LH), follicle stimulating hormone (FSH) and thyroid stimulating hormone (TSH)]. The beta subunit is unique to hCG. Concentrations of hCG rise exponentially after conception reaching a peak at about 9–12 weeks, then falling to reach a plateau at about 20 weeks. Some of the beta subunit (less than 1% of the intact dimeric hCG) is free in the blood and this molecule can be measured by the laboratory as a distinct entity. Free beta-hCG levels decline more quickly than total beta-hCG (by about 10 to 30 percent and 5 to 10 percent per week, respectively).

Alpha Fetoprotein (AFP)

AFP is a glycoprotein of 591 amino acids produced by the yolk sac and the fetal liver. Its level in fetal serum increases until the end of the first trimester and then gradually decreases. Concentrations are much lower in maternal serum but they continue to rise until about week 32. in the early second trimester, AFP increases by 15 to 20 percent per week.

Unconjugated Estriol (uE3)

Estriol is one of the three main steroid hormones produced by the feto-placental unit during pregnancy. It is made in the placenta from the 16-hydroxydehydroepiandrosterone produced by the fetal liver. Once in the maternal circulation most of the estriol undergoes conjugation with glucuronides or sulphate but about 10% remains as the unconjugated form. In the early second trimester estriol increases by 20 to 25 percent per week.

Inhibin-A (Inh A)

Inhibin-A is a dimeric molecule comprised of an alpha and a beta polypeptide chain linked by a disulphide bridge. (A similar molecule called inhibin-B has the same alpha chain but a different beta chain.) It is produced by the corpus luteum and the placenta during pregnancy with levels increasing during the first trimester, then declining to reach a plateau in the second trimester before increasing again in the third trimester. Inhibin A exhibits a shallow U-shaped curve with its nadir at 17 weeks of gestation.

5. Quantifying Analyte Levels

Analyte values are initially measured as standardized mass units. Each mass value is converted to the screened woman's gestation-specific multiple of the median (MoM), since the serum levels of each analyte are constantly changing during the periods of gestation in which screening is done. Patient data are usually expressed as MoM values after population-based medians are established, thereby eliminating the effects of assay differences and gestational age. Multiples of the median also provide a relatively simple way to compare an individual to the entire population being screened.

Changes in Serum Marker Levels[3]

Figure 4.1 shows changes in markers in each of the aneuploidies.

The Screening Protocols

A combination of ultrasound and biochemical markers have been used to achieve the maximal detection rate with a minimum false positivity. Table 4.1 shows the commonly available screening protocols.

	NT	CRL	FHR	β-hCG	PAPP-A
Trisomy 21	↑ 2.5	↔	↓	↑ 2.2	↑ 0.5
Trisomy 18	↑ 3.5	↔	↓	↑ 0.3	↑ 0.2
Trisomy 13	↑ 2.5	↔	↓	↑ 0.5	↑ 0.3
Turner's	↑ 7.0	↔	↓	↔	↑ 0.5
Triploidy	↑ 2.5	↔	↓	↑ 8.5	↑ 0.8

Fig. 4.1: Changes in markers

Interpretation of Prenatal Screening Tests

The purpose of prenatal tests is to identify the population at high risk for aneuploidy. The results of the screening test however should be evaluated using a patient specific risk based assessment. A woman's priori risk is determined based on her chronological age at the estimated date of delivery and history of previous Down syndrome pregnancy. This risk is then increased (or decreased) by a factor called the "likelihood ratio" (LR).

The LR is determined by comparing each of her serum marker MoM values with the reference distributions, after accounting for the degree of independence between each pair of markers (measured as an R value after log transformation of the MoMs). The final reported risk is her calculated patient-specific risk of having a fetus affected by Down syndrome in that pregnancy.

Table 4.1	The screening protocol	
Test	Markers	Time
Combined test	1. PAPP-A and beta-hCG 2. NT scan	• 10 to 13.6 weeks (with free beta-hCG) • 11 to 13.6 weeks (with total beta-hCG)
Quadruple marker test	1. AFP 2. uE3 3. beta-hCG 4. Inh A	 • 15 to 18 weeks
Full integrated test	NT and PAPP-A:10–13 weeks AFP, uE3, hCG, inh A: 15–18 weeks.	First and second trimester USG and biochemistry integrated to give a risk score
Serum integrated test	PAPP-A, AFP, uE3, beta-hCG, inh A	First and second trimester biochemistry integrated to give a risk score
Step-wise sequential testing	First trimester portion of integrated screen Offer CVS if high risk (e.g. ≥1 in 50) Low risk: Second trimester portion of the integrated test.	

hCG: Human chorionic gonadotropin

inh A: Inhibin A

PAPP-A: Pregnancy associated plasma protein A

AFP: Alpha fetoprotein

uE3: Unconjugated estriol

Screen-positive First-trimester Combined Test Results

A screen-positive test result indicates that the woman's risk of having a child with Down syndrome is equal to, or exceeds, a specific cut-off level that was predetermined by the laboratory based on the performance characteristics of the chosen screening test. A typical cut-off for the combined test is the term risk of Down syndrome of ≥ 1 in 250–300. This is associated with a false positive rate (FPR) of about 5 percent.

Screen-positive Integrated Test Results

A positive test result means the patient's risk of having a baby with Down syndrome is greater than a specified cut-off level.

A typical cut-off for the integrated test is a midpregnancy risk of Down syndrome of ≥1 in 100. This cut-off is associated with a FPR of 1 to 2 percent and odds of Down syndrome among the screen-positive women is about 1 in 5 to 1 in 10.

Screen-negative First-trimester or Integrated Test Results

A negative test result means the patient's risk of having a baby with Down syndrome is less than a specified cut-off level; it does not exclude the possibility of Down syndrome. With regard to Down syndrome screening, no further testing is recommended.

Efficacy of Screening Methods

Figure 4.2 shows the detection rate of each of the screening methods given a fixed false positive rate of 5%. Increasing the detection rate will increase the false positivity of the tests. By using a combination of first and second trimester markers like the integrated or serum integrated tests, detection rate is increased at a remarkably lower false positive rates.

Desirable Screening Protocols in Different Clinical Scenarios[7,8]

Antenatal Booking in First Trimester Low Risk Women

Flowchart 4.1

Women who Present in Second Trimester

Flowchart 4.2

Women at High Risk of Aneuploidies

- Maternal age at delivery 35 or more
- Fetal ultrasonographic findings indicating an increased risk of aneuploidy

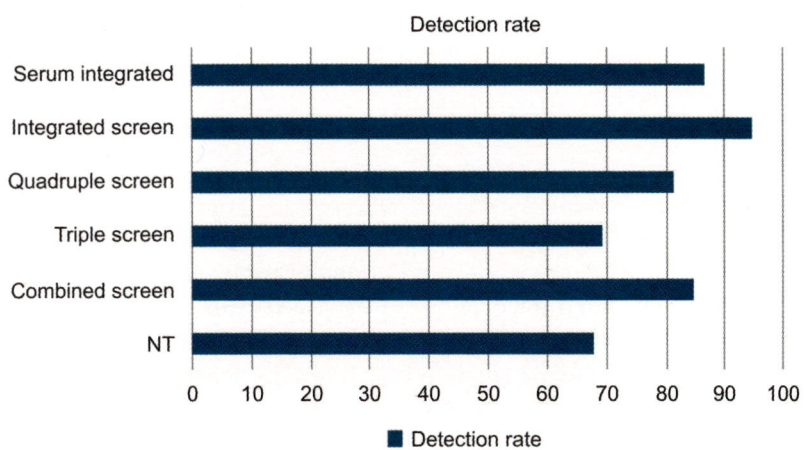

Fig. 4.2: Detection rates of various screening protocols for a fixed false positive rate of 5%[5,6]

Flowchart 4.1: Antenatal booking in first trimester low risk women

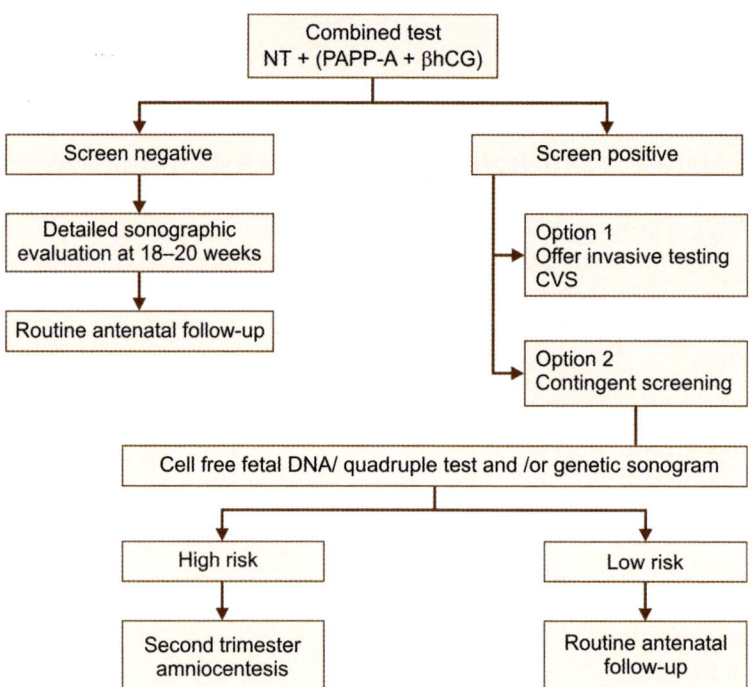

Flowchart 4.2: Women who present in second trimester

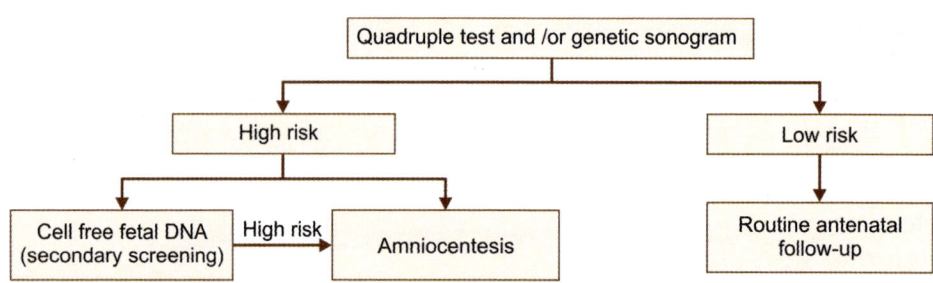

- Personal or family history of prior pregnancy with a trisomy
- Parental balanced robertsonian translocation with increased risk of T21 ot T13

Non-invasive prenatal testing using cell free DNA[9] can be offered as first line screening for aneuploidies in these conditions. Some couples would like to go for definitive invasive testing in such scenarios as well. Due consideration should be given to the couples' perspective during the counselling and appropriate test can be offered.

Even in women who want to undergo definitive invasive tests for aneuploidies, the routine NT scan in the first and detailed anomaly scan in the second trimester and trimester specific biochemical screen can be in place as they provide information on conditions other than aneuploidies also.

Factors affecting Biochemical Markers

Vaginal Bleeding

Vaginal bleeding does not significantly alter the levels of biochemical markers.

Twins[2]

Vanishing twins: Vanishing twin can contribute to biochemical analyte levels for many weeks. Hence, the risk calculation based on the maternal age and nuchal translucency only (i.e. without biochemistry).

Combined test in twins: For women screened using the combined test, where a dichorionic twin pregnancy is identified the risks will be reported for each fetus. In a monochorionic twin pregnancy both fetuses are either affected or unaffected so the risk will be the same.

Quadruple screening in twin pregnancies: This can be offered to women who present for the first time in the second trimester or where the nuchal translucency (NT) could not be measured in the first trimester.

Monochorionic twins

- The risk of a T21 birth from a monochorionic pregnancy is lower than that from a singleton pregnancy due to a higher fetal loss rate amongst affected pregnancies
- The performance of screening in monochorionic twins is comparable to that of

singletons: A detection rate is 80% for a standardised screen positive rate of 3%

Dichorionic twins

- The risk of a T21 birth of at least one baby from a dichorionic twin pregnancy is higher than that from a singleton pregnancy
- In dichorionic twins, where one is affected and the other unaffected, the performance is poorer due to the markers being less discriminatory but is better than using maternal age only, where the detection rate is only 30% for a 5% screen positive rate
- In dichorionic twins the detection rate is 40–50% for a standardised screen positive rate of 3%

Other clinical conditions

Table 4.2 shows alterations in biochemical markers in common clinical conditions.

SUMMARY

1. Prenatal screening for Down syndrome should be offered to all pregnant women.
2. A diagnostic test rather than screening is a reasonable choice for women at high risk of Down syndrome.
3. Prenatal screening programs based on maternal serum and ultrasound testing can detect up to 95 percent of pregnancies affected by Down syndrome with a false positive rate of 5 percent.
4. The first trimester combined test is desirable for early aneuploidy screening.
5. The full integrated test has the highest detection rate for Down syndrome, the

Table 4.2	Alterations in biochemical markers in common clinical conditions				
	PAPP-A	β-hCG	AFP	uE3	Inhibin-A
Diabetes[10]			Decreased (20%)	Decreased (5–10%)	
IVF[11]	Decreased (10–20%)	Increased		Decreased	Increased
Smoking[12]	Decreased	Decreased	Increased	Decreased	Increased
Increasing maternal weight[13]	Decreased	Decreased	Decreased	Decreased	Decreased

lowest rate of procedure-related losses per woman screened.

6. Screen positive women should be further evaluated by diagnostic tests for confirmation.

REFERENCES

1. Sherman SL, Allen EG, Bean LH, Freeman SB. Epidemiology of Down syndrome. Ment Retard Dev Disabil Res Rev. 2007;13(3):221–7.

2. NHS Down's, Edwards' and Patau's syndromes screening programme: Handbook for Laboratories, version 1.0, NHS Fetal Anomaly Screening Programme (Public Health England) April 2015.

3. Palomaki GE, Messerlian G, Canick JA. A summary analysis of Down syndrome markers in the late first trimester. In: Advances in Clinical Chemistry, Makowski G (Ed), Academic Press, 2007. Vol 43, p.177.

4. Norton ME, Jelliffe-Pawlowski LL, Currier RJ. Chromosome abnormalities detected by current prenatal screening and noninvasive prenatal testing. Obstet Gynecol 2014; 124:979.

5. Wald NJ, Rodeck C, Hackshaw AK, et al. First and second trimester antenatal screening for Down's syndrome: the results of the Serum, Urine and Ultrasound Screening Study (SURUSS). Health Technol Assess 2003; 7:1.

6. ACOG Committee on Practice Bulletins. ACOG Practice Bulletin No. 77: screening for fetal chromosomal abnormalities. Obstet Gynecol 2007; 109:217.

7. Malone FD, Canick JA, Ball RH, et al. First-trimester or second-trimester screening, or both, for Down's syndrome. N Engl J Med 2005; 353:2001.

8. Malone FD, Canick JA, Ball RH, et al. First-trimester or second-trimester screening, or both, for Down's syndrome. N Engl J Med 2005; 353:2001.

9. Committee Opinion No. 640: Cell-Free DNA Screening For Fetal Aneuploidy. Obstet Gynecol 2015; 126:e31.

10. Spencer K, Cicero S, Atzei A, et al. The influence of maternal insulin-dependent diabetes on fetal nuchal translucency thickness and first-trimester maternal serum biochemical markers of aneuploidy. Prenat Diagn 2005; 25:927.

11. Wald NJ, White N, Morris JK, et al. Serum markers for Down's syndrome in women who have had in vitro fertilisation: implications for antenatal screening. Br J Obstet Gynaecol 1999; 106:1304.

12. Bartels I, Hoppe-Sievert B, Bockel B, et al. Adjustment formulae for maternal serum alpha-fetoprotein, human chorionic gonadotropin, and unconjugated oestriol to maternal weight and smoking. Prenat Diagn 1993; 13:123.

13. Huang T, Meschino WS, Okun N, et al. The impact of maternal weight discrepancies on prenatal screening results for Down syndrome. Prenat Diagn 2013; 33:471.

Role of Invasive Tests (Amniocentesis, CVS and Cordocentesis) in Perinatology

Sachin Nichite, Pragya Tripathi Nichite

INTRODUCTION

Prenatal diagnosis is the science of identifying structural or functional abnormalities in the fetus. Diagnostic prenatal testing is mostly done with non-invasive techniques like ultrasound and biochemical markers. But in cases where these non-invasive tests show congenital anomalies, positive aneuploidy screening or previous positive history etc., invasive testing to obtain fetal cells for direct chromosomal/mutation testing may be necessary. Most commonly used procedures are: Amniocentesis, CVS (chorionic villus sampling) and cordocentesis/fetal blood sampling (FBS). The choice of test depends upon various factors like gestation age, expertise and indication.

PRE-TEST COUNSELLING

For any prenatal test, pre-test/pre-procedure counselling is vital and mandatory.

Following points to be covered and documented:

- Indication for offering invasive test
- A few lines about screening test which is done if applicable
- All possible options of invasive tests and difference between them in terms of accuracy, timing and complications
- Timing of procedure
- Simplified version of events during procedure
- Post procedure observation and monitoring
- Accuracy and limitations of particular test being performed
- Cost of procedure and investigations
- Indications of seeking medical advice following the test
- National and local estimated risks of procedure related pregnancy loss.

PRE-PROCEDURE CONSIDERATIONS

1. Rh-status of the mother
2. Universal maternal screening for HIV, HCV, HBV is not recommended as the risk of transmission is negligible
3. Antibiotic prophylaxis before procedure is currently not recommended although advised by many experts[28]
4. Ultrasound before procedure to check for Gestation age, number of foetuses, viability, amniotic fluid and placental location
5. Main principles of asepsis need to be followed
6. Local anaesthesia is not recommended for amniocentesis. In CVS, it can be used to reduce discomfort of the patient and before FBS may be considered in order to reduce risk of maternal movements.

AMNIOCENTESIS

Amniocentesis refers to trans-abdominal aspiration of amniotic fluid from the uterine cavity.

Indications

1. Positive aneuploidy screening test
2. Increased risk of fetal chromosomal/ genetic pathology based on previous obstetrical or family history.
3. Single or multiple major congenital anomalies
4. Maternal transmittable infectious disease.

Timing

All scientific and professional bodies currently recommend that amniocentesis should be performed at or beyond 15 + 0 weeks of gestation.[1, 10, 15]

Early Amniocentesis

Early amniocentesis should not be performed (before 15 weeks, Fig. 5.1) as multicentre RCT showed that early amniocentesis was associated with significantly higher rate of fetal loss, fetal talipes and post procedure amniotic fluid leakage.[8, 9]

Important aspects of Technique

- A 20–22 G needle to be inserted trans abdominally under continuous USG guidance[1–4]
- Entry into amnion should be firm to prevent tenting of amniotic fluid[2]

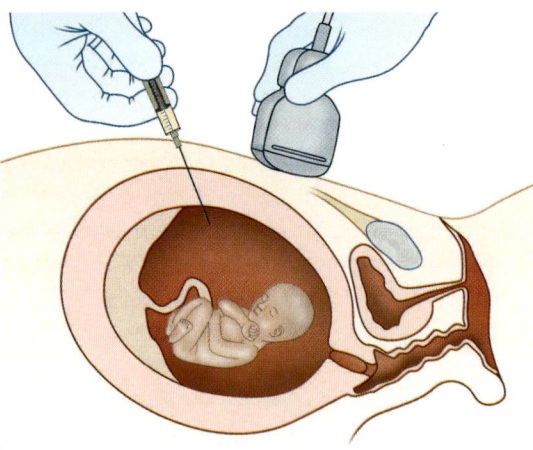

Fig. 5.1: Early amniocentesis

- Needle entry through the placental cord insertion site must be avoided and, if technically feasible, avoidance of the placenta is preferable (especially in Rh negative woman)[1–7]
- To minimise maternal contamination, first 2 ml of fluid should be discarded[15]
- Approximately 20–30 ml fluid (depending upon indication) is aspirated.

Complications

- **Fetal loss:** The additional risk of fetal loss in comparison with controls has been reported to vary from 0.1 to 1%, with recent reports being closer to the lower limit[11,12]
- **Amniotic fluid leakage:** The risk of membrane rupture after amniocentesis is 1–2%.[13–15] But the prognosis in these cases may be better than that in cases of spontaneous PPROM[16]
- **Chorioamnionitis:** The risk of chorio-amnionitis and uterine infection after genetic amniocentesis is low (<0.1%)[15]
- **Needle injury:** In modern techniques, occurrence of needle injury to the fetus is extremely rare[15]
- Serious maternal complications like sepsis are events.

Risk Factors for Complications

- Multiple attempts (>3 attempts)
- Blood stained or discolored fluid[10,17]
- Presence of fetal anomalies[10]

CHORIONIC VILLUS SAMPLING (CVS)

CVS is withdrawl of trophoblastic cells from placenta (Fig. 5.2).

Indications

1. Positive aneuploidy screening test
2. Increased risk of fetal chromosomal/ genetic pathology based on previous obstetrical or family history
3. Single or multiple major congenital anomalies.

Timings

CVS should be performed after 10 + 0 gestational weeks.[1,15]

Important aspects of Technique

- Needle should be inserted into placenta under continuous USG guidance
- Two techniques
  - Free hand or using biopsy adapter
  - The choice should be made according to operator experience or preference
- Access to placenta may be trans-abdominal or trans-cervical.
  Studies show that fetal loss and successful sampling rates were similar between two methods.[18]
- Trans-abdominal
  i. A needle of 17–20 G may be used[19]
  ii. Once the needle has reached the target, between 1 and 10 back-and-forth movements are performed, while the vacuum is maintained.
- Trans-cervical biopsy forceps are inserted trans-vaginally through cervical canal to the trophoblastic area or catheter with plastic or metal stylet under syringe aspiration may be used.
- The amount of villi obtained must be checked visually. A minimum amount of 5 mg villi in each sample is required to achieve a valid result.[2]

Complications

- **Fetal loss:** The additional risk of fetal loss has been reported to vary between 0.2% and 2%[1, 11]
- **Vaginal bleeding:** Reported to occur in 10% of cases and occurs more frequently after trans-cervical approach[22, 23]
- **Limp reduction/Oro mandibular hypoplasia:** The limbs or mandible seem to be more susceptible to vascular disruption before 10 weeks[2, 20, 21]

Risk Factors for Complications

- >2 attempts of needle insertion
- Heavy bleeding during procedure
- Gestation age <10 weeks
- Fetal structural anomalies and increased NT
- Lower levels of PAPP-A

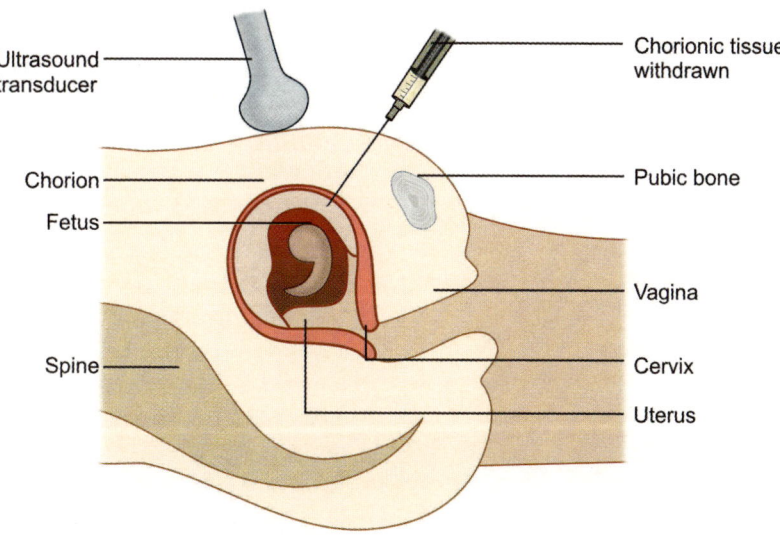

Fig. 5.2: CVS withdrawal

FETAL BLOOD SAMPLING/CORDOCENTESIS

The term cordocentesis refers to the ultra-sound guided puncture of the umbilical cord (umbilical vein) for either diagnostic (FBS) or therapeutic (intrauterine transfusion) purposes (Fig. 15.3).

Indications

In current practice, the following indications have become extremely rare and replaced by amniocentesis/CVS—karyotype, genetic testing, infections and metabolite/hormone studies.

Commonest indications nowadays are investigation for chromosomal mosaicism after amniocentesis or haematological assessment of the fetus (quantification of fetal anaemia or platelet /lymphocyte count).

Timing

FBS should be performed beyond 18 + 0 weeks.[24]

Important aspects of Technique

- 20–22 G needle is introduced trans abdominally under continuous guidance and inserted into umbilical vein.
- If anterior placenta, puncture of the cord at the level of placental insertion and if the placenta is posterior, a free loop of the cord or intra-abdominal portion of the umbilical vein is sampled.
- Care should be taken to avoid the umbilical arteries
- Aspiration by syringe is attempted by an assistant or operator until blood is obtained.

Complications

The risk of fetal loss after FBS is between 1 and 2%.[25–27]

Risk Factors for Fetal Loss

- Fetal anomalies
- FGR
- Gestation age <24 weeks

INVASIVE PROCEDURES IN MULTIPLE PREGNANCIES

Amniocentesis in Twins

In dichorionic twins
- Sampling of both amniotic sacs is recommended.
- Two techniques—two puncture technique and single puncture technique
- The risk of fetal loss has not been shown to be increased with 2 puncture compared with single puncture.[29]

In monochorionic diamniotic twins
Sampling of single sac is warranted when chorionicity has been clearly determined and fetal growth and anatomy is concordant.[30]

CVS in Twins

In dichorionic twins
- Either two separate punctures or a single puncture technique, sampling the two placentae in sequence may be performed.
- To reduce risk of unreliable or inaccurate results, placental sampling near cord insertion and avoidance of area around the dividing membrane is recommended.

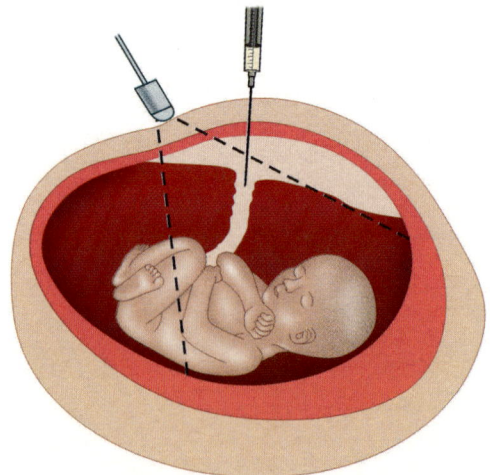

Fig. 5.3: Fetal blood sampling

In monochorionic twins

- A single sampling approach around amniotic equator is warranted.
- A shift to amniocentesis with two sampling approach must be considered after IVF or in case of discordant anomaly/growth.[30]

Post-procedure Instructions

- Bedrest or limiting physical activity for 12–24 hr is optional but usually advised.
- Use of paracetamol may be considered
- Administration of progesterone or tocolytic drugs has not been demonstrated to yield clear benefit though commonly advised
- Prophylactic anti-D immunoglobulin should be given to non-sensitized woman within 72 hrs. post-procedure unless the alleged father of the fetus is Rh negative.

REFERENCES

1. Royal College of Obstetricians & Gynaecologists. Amniocentesis and Chorionic Villus Sampling. Green-top Guideline No. 8, June 2010.

2. Wilson RD, Davies G, Gagnon A, Desilets V, Reid GJ, Summers A, Wyatt P, Allen VM, Langlois S; Genetics Committee of the Society of Obstetricians and Gynaecologists of Canada. Amended Canadian guideline for prenatal diagnosis (2005) change to 2005-techniques for prenatal diagnosis. J Obstet Gynaecol Can 2005; 27: 1048–1062.

3. Tabor A, Alfirevic Z. Update on procedure-related risks for prenatal diagnosis techniques. Fetal Diagn Ther 2010; 27: 1–7.

4. Cruz-Lemini M, Parra-Saavedra M, Borobio V, Bennasar M, Gonc´e A, Mart´inez JM, Borrell A. How to perform an amniocentesis. Ultrasound Obstet Gynecol 2014; 44: 727–731.

5. Athanasiadis AP, Pantazis K, Goulis DG, Chatzigeorgiou K, Vaitsi V, Assimakopoulos E, Tzevelekis F, Tsalikis T, Bontis JN. Comparison between 20G and 22G needle for second trimester amniocentesis in terms of technical aspects and short-term complications. Prenat Diagn 2009; 29: 761–765.

6. Uludag S, Aydin Y, Ibrahimova F, Madazli R, Sen C. Comparison of complications in second trimester amniocentesis.

7. Tabor A, Philip J, Madsen M, Bang J, Obel EB, Nørgaard-Pedersen B. Randomised controlled trial of genetic amniocentesis in 4606 low-risk women. Lancet 1986; 1: 1287–1293.

8. Randomised trial to assess safety and fetal outcome of early and midtrimester amniocentesis. The Canadian Early and Mid-trimester Amniocentesis Trial (CEMAT) Group. Lancet 1998; 351: 242–247.

9. Farrell SA, Summers AM, Dallaire L, Singer J, Johnson JA, Wilson RD. Club foot, an adverse outcome of early amniocentesis: disruption or deformation? CEMAT. Canadian Early and Mid-Trimester Amniocentesis Trial. J Med Genet 1999; 36: 843–846.

10. Kahler C, Gembruch U, Heling KS, Henrich W, Schramm T; DEGUM. [DEGUM guidelines for amniocentesis and chorionic villus sampling]. Ultraschall Med 2013; 34: 435–440.

11. Akolekar R, Beta J, Picciarelli G, Ogilvie C, D'Antonio F. Procedure-related risk of miscarriage following amniocentesis and chorionic villus sampling: a systematic review and meta-analysis. Ultrasound Obstet Gynecol 2015; 45: 16–26.

12. Wulff CB, Gerds TA, Rode L, Ekelund CK, Petersen OB, Tabor A; Danish Fetal Medicine Study Group. The risk of fetal loss associated with invasive testing following combined first trimester risk screening for Down syndrome—a national cohort of 147 987 singleton pregnancies. Ultrasound Obstet Gynecol 2016; 47: 38–44.

13. Wilson RD, Johnson J, Windrim R, Dansereau J, Singer J, Winsor EJ, Kalousek D. The early amniocentesis study: a randomized clinical trial of early amniocentesis and midtrimester amniocentesis. II. Evaluation of procedure details and neonatal congenital anomalies. Fetal Diagn Ther 1997; 12: 97–101.

14. Philip J, Silver RK, Wilson RD, Thom EA, Zachary JM, Mohide P, Mahoney MJ, Simpson JL, Platt LD, Pergament E, Hershey D, Filkins K, Johnson A, Shulman LP, Bang J, MacGregor S, Smith JR, Shaw D, Wapner RJ, Jackson LG. Late first-trimester invasive prenatal diagnosis: results of an international randomized trial; NICHD EATA Trial Group. Obstet Gynecol 2004; 103: 1164–1173.

15. American College of Obstetricians and Gynecologists. ACOG Practice Bulletin No. 88, December 2007. Invasive prenatal testing for aneuploidy. Obstet Gynecol 2007; 110: 1459–1467.

16. Borgida AF, Mills AA, Feldman DM, Rodis JF, Egan JF. Outcome of pregnancies complicated by ruptured membranes after genetic amniocentesis. Am J Obstet Gynecol 2000; 183: 937–939.

17. Hess LW, Anderson RL, Golbus MS. Significance of opaque discolored amniotic fluid at second-trimester amniocentesis. Obstet Gynecol 1986; 67: 44–46.

18. Jackson LG, Zachary JM, Fowler SE, Desnick RJ, Golbus MS, Ledbetter DH, Mahoney MJ, Pergament E, Simpson JL, Black S, et al. A randomized comparison of transcervical and transabdominal chorionic-villus sampling. The US National Institute of Child Health and Human Development Chorionic-Villus Sampling and Amniocentesis Study Group. N Engl J Med 1992; 327: 594–598.

19. Carlin AJ, Alfirevic Z. Techniques for chorionic villus sampling and amniocentesis: a survey of practice in specialist UK centres. Prenat Diagn 2008; 28: 914–919.

20. Mastroiacovo P, Botto LD, Cavalcanti DP, Lalatta F, Selicorni A, Tozzi AE, Baronciani D, Cigolotti AC, Giordano S, Petroni F, et al. Limb anomalies following chorionic villus sampling: a registry based case-control study. Am J Med Genet 1992; 44: 856–864.

21. Botto LD, Olney RS, Mastroiacovo P, Khoury MJ, MooreCA, Alo CJ, Costa P, Edmonds LD, Flood TJ, Harris JA, Howe HL, Olsen CL, Panny SR, Shaw GM. Chorionic villus sampling and transverse digital deficiencies: evidence for anatomic and gestational-age specificity of the digital deficiencies in two studies. Am J Med Genet 1996; 62: 173–178.

22. Brambati B, Lanzani A, Tului L. Transabdominal and transcervical chorionic villus sampling:

23. Papp C, Beke A, Mezei G, T' oth-P' al E, Papp Z. Chorionic villus sampling: a 15-year experience. Fetal Diagn Ther 2002; 17: 218–227.

24. Society for Maternal-Fetal Medicine (SMFM), Berry SM, Stone J, Norton ME, Johnson D, Berghella V. Fetal blood sampling. Am J Obstet Gynecol 2013 Sep; 209: 170–180.

25. Tongsong T, Wanapirak C, Kunavikatikul C, Sirirchotiyakul S, Piyamongkol W, Chanprapaph P. Fetal loss rate associated with cordocentesis at midgestation. Am J Obstet Gynecol 2001; 184: 719–723.

26. Maxwell DJ, Johnson P, Hurley P, Neales K, Allan L, Knott P. Fetal blood sampling and pregnancy loss in relation to indication. Br J Obstet Gynaecol 1991; 98: 892–897.

27. Antsaklis A, Daskalakis G, Papantoniou N, Michalas S. Fetal blood sampling-indication-related losses. Prenat Diagn 1998; 18: 934–940.

28. Mujezinovic F, Alfirevic Z. Technique modifications for reducing the risks from amniocentesis or chorionic villus sampling. Cochrane Database Syst Rev 2012; 8: CD008678.

29. Simonazzi G, Curti A, Farina A, Pilu G, Bovicelli L, Rizzo N. Amniocentesis and chorionic villus sampling in twin gestations: which is the best sampling technique? Am J Obstet Gynecol 2010; 202: 365.e1–5.

30. Audibert F, Gagnon A. Genetics Committee of the Society of Obstetricians and Gynaecologists of Canada; Prenatal Diagnosis Committee of the Canadian College of Medical Geneticists. Prenatal screening for and diagnosis of aneuploidy in twin pregnancies. J Obstet Gynaecol Can 2011; 33: 754–767.

efficiency and risk evaluation of 2,411 cases. Am J Med Genet 1990; 35: 160–164.

Ultrasound and Aneuploidy

Maimoona Ahmed, Shantala Vadeyar

INTRODUCTION

Aneuploidy (abnormal number of chromosomes) is the most common genetic abnormality detected on ultrasonography and its antenatal detection is one of the major goals of any prenatal screening program. The commonest aneuploidies detected on ultrasound include Trisomy 21 (Down syndrome), Trisomy 18 (Edward's syndrome), Trisomy 13 (Patau's syndrome), Triploidy and Monosomy X (Turner's syndrome). The following chapter provides a comprehensive description of the abnormal ultrasound findings which are more commonly seen associated with fetal aneuploidies.

ULTRASOUND SOFT MARKERS

Soft markers are ultrasound findings that are often associated with normal fetuses (i.e. normal variants), usually have no clinical sequelae, and are transient, resolving with advancing gestation or after birth.[1] However, they have been shown to carry an increased risk for fetal aneuploidy.[2] Both first trimester and second trimester soft markers have been described and extensively studied.

First Trimester Ultrasound Markers of Aneuploidy—the 11–13 + 6 Weeks Scan

The first trimester scan is a very important milestone that indicates the possibility of fetal aneuploidy by the assessment of various parameters given below. This scan provides information to the patient at a time when reproductive choices can be made with a great deal of privacy. Also it is clear that most couples want to be informed regarding the status of their fetus as early in pregnancy as possible.[3]

Nuchal Translucency (NT)

In 1992, Nicolaides et al introduced the term 'nuchal translucency', defined as the thickness of the translucent space between the skin and soft tissue overlying the fetal cervical spine. This method identifies about 75% of affected fetuses for a positive rate of about 5%.[4]

Criteria for measurement (Fig. 6.1)
- Gestation: 11–13 + 6 weeks
- CRL: 45–84 mm
- Fetal head and upper thorax occupy the whole screen
- Midsagittal plane with fetal spine down
- Fetal neck neutral
- The umbilical cord may be around the fetal neck in 5–10% cases and this finding may produce false increased NT. In such cases measure NT above and below the cord and take average.

Every NT measurement represents a likelihood ratio which is multiplied by the a priori maternal and gestational age-related

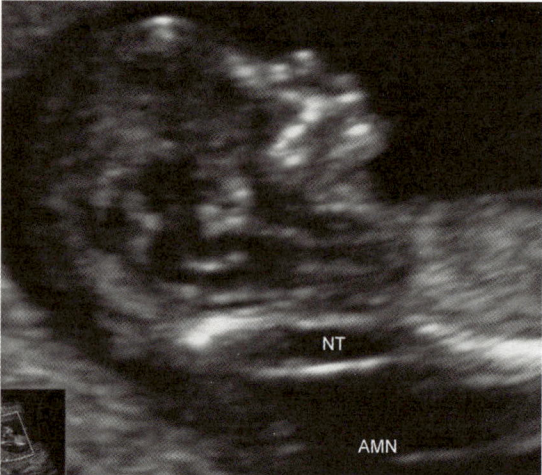

Fig. 6.1: Nuchal translucency

risk to calculate a new risk of the fetus being affected by aneuploidy. Thus, it is a screening test for fetal aneuploidy.

Increased NT (Fig. 6.2)

- Associated with an increased risk of fetal aneuploidy, structural anomalies, genetic syndromes, and adverse outcome.
- Fetal NT increases with CRL and therefore it is essential to take gestational age into account when determining whether a given NT thickness is increased.
- The risk increases as the NT increases.[5] To confirm aneuploidy with karyotype.

- Majority of the babies survive and develop normally.
- Persistence of increased NT at 14–16 weeks and evolution to nuchal edema or hydrops fetalis at 20–22 weeks—congenital infection or a genetic syndrome, e.g. Treacher Collins syndrome, DiGeorge syndrome, Joubert syndrome.
- There is a 10% risk of perinatal death or a live birth with a genetic syndrome that could not be diagnosed prenatally.
- The risk of neurodevelopmental delay in the survivors is around 3–5%.
- In the chromosomally normal group the parents should be counselled that even if the NT is more than 6.5 mm there is a one in three chance that the pregnancy would result in a live birth with no major defects.[6]
- The association between increased NT and a wide range of structural abnormalities and genetic syndromes constitutes an indication for detailed follow-up scans, including fetal echocardiography.[7]

NT ultrasound is an operator-dependent modality. Quality control and training in USG with registration and if possible grading of sonographers should be done in a specialized field like this to ensure high detection rate and standardization of practice. The unavailability of high end USG units at most of the

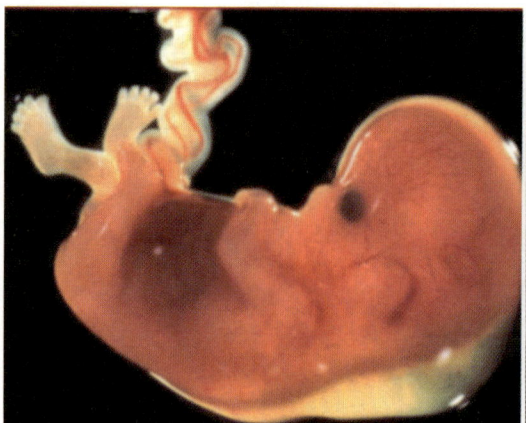
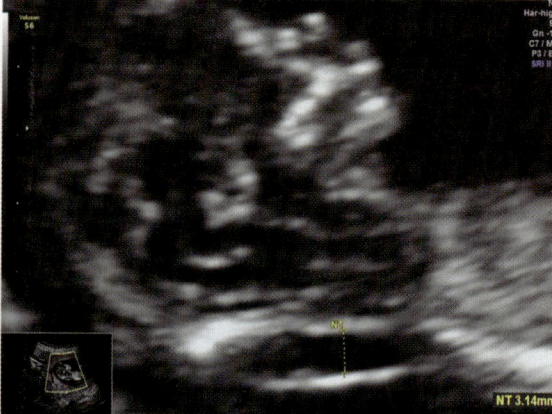

Fig. 6.2: Increased NT

community and government healthcare set-up is an important limiting factor.[8]

Nicholaides KH et al have summarized and compared the various first trimester screening techniques for aneuploidy in Table 6.1 with their detection rates.[4]

Fetal Heart Rate (FHR)

Criteria for measurement (Fig. 6.3)

- Transverse or longitudinal section of heart.
- Spectral/pulse wave Doppler: 6 to 10 cardiac cycles during fetal quiescence.
- FHR-calculated by US machine software.

FHR and aneuploidy (Fig. 6.4)

- Trisomy 13 and Turner's syndrome are associated with tachycardia, in trisomy 18 and triploidy there is a tendency for bradycardia. In trisomy 21 there is a mild tachycardia.[9]
- Tachycardia in Turner's syndrome and trisomy 13 may be due to delay in functional maturation of the parasympathetic system,

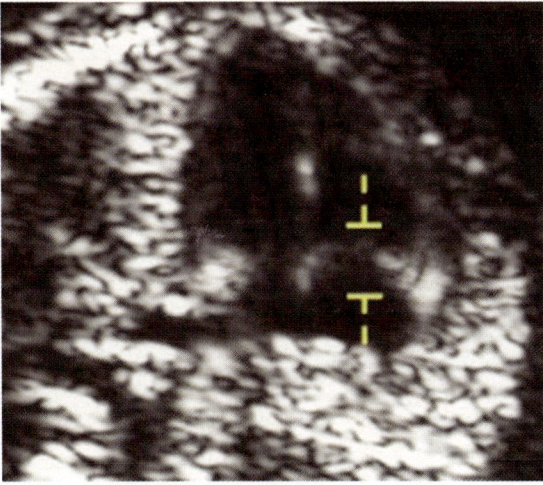

Fig. 6.3: Criteria for measuring FHR

resulting in delay in the physiological decrease in heart rate after 9 weeks.
- The bradycardia of trisomy 18 may be related to the fact that in these fetuses there is early onset severe growth restriction and development delay as compared to other aneuploidies.

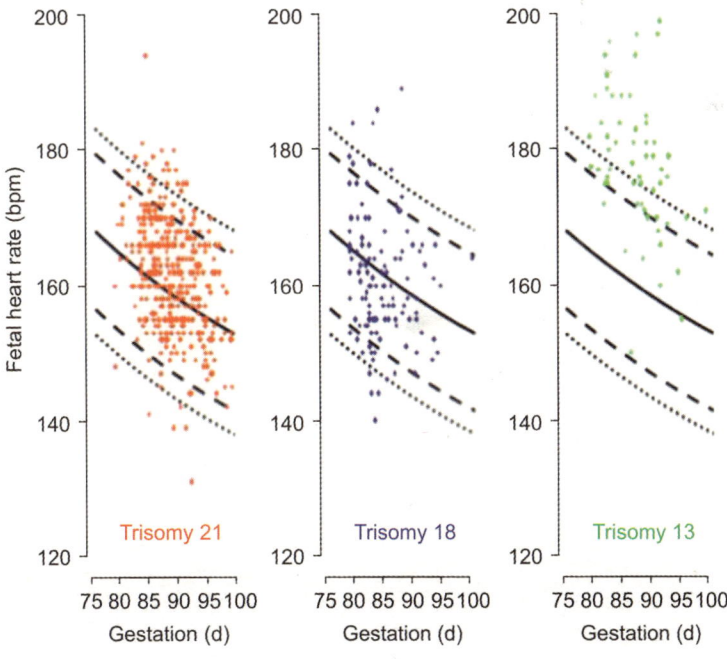

Fig. 6.4: FHR and aneuploidy—fetal medicine foundation

Table 6.1	Screening techniques and detection rate	
Method of screening	*Detection rate (%)*	*False-positive rate (%)*
MA	30	5
First trimester		
MA + fetal NT	75–80	5
MA + serum free β-hCG and PAPP-A	60–70	5
MA + NT + free β-hCG and PAPP-A (combined test)	85–95	5
Combined test + nasal bone or tricuspid flow or ductus venosus flow	93–96	2.5
Second trimester		
MA + serum AFP, hCG (double test)	55–60	5
MA + serum AFP, free β-hCG (double test)	60–65	5
MA + serum AFP, hCG, uE3 (triple test)	60–65	5
MA + serum AFP, free β-hCG, uE3 (triple test)	65–70	5
MA + serum AFP, hCG, uE3, inhibin-A (quadruple test)	65–70	5
MA + serum AFP, free β-hCG, uE2, inhibin-A (quadruple test)		70–75 5
MA + NT + PAPP-A (11–13 weeks) + quadruple test	90–94	5

MA, maternal age; NT, nuchal translucency; β-hCG, β-human chorionic gonadotrophin; PAPP-A, pregnancy-associated plasma protein-A

- Triploidy is associated with high rate of early intrauterine death and the bradycardia may represent a pre-terminal event.[10]

Newer first trimester markers: The newer markers such as nasal bone, tricuspid regurgitation and ductus venosus Doppler increase the performance of aneuploidy screening from 90% to 94% with a decrease in the false positives from 3% to 2.5%. The new markers can be assessed in all patients or can be offered to 15% of the patients with an intermediate risk (1 in 51 to 1 in 1000) after combined screening.[11] (Fig. 6.5)

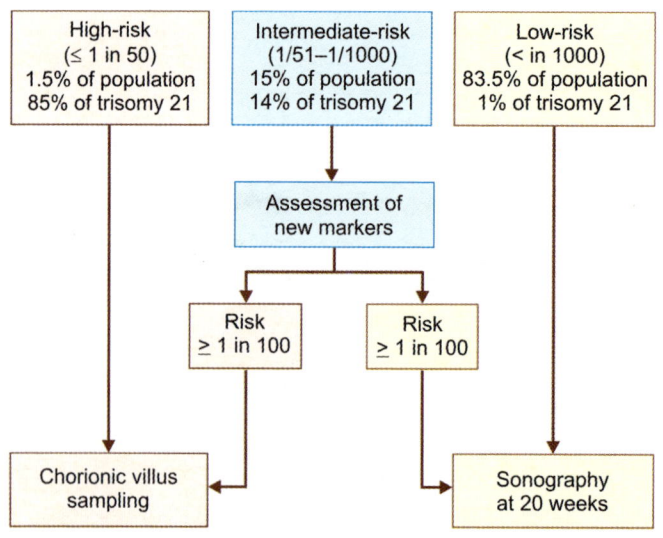

Fig. 6.5: Newer first trimester markers and screening

Absent Nasal Bone

Criteria for measurement (Fig. 6.6)

- *Gestation:* 11–13 + 6 weeks
- CRL: 45–84 mm
- Head–thorax occupy the whole screen
- Exact mid sagittal plane
- USG transducer should be parallel to direction of nose and palate
- Nasal bone is said to be present if it is seen more echogenic than the skin.
- *It is said to be absent if:* 1. not visible, 2. echogenicity same as that of the skin and 3. echogenicity less than the skin.

Absent nasal bone and aneuploidy (Fig. 6.7)

Several studies have demonstrated a high association between absent nasal bone at 11 – 13 + 6 weeks and trisomy 21 as well as other chromosomal abnormalities.[12] The absence of fetal nasal bone is independent of NT size and the two ultrasound screening methods could be combined into one modality, with a predicted sensitivity of 85% for a 1% false-positive rate.[13]

The frequency of absent nasal bone in euploid, trisomy 13, trisomy 18, and trisomy 21 was 2.5, 45, 53, and 60 percent, respectively.[4] However, the frequency of absent nasal bone

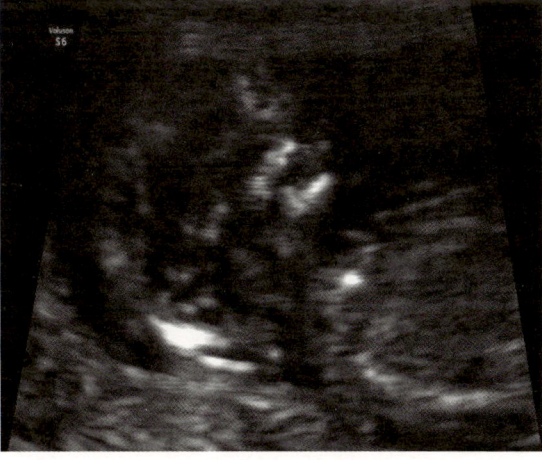

Fig. 6.7: Absent nasal bone

is higher in the Asian and Afro-Caribbean population as compared with Caucasians.

Tricuspid Regurgitation (TR)

Criteria for measurement (Fig. 6.8)

- GA: 11–13 + 6 weeks
- CRL: 45–84 mm
- Fetus not moving
- Fetal thorax occupy the whole screen
- Apical 4-chamber view
- Pulsed Doppler sample: 2–3 mm across tricuspid valve
- The insonation angle <300 from the direction of inter-ventricular septum

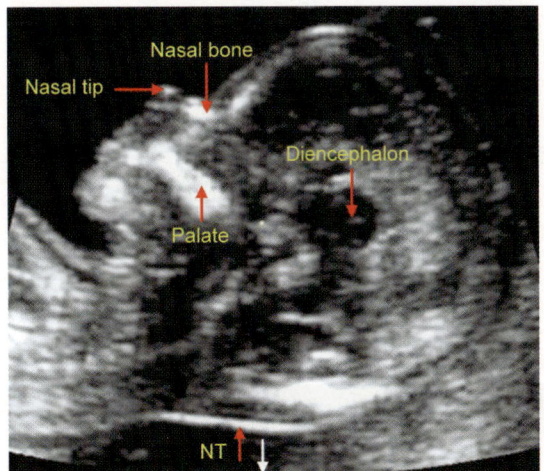

Fig. 6.6: Criteria for nasal bone assessment

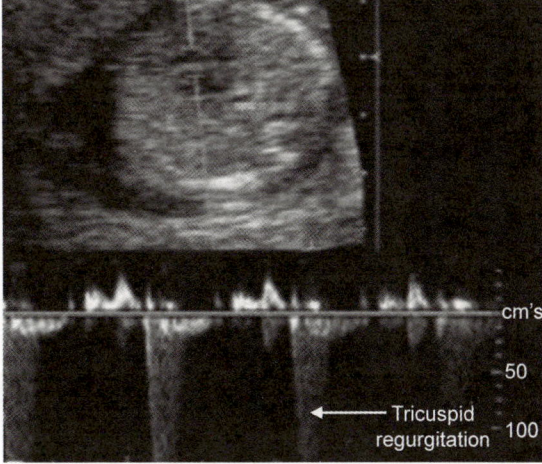

Fig. 6.8: Tricuspid regurgitation

- Sweep speed should be high 2–3 cm/sec
- TR—approximately half of systole, velocity more than 60 cm/sec

Between 11 and 13 + 6 weeks gestation, the prevalence of TR in fetuses with trisomy 21 is 55% as compared to 1% in chromosomally normal fetuses. The prevalence of TR in trisomy 18 and 13 and monosomy is 33%, 30% and 38% respectively.[14] Also along with aneuploidy, TR can be an early sign of cardiac defects. Hence, a detailed cardiac evaluation (fetal echocardiography) is warranted at 16–19 weeks.

Ductus Venosus (DV) Doppler

The ductus venosus (DV) is a fetal venous structure which connects the hepatic portion of the umbilical vein and the inferior vena cava. Approximately, 50% of the oxygenated blood returning from the placenta in the umbilical vein flows through DV.

Criteria for measurement

- GA 11–13 + 6 wks, CRL 45–84 mm
- *Gate:* 0.5–1 mm, insonation angle <30
- Low filter: 50–70 Hz, sweep speed: 2 cm/s

- Color flow mapping demonstrating the high velocity with aliasing at the narrow entrance of the DV confirms its identification.
- Normal waveform has 'S', 'D' and positive 'a' wave (Fig. 6.9).
- Abnormal waveform is associated with aneuploidy, cardiac defects and even fetal death [AG 1999].

Abnormal DV/reversed a wave (Fig. 6.10)

Between 11 and 13 + 6 weeks gestation, the prevalence of a wave reversal in trisomy 21 is 60% as compared to 3% in chromosomally normal fetuses. The prevalence of a wave reversal in trisomies 18 and 13 and monosomy X is 58%, 55% and 75% respectively.[15]

Frontomaxillary Facial Angle (FMF)

Another classical dysmorphic feature of individuals with trisomy 21 is a relatively flat face. This was measured objectively as the frontomaxillary facial angle.

Criteria for measurement (Fig. 6.11)

- Gestation: 11–13 + 6 weeks; CRL: 45–84 mm
- Head and thorax occupy the whole screen

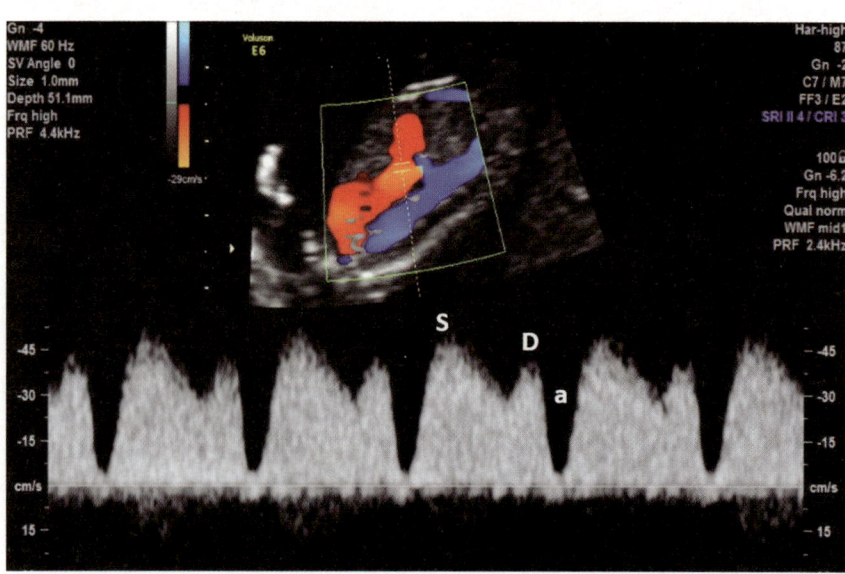

Fig. 6.9: Normal ductus venosus waveform

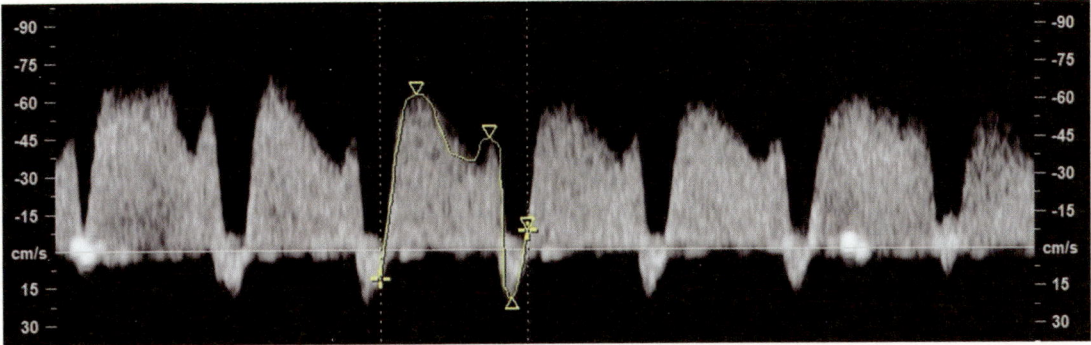

Fig. 6.10: Reversed a wave

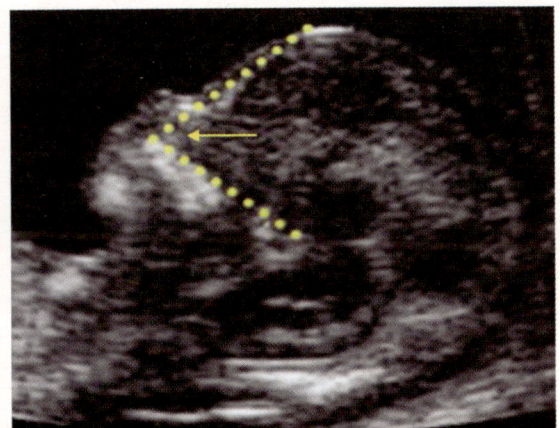

Fig. 6.11: FMF angle

- Exact mid-sagittal plane
- USG transducer should be parallel to direction of nose and palate
- Measured between
  i. Line along the upper surface of the palate and
  ii. Line which traverses the upper corner of the anterior aspect of the maxilla extending to the external surface of the forehead.

Frontomaxillary angle measurements in the first trimester have demonstrated that 70% of fetuses with trisomy 21 have an FMF measurement above the 95th centile of the normal population. However, the level of difficulty in obtaining the proper image has limited the use of FMF angle in general screening.[16]

Fetal Anomalies and Risk of Aneuploidy

Several fetal anomalies have been shown to significantly increase the risk of aneuploidy in the first trimester.[17]

Anomaly	Risk	Aneuploidy
Holoprosencephaly	1:2	Trisomy 13
Diaphragmatic hernia	1:4	Trisomy 18
Atrioventricular septal defect	1:2	Trisomy 21
Omphalocele	1:4:10	Trisomy 18
Megacystitis (bladder	1:10	Trisomy 13
length ≥ 7 mm)		and 18

Early fetal evaluation in the first trimester therefore, should include a complete structural survey as well as screening for soft markers indicative of aneuploidy.

Second Trimester Ultrasound Markers of Aneuploidy

Fetuses with aneuploidy have a variety of physical features that are potentially detectable by prenatal sonography between 15 and 20 weeks of gestation. A few of these features represent actual structural abnormalities that have clinical consequences regardless of karyotype while most are soft markers.[18] Table 6.2 enumerates these findings.

The screening ultrasound at 16 to 20 weeks should evaluate soft markers like thickened nuchal fold, echogenic bowel, mild ventriculomegaly, echogenic focus in the heart, and choroid plexus cyst. Identification of soft markers for fetal aneuploidy requires correla-

Table 6.2	Ultrasound markers of aneuploidy in second trimester		
	Structural anomalies (SA)	*Soft markers (SM)*	*IUGR*
Trisomy 21	Cardiac abnormalities	Nuchal fold thickening	√
	Duodenal atresia	Ventriculomegaly	
	Brachycephaly	Short femur or humerus	
	Hydrocephalus	Hypoplastic nose	
	Clinodactyly	Echogenic bowel	
	Cystic hygroma and hydrops	Pyelectasis	
Trisomy 18	Cardiac abnormalities	Choroid plexus cysts	√
	Esophageal atresia	Enlarged cisterna magna	
	Strawberry-shaped head	Ventriculomegaly	
	Diaphragmatic hernia	Short femur or humerus	
	Omphalocele	Hypoplastic nose	
	Meningomyelocoele	Echogenic bowel	
	Agenesis corpus callosum	Pyelectasis	
	Facial clefting	Single umbilical artery	
	Talipes		
	Rocker-bottom foot		
	Radial aplasia		
	Overlapping digits		
	Umbilical cord cyst		
	Cystic hygroma and hydrops		
Trisomy 13	Cardiac abnormalities	Echogenic intracardiac foci	√
	Diaphragmatic hernia	Enlarged cisterna magna	
	Omphalocele	Ventriculomegaly	
	Holoprosencephaly	Pyelectasis	
	Facial clefting	Single umbilical artery	
	Cyclopia		
	Agenesis corpus callosum		
	Rocker-bottom foot		
	Polydactyly		
	Talipes		
	Cystic hygroma and hydrops		

tion with other risk factors, including history, maternal age, and maternal serum testing results.[19]

Increased Nuchal Fold Thickness (NFT) (Fig. 6.12)

- The nuchal fold is the skin thickness in the posterior aspect of the fetal neck.
- Measurement is obtained in a transverse section of the fetal head at the level of the cavum septum pellucidum and thalami, angled posteriorly to include the cerebellum.

- Taken from the outer edge of the occiput bone to the outer skin limit directly in the midline.
- A measurement of 6 mm is considered significant between 18 and 24 weeks and a measurement of 5 mm is considered significant at 16 to 18 weeks.
- A meta-analysis reviewed the performance of a thick nuchal fold at 6 mm or greater and showed that the risk for Down syndrome increased by approximately 17-fold (CI 8–35).

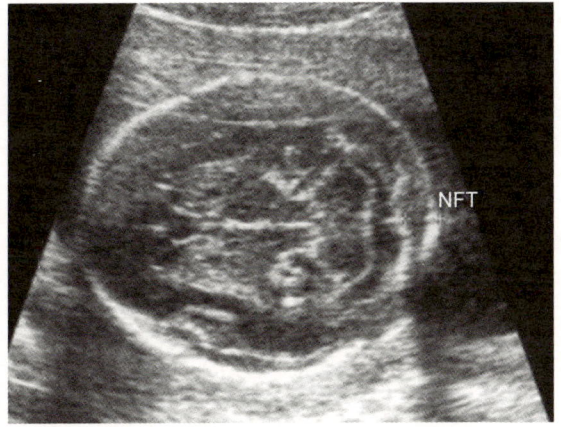

Fig. 6.12: Nuchal fold thickness

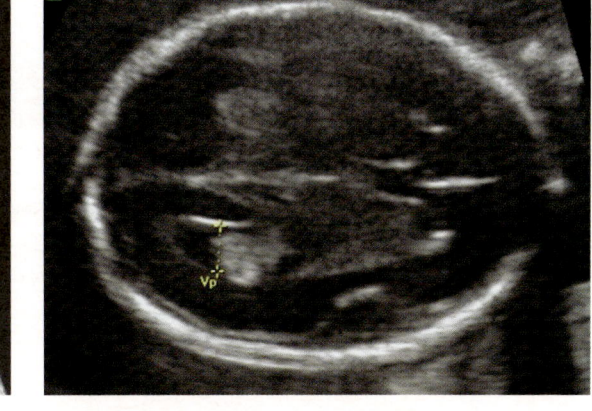

Fig. 6.13: Measurement of lateral ventricle

- Also can be associated with single gene abnormalities (e.g. Noonan syndrome, multiple pterygium syndrome), skeletal dysplasias and congenital cardiac defects.[20]

Mild Ventriculomegaly (MVM)

- Midline structures should be equidistant from the proximal and distal calvarium margins, cavum septum pellucidi anterior landmark, ambient cistern posterior landmark (Fig. 6.13).
- Measurement taken opposite internal parieto-occipital sulcus, perpendicular to the long axis, along the inner margins of the medial and lateral walls, at the level of the glomus of the choroid plexus.
- MVM is defined as measurements of 10 to 15 mm.
- In isolated MVM, the incidence of abnormal fetal karyotype is estimated at 3.8% (0 to 28.6%).
- Also can be associated with central nervous system pathology.
- A detailed anatomic evaluation looking for additional malformations or soft markers, laboratory investigation for the presence of congenital infection or fetal aneuploidy and MRI as a potential additional imaging technique.
- Neonatal assessment and follow-up.[19]

Absent or Hypoplastic Nasal Bones (Fig. 6.14)

- A midsagittal view of the fetal head, identifying the nasal bone, lips, maxilla, and mandible.
- Various criteria for detection of aneuploidy: Completely absent or hypoplasia with nasal bone length <2.5 mm or < 0.75 MoMs for the gestational age.[21]
- The presence of normal fetal nasal bone implies a 7-fold reduction in the risk of trisomy 21.

Echogenic Bowel (Fig. 6.15)

- Grade 0—normal (< liver), Grade 1—increased echogenicity (more than liver but less than bone), Grade 2—echogenicity equal to bone, Grade 3—greater than bone
- No further investigations are required for grade 1 echogenic bowel.
- Grade 2 and 3 fetal echogenic bowel is associated with both chromosomal and non-chromosomal abnormalities like bowel pathology, congenital infections and IUGR.[19]

Mild Pyelectasis (Fig. 6.16)

- Mild pyelectasis is defined as a hypoechoic spherical or elliptical space within the renal pelvis that measures 5 mm in the second trimester and 7 mm in the third trimester.

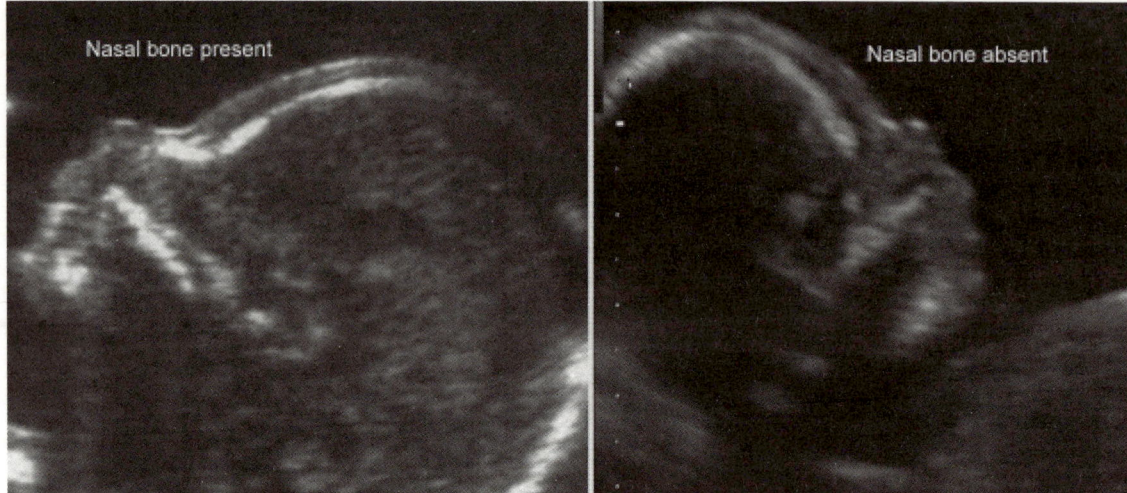

Fig. 6.14: Absent nasal bone (2nd trimester)

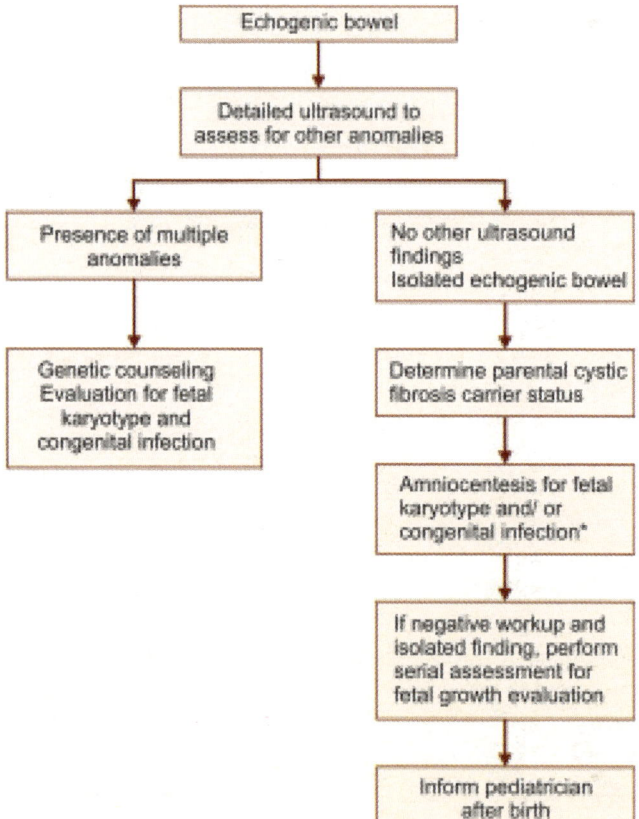

Fig. 6.15: Echogenic bowel algorithm. * TORCH infections like CMV, Toxoplasma, Rubella

- Measurement is taken on a transverse section through the fetal renal pelvis using the maximum anterior-to-posterior measurement.
- Isolated mild pyelectasis does not require fetal karyotyping.
- Follow-up scans for assessment of renal pelvis and post-natal evaluation is needed.[19]

Echogenic Focus in Heart (Fig. 6.17)

- Echogenic intracardiac focus (EICF) is defined as a focus of echogenicity comparable to bone, in the region of the papillary muscle in either or both ventricles of the fetal heart.
- Eighty-eight percent are only in the left ventricle, 5% are only in the right, and 7% are biventricular.
- Isolated EIF does not require any further investigation, it is commoner in Asian populations.
- Right-sided, biventricular, multiple, particularly conspicuous, or non-isolated EICF should be offered referral for expert review and possible karyotyping.[19]

Choroid Plexus Cyst (Fig. 6.18)

- Choroid plexus cysts (CPCs) are sono-graphically discrete, small cysts (≥ 3 mm) found in the choroid plexus within the

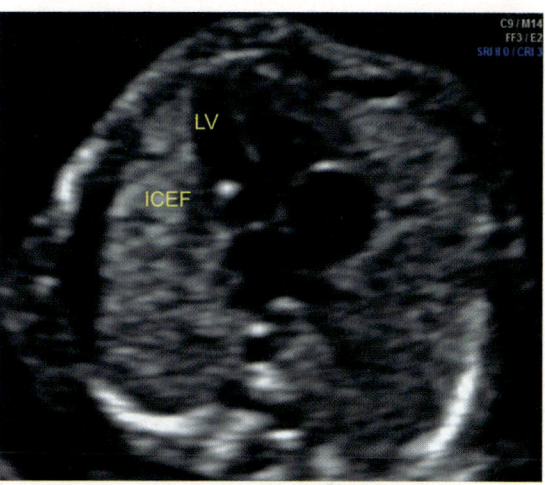

Fig. 6.17: Intracardiac echogenic focus

lateral cerebral ventricles of the developing fetus at 14 to 24 weeks' gestation.
- Size of CPCs is not of clinical relevance. Can be unilateral or bilateral.
- The incidence of CPCs is 50% in fetuses with trisomy 18, however, only 10% of fetuses with trisomy 18 will have CPCs as the only identifiable sonographic marker.
- The presence of CPCs in chromosomally normal fetuses is not associated with other fetal abnormalities or abnormal postnatal development.[19]

Short Femur and Short Humerus

- Isolated finding of short femur or short humerus has a low association with aneuploidy.
- However, such a finding should prompt measurement of all long bones to rule out skeletal dysplasia.

Meta-analysis of Second Trimester Markers for Detection of Trisomy 21 (Fig. 6.19)

In a meta-analysis by M Agathokleous et al, the authors summarized the screening performance of second trimester markers for detection of trisomy 21. Each individual marker was assigned a likelihood ratio (LR) and the background risk was multiplied by the LR as per Bayes' theorem. The presence of

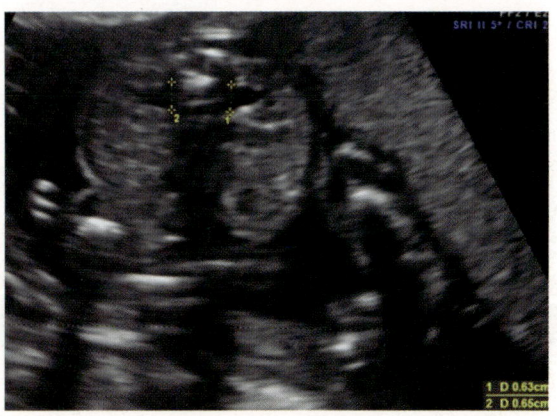

Fig. 6.16: Pyelectasis

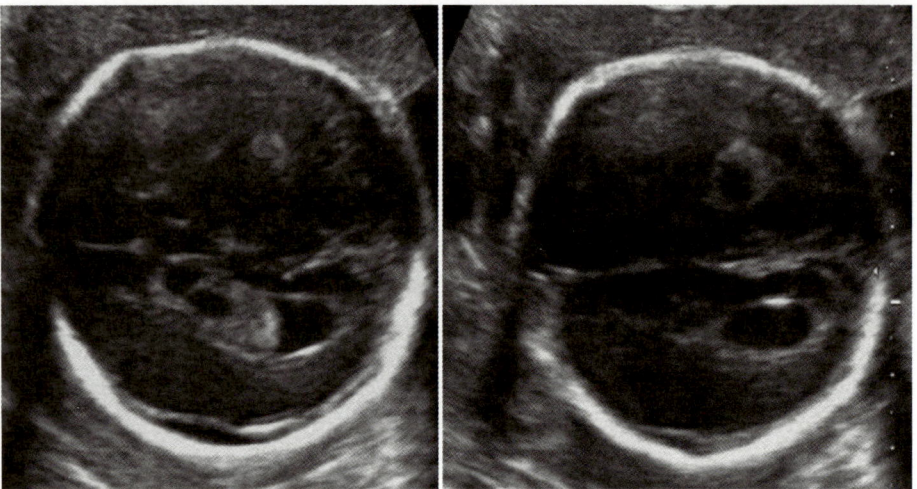

Fig. 6.18: Choroid plexus cysts

sonographic markers increased and their absence decreased the risk of T 21. Absence of all soft markers led to a 7.7-fold reduction in the risk.[22]

CONCLUSION

Advances in ultrasound technology and an improved understanding of early fetal development have made prenatal detection of

Meta-analysis of second trimester markers for trisomy 21 Agathokleous et al., UOG 2013					
Marker	DR	FPR	LR +ve	LR −ve	Isolated marker
Cardiac echogenic focus	24.4	3.9	5.8	0.80	0.95
Ventriculomegaly	7.5	0.2	27.5	0.94	3.81
Increased nuchal fold	26.0	1.0	23.3	0.80	3.79
Echogenic bowel	16.7	1.1	11.4	0.90	1.65
Mild hydronephrosis	13.9	1.7	7.6	0.92	1.08
Short humerus	30.3	4.6	4.8	0.74	0.78
Short femur	27.7	6.4	3.7	0.80	0.61
ARSA	30.7	1.5	21.5	0.71	3.94
Absent or hypoplastic NB	59.8	2.8	23.3	0.46	6.58

No markers LR 0.13 = 7.7 fold reduction
Meta-analysis 47 studies 1995–2012

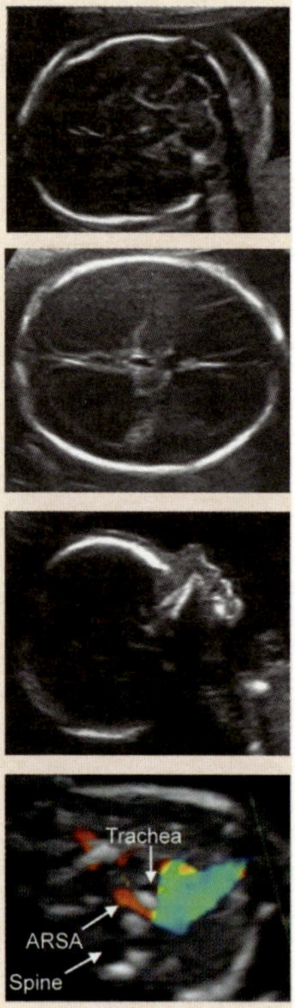

Trachea
ARSA
Spine

Fig. 6.19: Metaanalysis of soft markers for detection of trisomy 21

aneuploidy possible. An up-to-date knowledge concerning the various markers goes hand in hand with our ability to detect them, so as to provide the couple with the appropriate counselling and correct management. It is important to remember, however, that ultrasound is just a tool for identifying major anomalies and high risk situations. The confirmation of aneuploidy is only by invasive modalities like chorionic villous sampling and amniocentesis and laboratory methods of cytogenetics and molecular techniques.

REFERENCES

1. Bethune M. Literature review and suggested protocol for managing ultrasound soft markers for Down syndrome: thickened nuchal fold, echogenic bowel, shortened femur, shortened humerus, pyelectasis and absent or hypoplastic nasal bone. Australas Radiol 2007; 51:218.

2. Bethune M. Management options for echogenic intracardiac focus and choroid plexus cysts: a review including Australian Association of Obstetrical and Gynaecological Ultrasonologists consensus statement. Australas Radiol 2007; 51:324.

3. De Graaf Im, et al. Women's preference in Down's syndrome screening. Prenat Diagn 2001; 22:624–9.

4. NIcolaides KH. Screening for fetal aneuploidies at 11 to 13 weeks. Prenat Diag 2011;31:7–15.

5. Pandya PP, Brizot ML, Kuhn P, Snijders RJ, Nicolaides KH. First-trimester fetal nuchal translucency thickness and risk for trisomies. Obstet Gynecol 1994;84(3):420–3.

6. Bilardo CM, et al. Increased nuchal translucency thickness and normal karyotype: time for parental reassurance. Ultrasound Obstet Gynecol 2007 Jul; 30(1):11–8.

7. Souka AP, et al. Increased nuchal translucency with normal karyotype. Am J Obstet Gynecol 2005 Apr;192(4):1005–21.

8. Brigatti KW, Malone FD. First-trimester screening for aneuploidy. Obstet Gynecol Clin North Am 2004;31(1):v, 1–20.

9. Van Lith JMM, et al. Fetal heart rate in early pregnancy and chromosomal disorders. Br J Obstet Gynecol 1992;99: 741–4.

10. Liao AW, Snijders R, Geerts L, Spencer K, Nicolaides KH. Fetal heart rate in chromosomally abnormal fetuses. Ultrasound Obstet Gynecol 2000;16:610–3.

11. Sonek JD, et al. Additional first trimester ultrasound markers. Clin Lab Med 2010; 30:573–92.

12. Cicero S, Rembouskos G, Vandecruys H, Hogg M, Nicolaides KH. Likelihood ratio for trisomy 21 in fetuses with absent nasal bone at the 11–14 weeks scan. Ultrasound Obstet Gynecol 2004;23: 218–23.

13. Viora E, Masturzo B, Errante G, Sciarrone A, Bastonero S, Campogrande M. Ultrasound evaluation of fetal nasal bone at 11 to 14 weeks in a consecutive series of 1906 fetuses. Prenat Diagn 2003; 23(10):784–7.

14. Kagan KO, et al. Tricuspid regurgitation in screening for trisomies 21, 18 and 13 and Turner's syndrome at 11 to 13 + 6 weeks of gestation. Ultrasound Obstet Gynecol 2009;33: 18–22.

15. Maiz N, Valencia C, Kagan KO, et al. Ductus venosus Doppler in screening for trisomies 21, 18 and 13 and Turners syndrome at 11 to 13+6 weeks of gestation. Ultrasound Obstet Gynecol 2009;33: 512–7.

16. Borenstein M, Persico N, Kagan KO, et al. Frontomaxillary facial angle in screening for trisomy 21 at 11 to 13+6 weeks. Ultrasound Obstet Gynecol 2008;32: 5–11.

17. Snijders RJM, et al. Fetal abnormalities. In: Ultrasound markers for fetal defects. Snijders RJM, Nicolaides KH, Editors. Carnforth, UK: Parthenon Publishing; 1996. p.1–62.

18. Bryann Bromley, Benacerraf MD, et al. The genetic sonogram—a method of risk assessment for Down's syndrome in second trimester. J Ultrasound Med. 2002 Oct;21(10):1087–96.

19. SOGC clinical practice guidelines. Fetal soft markers in obstetric ultrasound. J Obstet Gynaecol Can 2005;27(6):592–612.

20. Smith-Blindman R, Hosmer W, Feldstein VA, Deeks JJ, Goldberg JD. Second trimester ultrasound to detect fetuses with Down syndrome: a meta-analysis. JAMA 2001;285(8): 1044–55.

21. Gianferrari, Elisa A, et al. Absent or short nasal bone length and the detection of Down syndrome in second trimester fetus. Obstetrics and Gynecology. 2007 Feb;109(2):371–75.

22. M Agathokleous, et al. Meta-analysis of second trimester markers for trisomy 21. Ultrasound Obstet Gynecol 2013; 41: 247–261.

Parental Concerns with Previous Affected Child

Vikas Bhatia

EDITORIAL NOTE

This is not an academic chapter—it is more a cry from the heart of a parent who has gone through the tragic loss of three newborns due to inborn errors of metabolism. We would like our esteemed doctors to read between the lines for messages which will help us in practice. There may be situations in which the obstetrician feels lost—referral to a fetal medicine consultant/geneticist may be appropriate. The need for appropriate counseling is highlighted by this personal narrative.

National Conference on Child Survival and Development in New Delhi, claims that of the roughly 26 million children born in India each year, 1.2 million die during the first four weeks. That is 30% of 3.9 million global neonatal deaths. According to the report, the current neonatal mortality rate (NMR) of 44 per 1,000 live births accounts for nearly two-thirds of all infant deaths (death before the age of one) and nearly half of under-five child deaths in India.[1]

PARENTAL CONCERN

For a couple the most mesmerizing moment of life is birth of their child. This feeling cannot be explained in words.

A newborn provides sense of responsibility to parents and completes the family, but yet it has exceptions too.

There are times when children suffer from many diseases or disorders, such as infection, genetic disorders, metabolic disorders and congenital disorders. Treatment and management of these newborns push their parents in deep pain and agony and also create uncertainties for future pregnancy.

This situation becomes worse when parents lose their baby without establishing a diagnosis. This is an intolerable and difficult to accept scenario that a newborn can have such life taking diseases or disorders. After facing such situation parents reach out to all possible solutions, so that they must not face this trauma in future.

INSTANCES OF PARENTAL ANXIETY

- At midnight a couple takes a newborn to hospital concerning that their child has not opened its eyes after five in the evening.
- A couple was worried about their child vomiting after every feed.
- Irregular heartbeat of a newborn which was very high some time and felt none at times, made its parents helpless.

Many of such reasons make parents run for solutions and many times no one has any answer!

Most important concern of these parents is the normal upbringing of their affected baby.

Many of the babies suffering from infections get to normal but some babies, especially those who are suffering from metabolic and genetic disorders, diagnosis itself becomes a tedious task. In our country about 16 lakh babies suffer from birth defect and/or genetic disorder. Here we suffer from lack of newborn screening and awareness towards these disorders. This lack of awareness also becomes a hindrance in management and treatment of these babies. Even parents of these newborns do not get appropriate doctor, hospital and facilities. For instance, for a parent of a baby suffering from metabolic or genetic disorder living in city like Jaipur, Nagpur, Nasik, Guntur, it is difficult to go to metro cities like Delhi, Mumbai, Hyderabad every time.

In a crisis situation, when a baby is admitted in NICU a specialist is needed. In a local hospital on unavailability of a specialist it becomes very difficult for the treating doctor to connect with the specialist doctor; even in this era of internet where video conferencing is very easily available.

REASONS FOR THE LOSS OF LIFE

- 35% Preterm birth
- 33% Infection
- 20% Birth asphyxia
- 12% malformation and other

Source: Liu et al, Lancet 2012 Statistical Report

6 out of every 100 suffer from birth defects.[2] Social look out: Parents have to face their social circle as well. Answering close relatives and folks become very difficult. Mostly they try to explain baby's problem without referring to any fact or scientific reason.

For example:

- *Najar lag gai*
- *Pichle janam ke karam*
- There is a problem in mother's milk
- Defective genes.

Few of these reasons are intolerable for parents. Whole family becomes helpless and clueless to answer and to cope with the situation.

PERSONAL EXPERIENCE

When I lost my first child in 1998 it seemed hard to believe that such things could happen to infants. However, after understanding the medical term "Sudden infant death syndrome" (SIDS) I had to make myself believe that it could happen. However, it could only be communicated through broken hearts in despair or shedding tears for dismantled life.

In 2000 when I had to go through the same stature of mind, it was clear that if the doctors used their presence of mind, then diagnosis could be done to get better outcome. Unfortunately I was too late to be aware about it. During the second child despite the history known to the gynecologist, her laid back attitude lead to the same situation again. Immediately the pediatrician convinced us to head to a major center in other city and referred to another doctor. My child was sick and in pain. I was quite nervous and not prepared to accept the same trauma the second time. Metabolic acidosis, ABG, lactic acid, ammonia these words were beating like drums in my head and were warning against the history repeating itself. On the other hand, the doctors at the hospital referred seemed to be knowledgeable but they never preferred to include me in all their process. Ironically in major cities we had never got to know about the doctor's visit while in small towns we could approach the doctors and pursue him. City doctors were quite blunt and professional, it was only in the last phase when one of the doctors who melted down and went out to inform us about our child's state; as she had foreseen and realized the critical stage. I had the guilt that it was me who had brought both mother and child here to save him but now when I witnessed the child's physical state—his lips had gone blue and nose bleeding. At that very time we wanted him to end his struggle rather than to want to let him survive. I was aghast but with no choice left. Next day we heard of the sad demise of our son. We were completely heartbroken as all our efforts were in vain and could not think of where our life would lead to.

There was an article published in India Today in 2000 that led me to Mumbai but yielded nothing even till 2004. Despite the indulgence of a team of experts from all parts of country (Mumbai, Delhi, Bangalore, Chennai, Cochin, Hyderabad), my inquisitiveness was regarding my child's exact diagnosis and my next step to that condition. That means once again I was back to square one. By that time I had become too keen to understand the case history that I once waited for 8 hours for the doctors to consult at Cochin. In yet another case I left the case file outside the doctor's residence when he was not available at home in Ahmedabad. I remained on toes for 5 consecutive years during which I learnt a lot about this profession. Here I found that the professionals did not take interest after a certain extent, there is lack of counseling and awareness, there's no heed to parent's enquiries, doctors hardly know metabolic cases, neither do doctors communicate. Nevertheless there were some praiseworthy points too, some doctors were wise enough to tackle and guide to some extent. I am in contact with those doctors till date. I cannot forget to mention a 16-year-old conversation with a doctor who told me to throw papers in some river and rather than fight with the God I have faith in and also question the doctor who had retained the report but had failed to produce a diagnosis report. It was only after 2005 when I grew to be techno savvy I started researching on metabolic errors.

I dared to take one more chance in 2009 and throughout the gestation period I struggled every day and night. The news of his arrival had mixed feelings; I geared up to bid a safe welcome of this newborn. I was pretty satisfied with the doctor under whose support and guidance we were. Nevertheless the error repeated we were very unfortunate like always as were fighting against a disorder that was yet fatal. We could not change impossible to possible. We had borne all that which could shatter anyone but we still fought for next 12 months and witnessed every expression of man outside sonography clinic. During this time I consulted about 4 gynecologists. One that was aware of the history, the other with whom we associated due to my known

metabolic expert and hospital where the baby had to be delivered. She was very blunt and inattentive and she initially denied taking up our case. By the grace of God finally I got to meet a gynecologist who was kind and less professional who knew her job well. I would like to take her support always, the pediatrician was also too cooperative, the teamwork was well woven but the disorder was once again fatal. Doctors were helpless and I was heartbroken. I gathered some courage and firmly thought to bring my baby home for once. Although I knew the little angel could not survive longer. As it was destined the boy passed away and took along with him everything that we owed he took from us our privilege of being at peace, our privilege of being a happy parent looking forward to welcome his bright future. I had to withstand this mental turmoil just to be by my wife and to support the thought that generated in my mind all these years struggle had made me even bolder and tougher. On that ground I send the reports to be diagnosed abroad the symptoms were quite similar to earlier we could have been lucky but all our efforts failed.

All these feelings heralded the forming of parents support group and with my own effort I want to provide not only ethical support to such parents but also to keep them better informed about IEMS. I want to represent the voice of those children who were too small to say what they had been through. My view on this campaign is quite clear that if we diagnose all this at early stage, then we could save the lives of many such children who die unnatural deaths. I would also want to provide a simple and effective diet which is beyond the reach of the common population. In India yet it would not be the food but their life saving component as the process of getting such food from other country is a herculean task.

NEW UPDATE ABOUT DIET

As a head of the organization MERD India, I am trying to campaign not only for newborn screening for treatable diseases but also awareness of inborn errors of metabolism; so that it can be of use to fight against such errors. I am very well aware of the fact that doctors do not hold parents at high esteem and that is why I wish to raise my voice for such parents so that newborn screening could be made mandatory in all states, as is present in Kerala and GMC Chandigarh. Through this book I wish to convey a message to all those who belong to this fraternity that if we detect any error through newborn screening, then we can diagnose the whole issue and through early treatment we could give a normal life to such children who have never had this advantage over living today when doctor and science are working hard to increase life expectancy. Diseases like inborn errors of metabolism are treated like gone cases and nothing is done about it, whereas the fact is that 20 out of 100 could be saved with modified diet/management/treatment and newborn screening and those 80 cases that are beyond any treatment. Such parents should at least be informed of **confirm diagnosis** so that they could think on the lines of parental diagnosis and other processes for the next time.

POSSIBLE WAY OUT FOR NEXT BIRTH

Selection of hospital and doctors: Selection of hospital, gynecologist and neonatologist is very crucial for parents who have passed through the trauma in previous baby. Parents should look for all the best possible facilities and specialization including gynecologist experience with best neonatal doctor.

Preconception test: If the previous baby has suffered/suffering from any chronic disease or genetic disorder, doctor should suggest preconception tests. This can have counseling and several tests. This situation is very horrifying as after all this parents are not sure if the baby they are expecting would be normal.

Prenatal test: After conception there are many tests possible including
- Ultrasonography

- Amniocentesis
- Chorionic villus sampling
- Fetal blood cells in maternal blood
- Maternal serum alpha-fetoprotein
- Maternal serum beta-HCG
- Maternal serum estriol

Even after all these tests couple is so afraid that they are not sure if they will have a healthy baby. After following every step given by gynecologist couple do not get peace of mind. Hence proper counseling before and after tests is needed.

Postnatal: After birth first concern of parents is whether the baby is looking normal, weight is okay; till the gynecologist and neonatologist do not assure the health of baby to be normal parents are worried.

Newborn screening: Parents who have some information of NBS from doctors, literature or internet are keen to go through this. Even if it is mandatory or not mandatory. As health of their baby is most important concern of parents and they do not want to go through any trauma again, hence parents of an affected child will be willing for the screening.

Neonatal care: Even after going through all this process parents are always worried about finding any sign of their previous experience in this baby. So counselling and confirmation of diagnosis are also very important.

SUMMARY

We witness more than **1.2 million** deaths of newborns in our country, thus we need medical fraternity to work as a team and provide the best possible solutions to parents in counseling, antenatal scanning, newborn screening and best neonatal care.

SUGGESTED READING

1. http://infochangeindia.org/children/news-scan/neonatal-death-rate-in-india-alarmingly-high-unicef-report.html
2. http://merdindia.com
3. http://nrhm.gov.in/nrhm-components/rmnch-a/child-health-immunization/rashtriya-bal-swasthya-karyakram-rbsk/background.html

Section

II

What an Obstetrician can do?

Preconceptional Care and Counseling

Pratima Mittal, Reeta Bansiwal

Preconception care refers to the process of identifying social, behavioral, environmental, and biomedical risks to a woman's fertility and pregnancy outcome and then reducing these risks through education, counseling, and appropriate intervention before conception.[1] Preconception intervention is more important than antenatal intervention for prevention of congenital anomalies since as many as 30 percent of pregnant women begin traditional prenatal care in the second trimester (>13 weeks of gestation), which is after the primary period of organogenesis. The Preconception Care Work Group of the Centers for Disease Control, USA recommends that preconception care should be an essential part of primary and preventive care.[2–3]

The three integral components of pre-pregnancy counseling are:
- Identification of risk factors related to pregnancy (screening).
- Patient education regarding pregnancy risks, management options and reproductive alternatives (information and counseling).
- Initiation of interventions, when possible, to provide optimum pregnancy outcome (interventions).

When and by Whom should Preconceptional Counseling be done?

Any health care provider—general practitioners, family physician, obstetricians and gyneco-logist and health providers at maternity hospitals can dispense preconceptional care and counseling during an encounter involving contraception, infertility, pregnancy testing, evaluation for sexually transmitted disease or vaginal infection, or periodic health examination, especially if the woman has pre-existing medical problems.

Addressing the preconceptional issues with the women in the reproductive age group at any point of their contact with the health care system is desirable because more than half the pregnancies are unplanned.

CHECKLIST FOR COMPREHENSIVE PRECONCEPTIONAL HEALTH PACKAGE[4]

Factors which facilitate the process of counseling are preconceptional care checklists; patient information sheets in local and simple language; posters and video films in the patient waiting areas; and availability of dedicated health care providers.

SCREENING

Screening of the couple through a detailed medical and family history, including genetic history of patient, spouse and family (Table 8.1); followed by examination is the first step in preconception care. The following needs to be screened:
- Hemoglobinopathies, e.g. sickle cell anemia, thalassemia if suspected by family history or examination.

Table 8.1	Genetic history of patient, spouse and family

History of congenital abnormalities
Neural tube defects
Heart defects
Cleft lip/palate
Any other

Chromosomal abnormalities
Down's syndrome
Mental retardation/learning disabilities (fragile X syndrome)

Advanced maternal or paternal age
Inherited diseases
Hemoglobinopathy
Muscular dystrophy
Thrombophilia
Cystic fibrosis
Huntington's chorea
Hemophilia
Metabolic disorders (e.g. phenylketonuria, diabetes)
Kidney disease
Deafness
Marfan syndrome
Any other

Ethnicity
Eastern European (Ashkenazi), Jews (Tay-Sachs, Canavan risk, etc)
French Canadian or Cajun (Tay-Sachs risk)
Mediterranean region (hemoglobinopathy risk)
Asia, including Southeast Asia and Western Pacific (hemoglobinopathy risk)
Africa and Middle East (hemoglobinopathy risk)
South America and Caribbean (hemoglobinopathy risk)
Caucasian (cystic fibrosis)

Consanguinity
Recurrent pregnancy loss, stillbirth, or early infant death
Maternal metabolic disorder

Adapted from: The preconception office visit, Joyce A Sackey, Up-to-date, 2015

- ABO and Rh(D) blood type to detect possibility of blood incompatibility between would-be couple.
- Infectious diseases like HIV, hepatitis B, C and syphilis as suggested by history and examination.
- Sexually transmitted diseases (other than syphilis, HIV and HBV) such as gonorrhea and chlamydia as indicated by history.
- Psychosocial and domestic issues.
- Rubella immunity status of females.
- Hemoglobin levels to identify anemia.
- Tests for diabetes, hypertension, etc. if directed by family history or examination.
- Estimation of BMI for identifying obesity/underweight.

INFORMATION AND COUNSELING

Imparting knowledge is the second step in preconception care, without which couple will not be ensured of preconception benefits.

- Imparting knowledge about reproductive biology and physiology of pregnancy.
- Information about different contraceptive options available for the couple.

  Avoiding unplanned and unwanted conception is an integral part of optimizing pregnancy outcomes and hence providing contraceptive advice when desired is an important step of preconceptional care. (Strength of recommendation: A; quality of evidence IV).[5]

- Timing of pregnancy

  The optimum biological age for pregnancy is between 20 and 35 years of age. Pregnancy at an earlier or later age entails poor outcomes for both the mother and the child. There is a 3 times increased risk of hypertension in women with advanced maternal age, besides the risk of placenta praevia, gestational diabetes, preterm birth and increased cesarean section rate, poor fetal outcomes in terms of increased risks of chromosomal anamolies, stillbirths, perinatal mortality, low birth weight and nursery admissions.[6] Teenage pregnancy has its own adverse effects in terms of abortions, anemia, eclampsia, preterm birth, operative deliveries and poor fetal outcomes.[7]

COUNSELING

Counseling has two aspects: General and specific.

General Counseling

- Regarding safe sex practices and behavior.
- Regarding ill effects of substance abuse, alcohol.
- About optimizing pre-pregnancy weight.
- About healthy lifestyle, nutritive diet and regular exercise.

Specific Counseling

Genetic counseling: Genetic counseling is the process by which patients at risk of a genetic disorder (identified by taking the genetic history—Table 8.1) are informed about the consequences of the disorder; the probability of developing and transmitting it; and the means by which this risk can be reduced. It aims at empowering the couple to take the appropriate decision and course of action in view of the risk.

- A thorough history for presence of high-risk factors needs to be taken to identify couples requiring genetic counseling (strength of recommendation: A; quality of evidence III).[5]
- The risk is assessed by taking detailed structured history of previous pregnancy outcomes, chromosomal disorders, genetic diseases in family and consanguinity. Consanguineous couples are counseled regarding increased risk of autosomal recessive disorders.
- A three generation pedigree chart is made in the cases with suggestive history.

The genetic counseling allows patients planning pregnancy to make informed reproductive decisions about adoption, surrogacy, use of donor sperm, *in vitro* fertilization after preimplantation genetic diagnosis, avoidance of pregnancy, and prenatal diagnosis.

INTERVENTIONS

It is the final step towards preconception care by providing appropriate care to cure the curable conditions and optimizing chronic conditions. The following care is recommended:

- Confirmatory tests if any of the screening test is positive
- *Immunization:* Strongly advisable vaccines are MMR and HBV vaccines and desirable are HPV, varicella, influenza, Tdap. Live vaccines (varicella, measles, mumps,

rubella) should be administered at least one month prior to pregnancy.

i. All women in the preconception period should be screened for rubella infection and vaccinated if non-immune (strength of recommendation: A; quality of evidence III).[5] The combination of MMR is preferred over rubella vaccine alone for the purpose of routine preconceptional vaccination (strength of recommendation: A; quality of evidence III).[5]

ii. Vaccination for hepatitis B should be done for eligible partners. Individuals, whose partner is HBsAg positive, are given a booster dose of hepatitis B vaccine if they have been vaccinated before.

iii. HPV vaccination is recommended for all women and girls (9–26 yrs) if not completed earlier (strength of recommendation: B; quality of evidence III).[5]

iv. Varicella immunization should be given to those who are unimmunized.

v. Influenza vaccine to those women who are at risk (becoming pregnant in influenza season). It can be given in pregnancy. Strength of recommendation: A; quality of evidence III).[5]

vi. Tetanus, diphtheria and pertusis (TdaP) can be offered if unimmunized and at risk of exposure to diptheria infection. It should be administered again during pregnancy in order to provide optimal protection to the baby during its first month of life (strength of recommendation: A; quality of evidence III).[5]

- *Genetic testing:* Offered in the preconceptional period when history suggestive. The types of testing could be (strength of recommendation: A; quality of evidence III):[5]

  i. Diagnostic work up
  ii. Carrier screening (e.g. thalassemia)
  iii. Prenatal testing (e.g. thalassemia, hemophilia, Duchenne muscular dystrophy)
  iv. Preimplantation testing (e.g. in couples with chromosomal translocations/X-linked disorders)
  v. Newborn screening

- *Referral:* Women with chronic disorders are referred to respective specialists so as to ensure good control of disease prior to pregnancy.

- *Folic acid supplementation:* All women of childbearing age should be recommended to take folic acid 0.4/0.5 mg daily, at least 1 month before conception to up to 3 months after conception to reduce the risk of neural tube defects (NTDs) (strength of recommendation: A; quality of evidence I).[5] Patients at high risk of NTDs (history of NTDs in women or their partners, or NTDs in previous pregnancy, hemolytic anemia, increased BMI (>30 kg/m^2), women with known MTHFR mutation, hemoglobinopathies (beta thalassemia), and medications affecting folate metabolism such as anticonvulsants), should be given 4 mg of folic acid at least 1 month before conception to up to 3 months after conception (strength of recommendation: A; quality of evidence I).[5]

- Iron supplementation in iron deficiency anemia.

- Treatment of STIs (gonorrhea, chlamydia, trichomonas, HIV) detected during screening.

- Cessation of smoking, alcohol and drugs before pregnancy.

- Limit caffeine consumption to less than 200 to 300 mg per day.

- Only cooked fish should be consumed and fish high in mercury content avoided.

- Glycemic control in women with diabetes— the American Diabetes Association recommends aiming for an A1C <7 percent prior to conception.[8]

- Optimizing weight in obese and lean.

- Replacement of teratogenic drugs like ACE inhibitors, ARBs, statins, lithium, valproic acid, streptomycin, tetracycline, methotrexate, etc. by safer alternatives a few months before pregnancy.

- Psychosocial issues needs to be tackled and discontinuation of substance abuse.

PRECONCEPTIONAL COUNSELING FOR WOMEN WITH MEDICAL DISORDERS

Various studies have shown improved outcome with preconceptional intervention in the following disorders.[9]

Diabetes: Preconceptional advice on diet, exercise and weight loss is crucial for good outcome. Women should be explained that with good glycaemic control the risks of miscarriage, congenital malformations, stillbirth and neonatal death is reduced. ADA recommends a HbA1C of <7% before planning pregnancy. Women with higher levels are at an increased risk of miscarriage and teratogenesis.

Hypertension: All women in the preconceptional period should be screened for hypertensive disorders especially those with previous hypertensive disorders in pregnancy, renal disease, autoimmune disorders or thrombophilias (strength of recommendation: A; quality of evidence IV).[5] ACE inhibitors should be stopped before pregnancy (fetal growth restriction, oligohydramnios, renal failure in fetus) and replaced by safer alternatives. Those with chronic hypertension should be assessed for end organ damage (heart, eye, kidney) and referred to the concerned specialist.

Asthma: Asthmatic women who are planning for conception should be advised about the probable asthma aggravation with pregnancy and need for achieving asthma control prior to conception with suitable pharmacotherapy (strength of recommendation: A; quality of evidence III).[5] Inhaled medications; both B2 agonists and steroids are safe in pregnancy and should be used for disease control (strength of recommendation: A; quality of evidence III).[5]

Thyroid disorders: Routine screening of thyroid function in women planning a pregnancy is advisable. Both in hypothyroidism and hyperthyroidism prompt referral to an endocrinologist is required.

Cardiac disease: All women should have at least a basic clinical cardiac assessment in the preconceptional period and referred to a cardiologist if required (strength of recommendation: A; quality of evidence IV).[5] Genetic counseling should be offered to women with congenital heart disease. (Strength of recommendation: A; quality of evidence III).[5]

Women should be advised strongly against pregnancy in following conditions (strength of recommendation: A; quality of evidence IV):[5]

- Severe pulmonary arterial hypertension of any cause
- Severe systemic ventricular dysfunction
- NYHA III–IV or LVEF <30%
- Previous peripartum cardiomyopathy with any residual impairment of left ventricular function
- Severe left heart obstruction
- Marfan syndrome with aorta dilated >40 mm
- Aortic dilatation > 50 mm in aortic disease associated with bicuspid aortic valve
- Native severe coarctation

Effective contraception like progesterone only contraceptives should be offered to the women who are not desirous of fertility at the present time or have been advised to postpone pregnancy for optimization of the cardiac condition.

Epilepsy: Women suffering from epilepsy should be counseled about the need to properly plan their pregnancies considering the risks of increased epileptic frequency in pregnancy, the potential effects of epilepsy and anticonvulsant drugs on pregnancy outcomes (strength of recommendation: A; quality of evidence III).[5] Women should preferably be placed on anticonvulsant monotherapy at the lowest effective dose to control seizures. It is recommended that the woman should be on a stable anticonvulsant regimen for at least 6 months (after dose modification or withdrawal) prior to conception (strength of recommendation: A; quality of evidence III).[5] Preconceptional folic acid (5 mg/day) is advised for women on anticonvulsants and to continue throughout pregnancy.

Chronic renal disease: Women should be informed that the pregnancy leads to aggravation of a renal disease, risk of hypertension (10% risk of fetal loss if pre-existing) and proteinuria. Renal disease during pregnancy is associated with risk of prematurity, growth restriction and deterioration in maternal renal function. Women with renal transplants should be asked to avoid pregnancy for a minimum of 2 years until renal function is optimized on a reduced amount of immunosuppressants.

Systemic lupus erythematosus: The prognosis is best when the disease has been quiescent for at least six months prior to pregnancy and the patient's underlying renal function is stable or near normal. Maternal medications may need to be changed because of potential fetal risks.

HIV infection: All HIV positive couples should be counseled to practice effective dual contraception (condom and/or hormonal contraception and/or intrauterine devices) to prevent unintended pregnancy until the viral load is suppressed below the limit of detection. The risk of transmission to the uninfected partner and the fetus should be discussed.

- The couple should be counseled about the need for initiation and continuation of combination antiretroviral therapy (cART) (strength of recommendation: A; quality of evidence IV)[5]
- For sero discordant couples, in whom the woman is HIV-positive, it is preferable to attempt home insemination with the partner's sperm during ovulation for 3 to 6 months before considering other methods. If the male partner is HIV positive, then a referral to a fertility specialist should be considered and an option of sperm washing with intrauterine insemination should be given (strength of recommendation: A; quality of evidence III).[5]

- Lifelong ART should be given to all HIV-infected pregnant and breastfeeding women (strength of recommendation: A; quality of evidence III).[5]

Psychosocial issues and substance abuse: Healthcare providers should screen for depression, anxiety and other psychotic disorders like mania and schizophrenia by a personal or family history and referral to psychiatrist accordingly. Couple needs to be told of untoward effects of substance abuse (nicotine, coccaine, alcohol, etc) on future pregnancy and discontinuation of same for the best pregnancy outcomes.

Thromboembolic diseases: These women are at higher risk of thromboembolic complications during pregnancy and adverse fetal outcomes. Women diagnosed to have APS can benefit from preconceptional aspirin and early initiation of prophylactic heparin therapy as soon as pregnancy is confirmed in the first trimester of pregnancy (strength of recommendation: A; quality of evidence I).[5]

PRECONCEPTIONAL COUNSELING FOR WOMEN WITH BAD OBSTETRIC HISTORY (BOH)

Women with BOH should be screened for the following and needs treatment accordingly:

i. Testing for antiphospholipid antibody syndrome (APS) with anticardiolipin antibodies (ACL), lupus anticoagulant (LAC), and anti-beta 2 glycoproteins (anti B2GP), in recurrent fetal demise in more than 10 weeks pregnancy, severe IUGR, severe oligohydramnios, early onset hypertension, eclampsia. If positive test needs to be repeated 12 weeks apart for confirmation.

ii. Karyotyping of parents in case of early first trimester abortions.

iii. Pelvic ultrasound and hysteroscopy to rule out uterine anomaly in cases of second trimester losses.

iv. VDRL to rule out syphilis

v. Blood sugar and thyroid profile

vi. Genetic counseling in 3 or more early trimester losses

vii. Folic acid supplementation for previous history of neural tube defects.

According to WHO, preconception care has a positive effect on a range of health outcomes and preconception care not only aims primarily at improving maternal and child health, but it also brings health benefits to the adolescents, women and men, irrespective of their plans to become parent.[10] The Preconception Care Work Group of the Centers for Disease Control, USA also recommends that preconception care should be an essential part of primary and preventive care.[2,11] Hence it is concluded that preconception counseling and screening is a very important intervention, which prepares a woman embarking into motherhood, physically as well as emotionally and hence ensures a good maternal and fetal outcome.

REFERENCES

1. Johnson K, Posner SF, Biermann J etal. Recommendations to improve Preconceptional Health and Health care. MMWR Recomm Rep 2006;55:1.

2. Atrash H, Jack BW, Johnson K et al. Where is "W"oman in MCH? Am J Obstet Gynecol 2008; 199:S259.

3. Jack BW, Atrash H, Coonrod DV et al. The clinical content of preconceptional care: an overview and preparation of this supplement. Am J Obstet Gynecol 2008; 199:S266.

4. Farahi N, Zolotor A. Recommendations for preconception Counseling and Care. Am Fam Physician 2013;88 (8):499–506.

5. FOGSI Good Clinical Practice recommendations on preconception care- India (2016).

6. Sohni V Dean, Zohra S Lassi, Ayesha M Imam, Zulfiqar A Bhutta. Preconception care: promoting reproductive planning. Reproductive Health 2014, 11(Suppl 3):52.

7. SH Mahavarkar. A Comparative study of teenage pregnancy. J obstet Gynecol. 2008 Aug;28(6):604–7.

8. American Diabetes Association. (12) Management of diabetes in pregnancy. Diabetes Care 2015; 38 Suppl: S77.

9. Joyce A Sackey The preconception office visit, Up-to-date, 2015, Oct 09, 2015.

10. Meeting to develop a global consensus on preconception care to reduce maternal and childhood mortality and morbidity. Geneva, World Health Organization, 2013.

11. Korenbrot CC1, Steinberg A, Bender C, Newberry S. Preconception care: a systematic review. Matern Child Health J. 2002 Jun;6(2):75–88.

Antenatal Care and Prenatal Counseling

Aparna Sharma, Juhi Bharti, Alka Kriplani

Antenatal care (ANC) is defined as care provided by health care professionals to pregnant women in order to attain the best possible outcome for both mother and baby. Every woman deserves quality antenatal care that gives them a positive pregnancy experience. WHO has defined positive pregnancy experience as "maintaining physical and sociocultural normality, maintaining healthy state for mother and baby, having effective transition to positive labor and birth and achieving positive motherhood (including maternal self-esteem, competence and autonomy)".[1] Women-centered approach has to be provided.

FREQUENCY OF ANTENATAL APPOINTMENTS

The latest World Health Organisation (WHO) recommendations have modified four visit ANC model to a minimum of eight contacts.[1] The word "contacts" is used to emphasize upon active participation of both the expectant mothers and health care providers. The recommended schedule is shown in Fig. 9.1. They have increased the number of visits in third trimester so as to detect and prevent high risk conditions like hypertensive disorders.

National Institute of Clinical Excellence (NICE) recommends minimum of 10 appointments for a nulliparous uncomplicated pregnancy and 7 for a parous women.[2] The frequency of appointments may vary depending on individual's requirements.

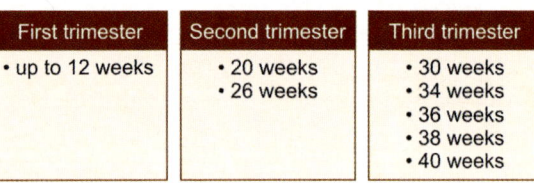

First trimester	Second trimester	Third trimester
• up to 12 weeks	• 20 weeks • 26 weeks	• 30 weeks • 34 weeks • 36 weeks • 38 weeks • 40 weeks

Fig. 9.1: WHO ANC model for an uncomplicated pregnancy[1]

PROTOCOLS OF ANTENATAL CARE

1. Initial assessment (clinical)
2. Initial assessment (laboratory)
3. Nutritional supplements and lifestyle advice
4. Immunization in pregnancy

1. Initial Clinical Assessment

1.1 *Height, weight, and body weight index (BMI):* All these should be measured on first visit. Pre-pregnancy BMI helps in deciding appropriate nutrition and any special care if needed as both under-nutrition and obesity are associated with poor pregnancy outcomes.

Measurement of weight at every visit to assess weight gain during pregnancy is not recommended except in certain clinical circumstances.

1.2 *Breast examination:* Traditionally breast examination was done to look for flat or inverted nipples to avoid problems during postnatal breastfeeding. However, it was not found to improve breastfeeding issues. Therefore, routine breast examination during pregnancy is not required.

1.3 *Blood pressure:* Blood pressure measurement on first visit detects women with pre-existing chronic hypertension. It should be measured at every antenatal visit and more frequently if at high risk of hypertensive disorders.

1.4 *Urinalysis:* Urinalysis is done to detect protein and sugar in urine by dipstick. Proteinuria of 1+ or more warrants further tests and treatment accordingly. Glycosuria of 2+ on one occasion or 1+ on two or more occasions requires oral glucose tolerance testing to rule out diabetes.

1.5 *Pelvic examination:* Routine pelvic examination should not be done as it does not predict accurate gestational age, preterm birth or cephalopelvic disproportion.

1.6 *Psychiatric screening:* Any history of mental illness in the past or present, including schizophrenia, bipolar mood disorder, etc. should be asked about. Family history of perinatal illness also is a risk factor. If positive history is present, woman should be asked if they wish to receive help regarding the same. For symptoms of depression, Edinburgh postnatal depression scale should be used.

2. Initial Assessment (Laboratory)

2.1 *Anemia:* Screening for anemia is done at first contact and repeated at 28 weeks. Treatment is done accordingly.

2.2 *Blood group:* ABO blood grouping and rhesus D status is advised at first contact. In cases of Rh-negative non-sensitised pregnancy, Anti-D is advised at 28 weeks to prevent isoimmunisation.

2.3 *Screening of infections*

    2.3.1. *HIV:* Screening for HIV should be offered to all women early in pregnancy.

WHO recommends provider-initiated testing and counselling (PITC) for HIV which includes pre-test counselling, obtaining consent with the mother having rights to refuse testing. ART is recommended for all HIV positive women irrespective of CD4 count to improve maternal health, mother to child transmission and prevent horizontal transmission to the partner. In case of generalised HIV epidemic, all HIV negative pregnant women need to be retested in third trimester, labor or postpartum. In case of concentrated HIV epidemic, retesting is indicated in a serodiscordant couple or vulnerable population group.

2.3.2. *Syphilis:* Testing should be done in the first trimester. Protocol is shown in Fig. 9.2.

2.3.3. *Hepatitis B:* Screening for Hepatitis B virus is offered so that timely post-natal interventions can be done to prevent mother to child transmission.

2.3.4. *Rubella:* If a woman is found to be susceptible to rubella, then vaccination should be offered in postnatal period for protection in future pregnancies.

2.3.5. *Tuberculosis:* In areas with high prevalence of tuberculosis (TB) (i.e. >100/100000 population), systematic screening for active TB in pregnant females should be considered. Early initiation of anti-tubercular treatment (ATT) is associated with improved fetomaternal outcomes.[1]

2.3.6. *Asymptomatic bacteriuria:* Midstream urine culture is recommended early in pregnancy to rule out asymptomatic bacteriuria. It prevents risk of pyelonephritis.

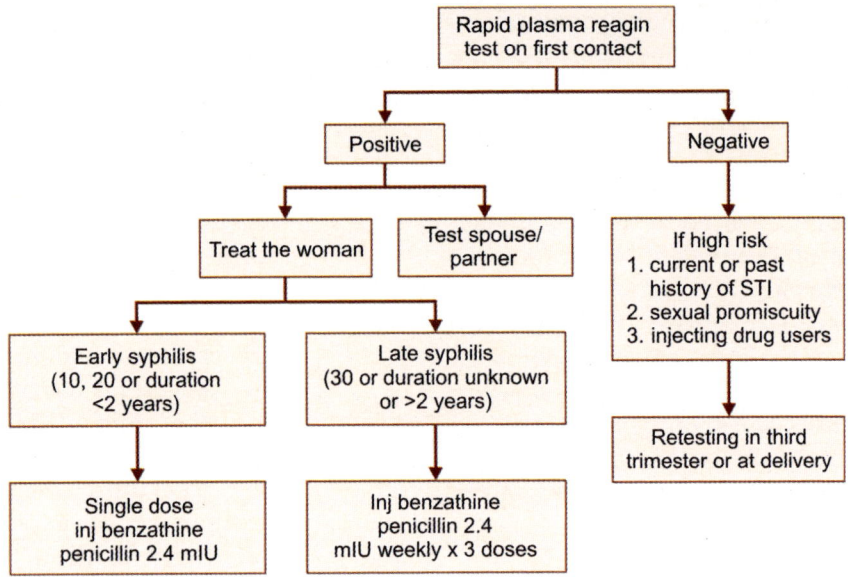

Fig. 9.2: Protocol of syphilis testing and management in pregnancy[3]

2.4 *Glucose tolerance test:* Indian women are at high risk of diabetes in pregnancy. Universal screening is recommended for all women. The screening for diabetes should start from the first visit by a blood sugar fasting and post prandial. In women who are at high risk of diabetes like BMI >30 kg/m^2, history of diabetes in previous pregnancy, previous baby ≥4.5 kg (macrosomia), family history of diabetes in first degree relative, family origin with increased risk of diabetes like South Asian, Middle Eastern, etc. a glucose tolerance test can be offered at the first visit itself. All women irrespective of their previous screening status should be offered glucose tolerance test with 75 gm glucose. The interpretation of the tests is shown in Table 9.1.

3. Nutritional Supplements and Lifestyle Advice

Pregnancy is a state of increased nutritional requirements to meet both maternal and fetal demands. A balanced intake of energy, proteins, vitamins and minerals is required to have an optimum outcome.

3.1 *Iron and folic acid:* Global prevalence of anemia in pregnant females is 38.2%, South East Asia and Africa being the most affected regions (48% and 46.3% respectively).[4] WHO recommends daily oral intake of 30 to 60 mg elemental iron and 400 µg of folic acid to prevent maternal anemia, preterm births, and puerperal sepsis. In cases where women are intolerant to oral iron due to side effects and prevalence of anemia in pregnant population is less than 20%, they can take intermittent oral iron and folic acid supplement for prevention of anemia. The intermittent dose recommended is 120 mg oral elemental iron and 2.8 mg folic acid weekly. If anemia is diagnosed, the dose is increased to 120 mg elemental iron and 400 µg of folic acid daily till hemoglobin level rises to normal value followed by standard prophylactic dose.[1]

Folate (vitamin B$_9$) deficiency can be a cause of fetal neural tube defects. Therefore, intake of folic acid in a dose of 400 µg per day up to 12 weeks of pregnancy is recommended.[2]

Table 9.1	Interpretation of glucose screening tests				
	Fasting mg/dL	1 hour mg/dL (75-g OGTT)	2 hour mg/dL (75-g OGTT)	HbA1C	Random
ADA DM/overt Diabetes (Any one criteria) (First trimester or anytime in pregnancy)	>126		200	>6.5%	200 mg/dL with symptoms of hyperglycemia
WHO/IADPSG DM (First trimester or anytime in pregnancy	≥126		≥200	≥6.5	
GDM (75-g OGTT)	92 mg/dL	180	153 mg/dL		

ADA: American Diabetes Association
DM: Diabetes Mellitus
IADPSG: International Association of Diabetes in Pregnancy Study Group
WHO: World Health Organisation
OGTT: Oral Glucose Tolerance Test 75 gm Glucose load

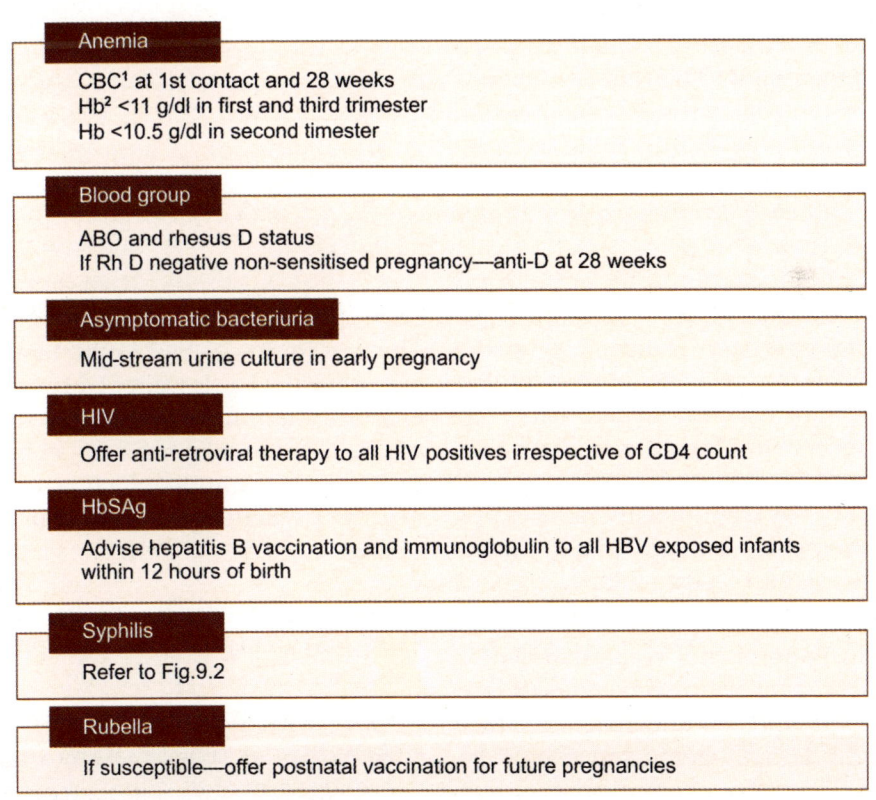

Anemia
CBC[1] at 1st contact and 28 weeks
Hb[2] <11 g/dl in first and third trimester
Hb <10.5 g/dl in second timester

Blood group
ABO and rhesus D status
If Rh D negative non-sensitised pregnancy—anti-D at 28 weeks

Asymptomatic bacteriuria
Mid-stream urine culture in early pregnancy

HIV
Offer anti-retroviral therapy to all HIV positives irrespective of CD4 count

HbSAg
Advise hepatitis B vaccination and immunoglobulin to all HBV exposed infants within 12 hours of birth

Syphilis
Refer to Fig.9.2

Rubella
If susceptible—offer postnatal vaccination for future pregnancies

[1]CBC: Complete blood count; [2]Hb: Hemoglobin

Fig. 9.3: Summary of laboratory assessment

3.2 *Calcium and Vitamin D:* Maternal hypocalcemia and vitamin D deficiency are found to be associated with adverse maternal and fetal outcomes like pre-eclampsia, low birth weight and neonatal hypocalcemia.[5]

Routine screening of vitamin D in pregnancy is not cost effective and hence not recommended. Cochrane meta-analysis finds that as far as vitamin D is concerned, evidence of routine supple-mentation is still unclear. Though it can reduce risk of pre-eclampsia and low birth weight baby, preterm births may get increased.[6] For women at high risk of vitamin D deficiency such as dark pigmented skin (e.g. South Asian origin) or limited sunlight exposure, RCOG recommends 1000 units/day. The doses are shown in Table 9.2. Pregnant women should be encouraged to increase sunlight exposure duration and adequate nutrition to maintain vitamin D levels.

If vitamin D deficiency is documented, the dose advised is 20000 IU cholecalciferol weekly or 10000 IU ergocalciferol twice weekly for 4–6 weeks followed by standard dose.[7]

WHO recommends daily calcium supplementation with 1.5–2 g/day oral elemental calcium from 20 weeks gestation.[1] It can be taken in three divided doses to improve compliance. This is in accordance with Cochrane review that suggests calcium supplementation in high doses ($\geq$1 g/day) significantly reduces risk of pre-eclampsia especially in population with low calcium dietary intake.

3.3 *Vitamin A:* Routine antenatal vitamin A supplementation is not recommended to improve pregnancy outcomes. According to NICE, vitamin A in the dose > 700 µg/day might be teratogenic and should be avoided. They also recommend avoiding consumption of dietary products rich in vitamin A like liver and liver products.[2]

3.4 *Other vitamins and micronutrients:* Zinc and other micronutrients are not recommended in pregnancy to improve maternal and neonatal outcomes. Similarly, vitamin B_6, C and E are also not recommended.

3.5 *Alcohol, smoking and substance abuse:* At the time of first contact and at every antenatal appointment, all women should be enquired regarding tobacco use, second hand smoke, alcohol consumption or any other substance abuse. Specific risks of these addictions have to be discussed. Smoking cessation support has to be provided. Women have to be informed that they should stop drinking alcohol particularly in first trimester.

3.6 *Exercises in pregnancy:* Pregnancy is an appropriate time for healthy lifestyle and behaviour modifications as these women are motivated and frequently come in contact with health care providers. Women who were previously active should continue with the exercises and previously inactive women should be encouraged to begin exercises from second trimester.[8]

What is the appropriate duration of exercise?

Thirty minutes of low to moderate intensity exercises like brisk walk (3-4 times per week) is appropriate for a pregnant woman. There are also set guidelines for target heart rate up to 160 beats/minute which is safe and good for ideal weight gain and improves blood glucose control.

Activities to be strictly avoided are:
- Certain yoga postures and supine positions that can cause hypotension.
- Contact sports (e.g. basketball)
- Activities at risk of falling (e.g. off-road cycling, downhill skiing, etc.)
- Scuba diving that may cause fetal birth defects and decompression diseases.[9]

Table 9.2	Nutritional supplements and their doses		
Nutrient	*Dose*		*Advantages*
Iron	30–60 mg elemental iron daily		*Decreases:*
	If intolerant: 120 mg weekly elemental iron		Maternal anemia
	Deficiency: 120 mg elemental iron daily		Puerperal sepsis
			LBW, preterm births
Folic acid	400 µg daily or 2.8 mg weekly		Decreases: Maternal anemia
			Fetal neural tube defects
Calcium	1.5–2 gram oral elemental calcium daily (from 20 weeks)		Decreases: Pre-eclampsia
Vitamin D (RCOG)	400 units daily (10 µg)		Decreased:
	If risk of pre-eclampsia	800 units + calcium	Pre-eclampsia, LBW
	If high risk of Vit D deficiency	1000 units	
	If documented evidence	20000 IU weekly × 4–6 weeks	
Vitamin B$_6$, C and E	Not recommended		
Zinc	Not recommended		

3.7 *Travel in pregnancy*

Air travel: Occasional air travel is safe in pregnancy in the absence of any known medical or obstetrical complication. The safest time to fly is before 36 weeks in uncomplicated singleton pregnancy and 32 weeks in uncomplicated twin pregnancy.

Car travel: While travelling by car, seat belts have to be used correctly. Three-point seat belt means using both shoulder strap and lap strap. The shoulder strap should lie over one shoulder and pass between the breasts. The lap strap should lie below the belly, over the hip bones.

4 Immunization

4.1 *Tetanus toxoid (TT) vaccination:* It is a must for all pregnant females for prevention of maternal and neonatal tetanus. The dose depends on previous vaccination exposure history and is shown in Tables 9.3 and 9.4.[10]

4.2 *Tetanus, diphtheria and pertussis (Tdap):* Centres for disease prevention and control (CDC) recommends single dose of Tdap vaccine to mothers in every pregnancy irrespective of previous vaccination history.[11] Though it is safe to be given at any point in pregnancy, but the optimal time recommended is between 27 weeks and 36 weeks. This is to achieve adequate maternal antibody formation and thus ensuring passive transfer of antibodies to neonate. This prevents infant morbidity and mortality due to pertussis (whooping cough). This should be given at least two weeks before delivery to allow maternal antibody response. If Tdap vaccine has not been given during pregnancy, she should receive it in immediate postpartum period.

In cases of unknown or incomplete tetanus vaccination history, they should receive three doses of tetanus and reduced diphtheria toxoids (Td). The dosing schedule to be followed is 0, 4 weeks and third dose after 6–12 months. The third dose of Td should be replaced by Tdap vaccine preferably in third trimester.

4.3 *Flu vaccine:* Centers for Disease control and Preventions' Advisory Committee on Immunization Practices (ACIP) and

Table 9.3	Immunization schedule for women without previous exposure to TT, Td or DTP	
Dose of TT or Td	When to give	Expected duration of protection
1	As early as possible	None
2	At minimum of 4 weeks after TT1	1–3 years
3	At minimum of 6 months after TT2 or in next pregnancy	At least 5 years
4	≥1 year after TT3 or in next pregnancy	At least 10 years
5	≥1 year after TT4 or in next pregnancy	For entire childbearing age and longer

Table 9.4	Immunization of women who were previously immunized during infancy, childhood or adolescence		
Age at last vaccination	Previous immunization (based on written records)	Recommended immunizations	
		At present contact/ pregnancy	Later (at intervals of at least one year)
Infancy	3 DTP	2 doses of TT/Td (at least 4 weeks interval between doses)	1 dose of TT/Td
Childhood	4 DTP	1 dose of TT/Td	1 dose of TT/Td
School age	3 DTP + 1DT/Td	1 dose of TT/Td	1 dose of TT/Td
School age	4 DTP + 1 DT/Td	1 dose of TT/Td	None
Adolescence	4 DTP + 1 DT at 4–6 years + 1 TT/Td at 14–16 years	None	None

ACOG recommend single shot of inactivated influenza vaccine to all females getting pregnant in flu season.[12] It can be taken at any gestation. Transplacentally acquired antibodies also protect the neonate from flu.

Conclusion: Antenatal period is an important opportunity to improve health of both mother and baby. It often serves as first contact of a woman with a health care provider and thus has the advantage of an overall health screening. It helps to reduce maternal and perinatal mortality. The summary of antenatal care for an uncomplicated pregnancy is given in Table 9.5.

REFERENCES

1. WHO recommendations on antenatal care for a positive pregnancy experience. Geneva: World Health Organisation 2016. [Online]. Available at: http://apps.who.int/iris/bitstream/10665/250796/1/9789241549912-eng.pdf

2. Antenatal care for uncomplicated pregnancy. National Institute of Clinical Excellence (NICE) 2008. [Online]. Available at: https://www.nice.org.uk/guidance/cg62/evidence.

3. Screening for Syphilis during pregnancy. Ministry of health and family welfare 2014. Available at: http://nrhm.gov.in/images/pdf/programmes/maternal-health/guidelines/Syphilis_Doc_Low-res_5th_Jan.pdf

4. The global prevalence of anaemia in 2011. Geneva: World Health Organization 2015 [Online]. Available at: http://apps.who.int/iris/bitstream/10665/177094/1/9789241564960_eng.pdf.

5. Aghajafari F, Nagulesapillai T, Ronksley PE, Tough SC, O' Beirne M, Rabi DM. Association between maternal serum 25-hydroxyvitamin D level and pregnancy and neonatal outcomes: systematic review and meta-analysis of observational studies. BMJ 2013; 346: f1169

6. Hofmeyr GJ, Lawrie TA, Atallah ÁN, Duley L, Torloni MR. Calcium supplementation during pregnancy for preventing hypertensive disorders and related problems. Cochrane Database of Systematic Reviews 2014,(6): CD001059. DOI: 10.1002/14651858.CD001059.pub4.

Table 9.5		Antenatal care in an uncomplicated pregnancy	
Number of visit	*Timing*	*Antenatal care*	*Danger signs*
First contact	Up to 12 weeks	Measure blood pressure (BP), height, weight, BMI Folic acid supplementation Advise all antenatal investigations • Blood group and rhesus D typing • Complete blood count (CBC) • Offer screening for HIV, Hepatitis B, syphilis, rubella susceptibility • Urine tests for routine microscopy and culture • Assess risk of diabetes and offer GTT accordingly • Offer first trimester screening (early USG and dual screen) Advise stoppage of smoking, alcohol or any substance abuse Offer TT immunisation Enquire about intimate partner violence and offer support	Bleeding Pain Syncopal attack Excessive vomiting Dehydration High fever Pain or burning during micturition Leg or calf pain
Second contact	20 weeks	Discuss results of all previous tests and their implications Measure BP, urine for protein Advise anomaly scan Discuss role of exercises, its benefits Supplement iron and calcium	Bleeding per vaginum Leaking per vaginum Pain abdomen High BP record
Third contact	26 weeks	Glucose tolerance test Measure BP, Urine for protein Measure symphysio fundal height (SFH) If Rh negative, advise anti-D after indirect Coombs' test	Bleeding per vaginum Leaking per vaginum Pain abdomen High BP record
Fourth contact	30 weeks	Measure BP, urine for protein Measure SFH Continue iron and calcium Repeat CBC	Bleeding per vaginum Leaking per vaginum Pain abdomen High BP record
Fifth contact	34 weeks	Measure BP, urine for protein Measure SFH Continue iron and calcium	Headache, epigastric pain, blurring of vision Decreased fetal movements Seizures
Sixth contact	36 weeks	Same as above + Birth preparedness Advise regarding signs of labour Information on breast feeding Confirm presentation Offer ECV if breech	
Seventh contact	38 weeks	Measure BP, urine for protein Birth and emergency plan Discuss epidural analgesia	
Eighth contact	40 weeks	Offer induction of labour at 41 weeks Postnatal care, pregnancy spacing	

7. Vitamin D in pregnancy. Scientific impact paper 43, RCOG 2014: 1–11.

8. Schmidt SM, Chari R, Davenport MH. Exercise During Pregnancy: Current Recommendations by Canadian Maternity Health Care Providers. J Obstet Gynaecol Can 2016; 38(2): 177–178.

9. Physical activity and exercise during pregnancy and the postpartum period. Committee Opinion No. 650. American College of Obstetricians and Gynecologists. Obstet Gynecol 2015; 126: e135-42.

10. Maternal immunization against tetanus. Geneva: World Health Organisation 2006. [Online]. Available at: http://www.who.int/reproductive health/publications/maternal_perinatal_health/immunization_tetanus.pdf

11. Update on immunization and pregnancy: tetanus, diphtheria, and pertussis vaccination. Committee Opinion No. 566. American College of Obstetricians and Gynecologists. Obstet Gynecol 2013;121: 1411-4.

12. Influenza vaccination during pregnancy. Committee Opinion No. 608. American College of Obstetricians and Gynecologists. Obstet Gynecol 2014; 124: 648–51.

Antenatal Care: Minimum Required

Pradnya Supe, Dheera Samadariya

Preconceptional Care

- Because health during pregnancy depends on health before pregnancy, antenatal care should ideally begin before conception. A comprehensive preconceptional program has the potential to reduce risks, promote healthy lifestyles, and bring about readiness for pregnancy.

- Folic acid supplementation should be started while planning pregnancy as most major malformations have already occurred in the first 8 to 10 weeks by the time pregnancy is confirmed in most cases. Recommended dose: 400 µg daily in all women and 4 mg/day in those with previous history of neural tube defect (NTD).

- Immunity to rubella should be checked and immunization provided to the non-immune.

- Pre-conceptional counseling is particularly helpful in women with medical ailments such as diabetes, epilepsy, asthma, depression, HIV for optimizing their medication and the medical condition to make the pregnancy safer for the mother and improve the perinatal outcome.

Antenatal Appointments (Schedule and Content)[1]

- The schedule below, which has been determined by the purpose of each appointment, presents the recommended number of antenatal care appointments for women who are healthy and whose pregnancies remain uncomplicated in the antenatal period: 10 appointments for nulliparous women and 7 for parous women.

- These appointments follow the woman's initial contact with a healthcare professional when she first presents with the pregnancy and from where she is referred into the maternity care system.

First Contact with a Healthcare Professional

- Give information (supported by written information and antenatal classes), with an opportunity to discuss issues and ask questions.

- Topics covered should include:
  - Folic acid supplementation, food hygiene, including how to reduce the risk of a food-acquired infection,
  - Lifestyle advice, including smoking cessation, recreational drug use and alcohol consumption;
  - All antenatal screening, including risks and benefits of the screening tests.

Booking Appointment (ideally by 10 weeks or the earlier the better)

- At the booking appointment, give the following information (supported by written information and antenatal classes), with an opportunity to discuss issues and ask questions.

- Topics covered should include:
  - How the baby develops during pregnancy, nutrition and diet, including vitamin D supplementation,
  - Exercise, including pelvic floor exercises,
  - Antenatal screening, including risks and benefits of the screening tests,
  - Pregnancy care pathway, place of birth,
  - Breastfeeding, including workshops, participant-led antenatal classes, maternity benefits.
- At this appointment:
  - Detailed history covered under 3 major headings—medical, obstetrical and family history. Complete details of all previous deliveries and abortions should be taken.
  - Identify women who may need additional care and plan pattern of care for the pregnancy.
  - Breast and thyroid examination
  - Hemoglobin estimation, CBC
  - Check blood group and rhesus D status. If patient's blood group is Rh negative—check husband's blood group. If he is Rh positive, patient's indirect Coombs' test (ICT) and Rh titres to be checked.
  - Offer screening for hemoglobinopathies, anemia, red-cell alloantibodies, hepatitis B virus, HIV, rubella susceptibility and syphilis (VDRL)
  - Offer screening for asymptomatic bacteriuria
  - Offering screening for Down's syndrome
  - Offer early ultrasound scan for gestational age assessment and early diagnosis of multiple pregnancy.
  - Offer ultrasound screening for structural anomalies
  - Measure height, weight and calculate body mass index (pre-pregnancy BMI is ideal)
  - Measure blood pressure and test urine for proteinuria
  - Offer screening for gestational diabetes and pre-eclampsia using risk factors.
  - Fasting and post-prandial blood sugars in select high-risk population in whom one suspects overt diabetes (obesity, h/o macrosomia or malformed baby, unexplained stillbirth, or GDM in previous pregnancy).
  - Diagnosis of GDM is by the glucose tolerance test with 75 g (WHO recommendation) or 100 g[2–4] One stage screening approach). Retesting is advised at 24–28 weeks if negative. Routine testing is done at 24–28 weeks.
  - Ask about any past or present severe mental illness or psychiatric treatment
  - Ask about mood to identify possible depression
  - Ask about the woman's occupation to identify potential risks.
  - At the booking appointment, for women who choose to have screening, the following tests should be arranged: Down's syndrome screening using 'combined test' at 11 weeks 0 days to 13 weeks 6 days (FTS, i.e. first trimester screening-NT scan with beta HCG and PAPPA)

16 Weeks

The next appointment should be scheduled at 16 weeks to:

- Review, discuss and record the results of all screening tests undertaken; reassess planned pattern of care for the pregnancy and identify women who need additional care.
- Investigate a hemoglobin level below 11 g/100 ml and consider iron supplementation if indicated
- Measure blood pressure and test urine for proteinuria
- All women should receive 2 injections of tetanus toxoid at an interval of 4–6 weeks

to prevent neonatal tetanus. The first is usually given at 16 weeks (can also be administered in the first trimester, especially in those who are likely to default in subsequent visits).

- *Iron and calcium supplementation:* WHO recommendation—60 mg of elemental iron with 400 µg folic acid continued up to 3 months postpartum[5]

  National Nutritional Anemia Control Programme: 100 mg elemental iron with 500 µg folic acid for a minimum 100 days up to 3–6 months postpartum[6]
- 1000 mg of calcium supplementation per day.
- Serum screening test (triple or quadruple) at 15 weeks 0 days to 20 weeks 0 days.

18 to 20 Weeks

- At 18 to 20 weeks, an ultrasound scan should be performed for the detection of structural anomalies.
- For a woman whose placenta is found to extend across the internal cervical os at this time, another scan at 32 weeks should be offered.

24–25 Weeks

At 25 weeks, another appointment should be scheduled for nulliparous women. At this appointment:

- Measure and plot symphysio-fundal height
- Measure blood pressure and test urine for proteinuria

28 Weeks

The next appointment for all pregnant women should occur at 28 weeks. At this appointment:

- Offer a second screening for anemia and atypical red-cell alloantibodies
- Investigate a hemoglobin level below 10.5 g/100 ml and consider iron supplementation, if indicated
- Repeat ICT in Rh negative women and offer anti-D prophylaxis 300 µg to rhesus-negative non-immunised women

- Measure blood pressure and test urine for proteinuria
- Measure and plot symphysis-fundal height

31 Weeks

Nulliparous women should have an appointment scheduled at 31 weeks to:

- Measure blood pressure and test urine for proteinuria
- Measure and plot symphysis-fundal height
- An interval growth ultrasound can be done at 31–32 weeks.
- Discuss and record the results of screening tests undertaken at 28 weeks; reassess planned pattern of care for the pregnancy and identify women who need additional care.

34 Weeks

At 34 weeks, all pregnant women should be seen again.

- Topics covered should include:
  - Preparation for labor and birth, including information about coping with pain in labor and the birth plan, recognition of active labor.
  - Counsel regarding labor analgesia and offer, if required.

At this Appointment

- Measure blood pressure and test urine for proteinuria
- Hemoglobin estimation to be repeated
- Measure and plot symphysis-fundal height

36 Weeks

At the 36-week appointment, all pregnant women should be seen again.

- Topics covered should include:
  - Breastfeeding information, including technique and good management practices that would help a woman succeed, such as detailed in the UNICEF Baby Friendly Initiative

– Care of the new baby, vitamin K prophylaxis and newborn screening tests
– Postnatal self-care, awareness of 'baby blues' and postnatal depression.

At this appointment:
- Measure blood pressure and test urine for proteinuria
- Measure and plot symphysis-fundal height
- Check position of baby
- For women whose babies are in the breech presentation, offer external cephalic version (ECV)
- Information on Cord Blood Banking can be provided.
- Routine private cord blood collection and banking is not recommended unless there is a medically indicated reason (RCOG and ACOG)[7–9]

37–38 Weeks

Another appointment at 38 weeks will allow for:
- Measurement of blood pressure and urine testing for proteinuria
- Measurement and plotting of symphysis-fundal height
- Hemoglobin estimation at term
- A full digital examination of the pelvis is undertaken, especially in primigravidae. The relationship of the fetal head to the pelvic brim becomes important as term approaches. In primigravidae, the head is usually engaged by 37 weeks. If it is not, rule out disproportion and look for other causes of a floating head.
- A term ultrasound may be done to look for the presentation, liquor, placental position, estimated fetal weight.

40 weeks

For nulliparous women, an appointment at 40 weeks should be scheduled to:
- Measure blood pressure and test urine for proteinuria
- Measure and plot symphysis-fundal height

41 Weeks

- For women who have not given birth by 41 weeks:
  - A membrane sweep should be offered
  - Induction of labor should be offered
- Blood pressure should be measured and urine tested for proteinuria
- Symphysis-fundal height should be measured and plotted

General

- Throughout the entire antenatal period, healthcare providers should remain alert to risk factors, signs or symptoms of conditions that may affect the health of the mother and baby, such as domestic violence, pre-eclampsia and diabetes.
- At each visit, give information, with an opportunity to discuss issues and ask questions, including discussion of the routine anomaly scan; offer verbal information supported by antenatal classes and written information.
- Provided pregnancy continues without any complications, subsequent visits can also be divided as: Every month till 28 weeks, every fortnight till 36 weeks, and then weekly until delivery.

Weight Gain

The average weight gain during pregnancy is about 10–12.5 kg. One-third of this weight increase is put on in the first 20 weeks, another third between the 20th and 30th week, and the remaining third between 30th week and term (Table 10.1).

Diet

- The first requisite is an adequate supply of fluid. It is the best and natural defense against constipation and helps to tread down stasis in the urinary tract with its resultant liability to infection.
- Diet should contain "roughage" foods as fruits and vegetables

Table 10.1	Recommended ranges of weight gain during Singleton gestations stratified by prepregnancy body mass index[a]			
Weight-for-height category			Recommended total weight gain	
Category	BMI		kg	lb
Low	< 19.8		12.5–18	28–40
Normal	19.8–26		11.5–16	25–35
High	26–29		7–11.5	15–25
Obese	>29		15	

[a]The range for twin pregnancy is 35 to 45 lb (16 to 20 kg). Young adolescents (< 2 years after menarche) and African–American women should strive for gains at the upper end of the range. Shorter women (< 62 in. or < 157 cm) should strive for gains at the lower end of the range. BMI = body mass index. (Weight in kg/height in m^2)

- Daily caloric intake increased by +300 kcal, protein intake needs to be increased by +15 g/day
- Pregnant women are prone to a deficient intake of iron and vitamin D, and iron deficiency anemia is still common.
- One litre of milk, or at least half litre, go far towards meeting the dietary demands of pregnancy, including 1000 mg of calcium, in addition to proteins and vitamins A and D.
- Adequate calcium supplementation may play a role in prevention of pre-eclampsia
- Salt encourages the urinary excretion of calcium so its intake should be minimized.
- Phosphorus is also necessary for the skeletal growth, supplied by eggs, cheese, milk, meat, liver and oatmeal
- Low serum zinc levels have been associated with an increased risk of congenital malformations, pre- and post-term labor and IUGR. Sources are milk, nuts, leafy and root vegetables, whole grains.
- Iodine deficiency can lead to cretinism and neonatal hypothyroidism. Iodized salt is advised.

Rest

- Rest is as important as exercise.
- A minimum of 8 hours in bed at night is recommended, and during the afternoon the expectant mother should be encouraged to lie down or at least put her feet up for one hour.

Exercise

- Being active in pregnancy helps in reducing backache, constipation and edema.
- Walking is the most natural and suitable and also the most preferred.
- Exercises that use multiple muscle groups increase the sensitivity of peripheral tissues to insulin and therefore particularly benefit type 2 diabetics. To this end, golf and swimming are also good sports.
- In general, it is safe to continue all forms of exercise to which the woman has been habituated before embarking on her pregnancy, including aerobic exercises.
- All exercises should be stopped short of fatigue and not continued to exertion.
- Exercises that involve prolonged periods of motionless standing and risk of abdominal trauma should be avoided.

Travel

- Long car journeys should be interrupted around every 2 hours in order to allow a change of position and the re-establishment of healthy circulation.
- Travel by air nowadays in well pressurized aircraft has no risk peculiar to pregnancy and is suitable for long journeys because of its freedom from fatigue. For very long

journeys, the same rules of stimulating the circulation apply.

Coitus

- Pregnant woman should be informed that sexual intercourse in pregnancy is not known to be associated with any adverse outcomes.
- However, many are of the opinion that there should abstinence in the first trimester and last 4 weeks of pregnancy.
- The use of lateral position in coitus prevents deeper penetration and this is worth remembering when counseling married women with mitral valve disease, where combination of tachycardia and supine position may cause acute pulmonary distress.

Dental Hygiene

- It is desirable to treat any focus of sepsis.
- Dental extractions under a general anesthetic are best undertaken in mid-pregnancy.

Antenatal Classes

- The primary objective is to create awareness and eliminate fear.
- These "mothers' meetings" are usually designed to include classes in mothercraft, physiotherapy including lessons in muscular relaxation and talks on the physiology of pregnancy and labor.
- Special attention is given to breastfeeding and antenatal preparation of breasts to ensure subsequent success.

REFERENCES

1. National Institute For Health and Care Excellence (NICE)(2014) Antenatal Care: NICE clinical guideline 62 (NG62).
2. O'Sullivan J, Mahan C, Charles D, et al. Screening criteria for high risk gestational diabetic patients. Am J Obstet Gynecol 1973; 116: 895.
3. National Diabetes Data Group. Classification and diagnosis of diabetes mellitus and other categories of glucose intolerance. Diabetes 1979; 28: 1979.
4. Carpenter MW, Coustan DR. Criteria for screening tests for gestational diabetes. Am J Obstet Gynecol 1982; 144: 768
5. Stoltzfus RJ, Dreyfuss ML. Guidelines for the use of iron supplements to prevent and treat iron deficiency anaemia. INACG. International life Series Institute Press, Washington,1998.
6. Indian Council of Medical Research. Supplementation trial in pregnant women with 60 mg,120 mg and 180 mg iron with 500 µg of folic acid. Indian Council of Medical Research, New Delhi, 1992,p.641.
7. Umbilical Cord Blood Banking. ACOG committee opinion 2015; number 648.
8. RCOG Scientific Advisory Committee Opinion Paper 2 umbilical cord blood banking (2006).
9. RCM Position Statement Commercial Cord Blood Collection (2011).

Antenatal Assessment of Fetal Wellbeing

Vandana Bansal

INTRODUCTION

Antenatal fetal surveillance is a domain in modern obstetrics which has attained great importance with the availability of high resolution ultrasound and Doppler, easy availability of non-stress test and better understanding of fetal physiology and disease. Significant improvement in maternal outcome and reduction in maternal mortality has allowed concentration on a second paradigm of obstetrical care, obtaining the best possible fetal outcome. Antenatal fetal surveillance offers a unique opportunity to timely detect morbid changes in the fetal status and implement intervention to avoid death or disability.

1 in 200 pregnancies end in a stillbirth and around 1 in 300 babies die in the first four weeks of life.[1] Stillbirths account for more than 55% of all perinatal mortality in the United States and could potentially be prevented by effective fetal surveillance tests.[2] 70 to 90% fetal deaths occur before the onset of labor. Approximately 30% of fetal deaths may be attributed to asphyxia (IUGR, prolonged gestation, maternal disorders), 15% to congenital malformations and chromosomal abnormalities, and 2% to infection. Just over 50% of stillbirths have no identifiable etiology. About half of these unexplained stillbirths may be associated with growth restriction.[1] The percentage of these unexplained stillbirth increase with advancing gestation. Antepartum late fetal death is the component of perinatal mortality that has shown greatest resistance to change over recent years.[3]

The aim of antenatal fetal surveillance is to ascertain fetal well-being as well as to determine fetuses which are compromised and at risk of intra or extra uterine fetal demise so that timely action can be taken so as to minimize perinatal morbidity and maximize their future potential. Timely action after an abnormal surveillance test may be either an immediate delivery or further specialized test. While each test on its own may not be diagnostic of fetal jeopardy, identifying pregnancy at-risk, combining various fetal surveillance methods and assessing changes in trends may improve predictive accuracy of these tests and aid clinical decision-making.

These tests should not only be sensitive enough to detect a compromise fetus but also specific so that they do not give an abnormal result for a healthy fetus. False positive results may lead to unnecessary intervention in a premature fetus increasing fetal and maternal morbidity and increasing parental anxiety. Antepartum fetal surveillance tests should help permit early delivery if it is considered safer than continued *in-utero* observation.

Although antenatal surveillance may identify suspected fetal compromise and gives the obstetrician time to intervene before progressive hypoxia leading to death or damage to the fetus, there are limitations to the generalized use of these tests of fetal well-being in low risk obstetric population. Unfortunately patients are not adequately informed about these limitations and often have unreasonable expectations that generate animosity towards the obstetrician and are a source of medico-legal problems. In the first place, there is no ideal single test that can detect all fetal problems. Secondly, none of the tests of fetal surveillance are absolutely specific and interventions due to a false positive test may lead to iatrogenic premature delivery. On many occasions the tests of fetal well-being are unable to timely detect fetal problem and intervention may be too late to be of any benefit. Finally acute catastrophic events like a placental abruption or a cord prolapse are not predicted by these tests of fetal surveillance and are not amenable to prevention.

At present only about 25% of stillbirths are preventable.[4] The most important preventable category is placental insufficiency secondary to abnormal placentation. Most tests of antepartum fetal surveillance have developed to determine the possibility of placental insufficiency and the fetal response to hypoxia. Prevention of stillbirth starts with identification of patients at high risk for placental insufficiency such as advanced maternal age, obesity, low socioeconomic status, cigarette smoking, previous poor obstetric history and maternal medical disorders (diabetes, hypertension, connective tissue disorders, thyroid dysfunction, cardiac, renal disease and severe asthma). These high-risk pregnancies are a small segment of obstetric population but are responsible for large majority of poor fetal outcome. High-risk identification remains the critical first step in antenatal surveillance.

NORMAL FETAL BEHAVIOR AND BIORHYTHMS THAT FORM THE BASIS OF ALL SURVEILLANCE TESTS

As obstetricians we are the primary physicians for the fetus and understanding its behavior, recognizing normal and abnormal response will aid in interpreting the tests of fetal well-being and decision-making.

Fetal biophysical activity is a reflection of an intact central nervous system (CNS). Normal fetal heart rate varies continuously due to vagal and sympathetic interaction to adjust cardiac output in order to meet the demands of fetal activity. Sympathetic impulses increase the fetal heart rate and generate acceleration while parasympathetic impulses decrease the heart rate. Acceleratory periods occur in response to movement and are superimposed on baseline variability. Baseline FHR is influenced by gestational age, maternal medications, fetal anomalies and fetal acidosis. Fetal sympathetic system matures earlier; hence the baseline heart rate is faster in preterm. As gestational age increases, FHR comes under parasympathetic predominance resulting in decreased baseline heart rate, increasing variability and responsiveness.

Maternal perception of fetal movement has been a traditional indicator of fetal well-being. The mother usually perceives 70 to 80% of these movements.[5] The periods of fetal activity last for about 40 minutes and the periods of rest about 20 minutes. Cycling between activity and quiescence occurs over a time span of approximately 40 minutes at term. Activity is highest in late evenings. Human fetal breathing movements occurs 30% of the time and gross body movements 10% of the time during the last 10 weeks of pregnancy. 90% of the gross body movements are accompanied by fetal heart accelerations in presence of an intact neurological coupling between fetal CNS and heart. Fetal breathing is inhibited by maternal smoking and narcotics and stimulated by maternal hyperglycemia.[6]

Individual fetal activity is regulated by specific centers in the central nervous system. Coordinated physiological activities such as breathing and movements require an intact non-hypoxic CNS. Individual centers are sensitive to varying degrees of fetal acidemia.

The presence of different fetal behavioral states is described in literature. These well-developed behavioral characteristics, similar to those observed in the neonate, are exhibited in the fetus in the third trimester. Criteria used for defining the four recognizable coordinated behavioral states in the human fetus as described by Nijhuis and colleagues[7] are as follows:

- State 1F (quiet sleep state) is a complete quiescent state, which can be regularly interrupted by brief gross body movements; fetal breathing if present is regular; eye movements absent; stable fetal heart rate with a narrow oscillatory bandwidth of less than ten beats per minutes. It is analogues to the non REM sleep.
- State 2F (active sleep state) includes frequent gross body movements, frequent eye movements, FHR accelerations superimposed on wider oscillation band-width (10–15 beats per minute); fetal breathing movements frequent and regular. This state is analogous to rapid eye movement (REM) in the neonate.
- State 3F (quiet awake state) includes continuous eye movements in the absence of body movements and no accelerations of the fetal heart rate. The existence of this state is disputed.
- State 4F (active awake state) is one of vigorous body movements with continuous eye movements and fetal heart rate accelerations on a wider oscillation band width of 25 beats per minutes; irregular fetal breathing movement present between episodes of gross body movement. This state corresponds to fetal wakefulness.

Fetus shifts from one state to another in a coordinated manner. In the third trimester it spends majority of its time in states 1F and 2F. Near term fetus spends 25% of its time in a quiet sleep state (state 1F) and 60–70% in an active sleep state (state 2F). Usually the fetus moves between 'sleep state' and 'awake state' periodically every 20 to 40 minutes. However, sleep cyclicity has been described as varying between 20 minutes to as long as 75 minutes.[8]

SIGNIFICANCE OF FETAL BEHAVIORAL STATES IN CLINICAL PRACTICE

While normal fetal behavior reliably equates with normal healthy fetus, abnormal fetal behavior may, in addition to hypoxemia, be due to other factors and awareness of these influences is essential before inappropriate conclusions regarding fetal welfare are made.

These sleep patterns are coupled with alterations in biophysical activities. In quiet sleep state (1F), rapid eye movements and repetitive mouthing movements are present but almost all other movements are absent, this may make interpretation of both NST and BPP falsely abnormal. Active sleep (2F state) provides excellent opportunity for monitoring due to clustering of activities. State 4F, a 'jogging fetus,' may show sustained acceleration in fetal heart and the return to baseline may be incorrectly interpreted as pathological tachycardia with deceleration.

Most obvious implication of fetal behavior states in clinical practice relates to interpretation of a NST to assess well-being. A reactive NST is observed most often during state 2F; state 4F show tracings with multiple large amplitude accelerations lasting for 30 seconds to several minutes. Tracing is usually nonreactive in 1F state. Distinguishing whether the fetal NST is nonreactive due to quiescent behavior or hypoxic is possible by either extending the recording time or by awakening the fetus by stimulation so as to alter the behavior state. A healthy fetus will eventually change to 2F or 4F state.

Thus, observation of normal biophysical activity indicates functional and non-asphyxiated fetal CNS. But quiet sleep state, a normal periodic fetal state and fetal asphyxia, a pathological condition can both result in nonreactive NST and abnormal biophysical profile.

The time taken to achieve a satisfactory NST or BPP is strongly dependent on whether the fetus is in state 1F (quiet sleep) or states 2F and 4F (active state), with the recommendation that biophysical recording should be extended to at least 40 minutes before the BPP is considered suspicious.[9] State 1F and the nonreactive NST seldom persist longer than 40 minutes. In either case, subsequent observation of normal activities confirms normality, whereas persistent absence of activities suggests an asphyxial etiology.

FETAL RESPONSE TO ASPHYXIA

Hypoxia produces effects on multiple organs systems. A compromised fetus reduces its activity in response to decreased oxygenation as seen by a decreased maternal perception of fetal movement.

The biophysical response to hypoxia can be categorized into two main categories according to their temporal relation to the insult: Acute and chronic response. Changes that are seen at the time of the insult and for some variable time after are termed acute fetal response. This includes loss of fetal breathing movement, tone, movement, heart rate variability and reactivity. As the fetus becomes hypoxic both the baseline variability and accelerations in response to movement decreases. With worsening hypoxia, the fetus has intermittent episodes of bradycardia.

Fetus has the unique ability to bring about reflex redistribution of its cardiac output away from organs systems not vital to fetal life such as lung, kidney, gut and towards vital organs like heart brain and placenta. The degree of manifestation of these signs varies depending on severity and chronicity of the asphyxia insult. Chronic hypoxia leads to decrease perfusion to fetal kidneys and lung causing

decreased urine and lung fluid production finally resulting in diminished amniotic fluid. The reduction in amniotic fluid volume in response to asphyxia insult takes some time to develop and is termed chronic response.

So, initially hypoxemia will cause loss of acute variables (heart rate reactivity-NST, breathing and gross body movement, tone), while the amniotic fluid may be normal. But repeated hypoxic insults with periods of recovery would cause oligohydramnios while acute phase variables may be normal. Progressive severity may finally cause both oligohydramnios and abnormal acute phase variables.

ANTENATAL TESTS OF FETAL WELL-BEING

The ideal fetal surveillance test should be quick and easy to perform and yield readily interpretable and reproducible results. It should clearly identify a sick fetus but in turn not give an abnormal result for a healthy fetus. In short, any fetal testing modality must address the critical distinction: 'nearly dead or just sleeping'.

Historically, tests of fetal well-being were introduced in 1970s and attempts to identify fetal disease were based on biochemical analysis of maternal biomarkers such as human placental lactogen, placental alkaline phosphatase, leucine aminopeptidase (oxytocinase), progesterone, estriol, estrone and alpha fetoprotein. The concentrations of each and the variation over time were related to clinical outcome. With time and cumulative clinical experience, these tests as markers of fetal asphyxia, were abandoned and replaced by more specific and direct fetal biophysical indices of fetal condition.

The methods used in modern era to detect and evaluate the severity of acute or chronic fetal hypoxia are biophysical in nature. Fetal biophysical activity is a reflection of an intact central nervous system (CNS) of the fetus. Obstetric practice has shifted away from reliance on nonspecific maternal clinical markers of potential fetal disease, such as fundal height measurements and maternal perception of fetal movement, towards more specific and direct examination of fetus. In clinical use antepartum fetal surveillance are mainly based on assessment of fetal heart rate patterns combined with ultrasonography and color Doppler. Although none of the tests alone are diagnostic of fetal jeopardy, appropriate combinations may help in clinical decision making. The tests used are:

- Daily fetal kick count (DFKC)
- Non-stress test (NST)
- Contraction stress test (CST) (obsolete and not discussed)
- Fetal biophysical profile (BPP)
- Modified biophysical profile (MBPP)
- Vibro-acoustic stimulation test (VAST)
- Umbilical, middle cerebral, uterine, venous Doppler

PREGNANCIES THAT REQUIRE FETAL SURVEILLANCE TESTS

Although monitoring of all pregnancies is necessary, the nature and frequency of tests used vary. There is no scientific evidence to support routine fetal surveillance tests to monitor low-risk pregnancies. In the low-risk population these tests show low sensitivity, specificity, positive and negative predictive values for fetal and neonatal morbidity and mortality, with low cost-benefit values when applied. Identifying pregnancies at high risk for fetal compromise and requiring surveillance would be the first step in management. High-risk pregnancy requiring surveillance are given in Table 11.1.[10]

Kontopoulos and Vintzielos classified the pathophysiological processes leading to fetal damage or death into seven categories that may be used in deciding on the optimal tests to be done. However, not all maternal fetal condition fit the above criteria and so this categorization may be used along with individualization of each case clinically. Maternal fetal condition and underlying pathophysiology leading to possible damage is given in Table 11.2.[11]

A summary of recommended antepartum fetal tests in specific conditions of pregnancy as described by S Arulkumaran is given in Table 11.3.[12]

TIMING OF FETAL SURVEILLANCE TESTS

The most important consideration in deciding when to begin antepartum testing is the prognosis for neonatal survival. The severity of maternal disease is another important consideration. In general, with the majority of high-risk pregnancies, most authorities

Table 11.1	High-risk pregnancy requiring surveillance
Obstetric high-risk factors	*Medical high-risk conditions*
Gestational hypertension, pre-eclampsia	Chronic hypertension
Gestational Diabetes	Diabetes
Prolonged pregnancy	Obesity
Cholestasis of pregnancy	Cardiac disease
Previous stillbirth, recurrent abortions, FGR, previous birth asphyxia, neonatal deaths	Renal disease
	Hepatic disorder
Rh isoimmunized pregnancy/hydrops	Thyroid disease
Extremes of maternal age	Thrombophillias
Fetal growth restriction	Systemic lupus erythematosus
Discordant twins	Autoimmune disorders
Oligohydramnios/polyhydramnios	Sepsis

Table 11.2	Maternal fetal condition and underlying pathophysiology
Pathophysiological process	*Maternal fetal conditions*
Decreased utero-placental blood flow	• Chronic hypertension • Pre-eclampsia • Collagen vascular disease • Renal disease • Most fetal growth restriction
Decreased gas exchange	• Postdate pregnancy • Some fetal growth restriction (>32–34 weeks)
Metabolic aberration	• Fetal hyperglycemia • Fetal hyperinsulinemia
Fetal sepsis	• PROM • Intra-amniotic infection • Maternal fever/subclinical chorioamnionitis
Fetal anemia	• Fetomaternal hemorrhage • Erythroblastosis fetalis • Parvovirus infection
Fetal heart failure	• Cardiac arrhythmia • Nonimmune hydrops • Placental chorioangioma • Aneurysm of the vein of galen
Umbilical cord accident	• Cord entanglement monoamniotic twins • Velamentous cord/non-coiled cord • Oligohydramnios

recommend that testing begin by 32 to 34 weeks. Pregnancies with severe complications might require testing at 26 to 28 weeks or earlier depending upon the ability of neonatal intensive unit to take care of premature babies, if subsequent intervention is needed for a positive test result.

The frequency of testing would depend on the nature of fetal or maternal disease, its severity and degree of worsening and the result of the tests undertaken and may vary from weekly to even daily.

FETAL MOVEMENT COUNT

Fetal movement is an indirect measure of fetal central nervous system integrity. Reduction in fetal kick count may signal decreased fetal oxygenation. Fetal movement follows a circardian rhythm with more movements in the late evenings. Movements increase in frequency and force till about 32–34 weeks.

About 5–15% of all pregnancies report with decreased fetal movements some time of gestation. Although keeping a daily fetal kick count is the simplest and least costly method and is often the only method of fetal surveillance used, studies have shown that women fail to keep movement counts that last for several hours. Fetal movement monitoring has a low positive predictive value (2–7%) in a low-risk population. The antenatal care in uncomplicated pregnancy, guideline by the National Institute for Health and Care Excellence, 2008 suggests that routine formal fetal movement counting should not be offered.[13]

Several fetal movement counting protocols have been used, neither the optimal number of movement nor the ideal duration for counting them have been defined. A change in the usual pattern of fetal movements may be more important than absolute numbers.

Table 11.3	Specific conditions and the recommended tests of fetal well-being
Conditions	**Recommended tests**
Decreased utero-placental blood flow • Fetal growth restriction • Hypertensive disorders • Vascular disease	• Fetal movement count, NST and AFI weekly for women with mild pre-eclampsia and twice weekly for severe pre-eclampsia • Serial ultrasound to monitor fetal growth and umbilical artery Doppler biweekly to every 2 weeks depending on severity
Decreased gas exchange • Prolonged pregnancy	• Dating ultrasound in first trimester • Modified biophysical profile (NST, amniotic fluid assessment) or biophysical profile
Metabolic aberration • Diabetes	• Regular maternal glucose monitoring • Ultrasound every 4 weeks till 32 weeks and 2 weekly thereafter • Doppler if vasculopathy • Biophysical profile not useful as falsely reassuring with hyperglycemia
Fetal sepsis • PROM • Maternal pyrexia • Suspected infection	• Amniotic fluid assessment • NST • Biophysical profile
Fetal anemia • Fetomaternal bleed • Fetal Parvovirus infection • Fetal red cell alloimmunization	• Ultrasound evaluation of Middle cerebral artery peak systolic velocity for fetal anemia, hydrops and fetal liver length. • Amniocentesis/cordocentesis
Fetal heart failure • Cardiac arrhythmia • Nonimmune hydrops	• M-mode echocardiography • Ultrasound and Doppler assessment specially for venous circulation
Obstetric cholestasis • NST twice weekly	• Ultrasound for growth, liquor and Doppler 1–2 weekly
Twin pregnancy	• Identification of chorionicity and correct dating by ultrasound • Serial ultrasound with adjunctive use of Doppler at 28, 32, 36 weeks • Earlier and more frequently in monochorionic twins

The method most commonly used is the "count to 10". Patients are instructed to begin counting fetal movements until they reach 10 movements. If 10 movements are noticed in 10 hours, or less, the fetus is in good health. If the mother notices less than 10 movements in 10 hours, she should have further evaluation.[14] ACOG practice bulletin considers 10 movements in 2 hours as reassuring.[15]

Another method used for daily fetal kick count is to instruct the women to count fetal movement for one hour, two to three times each day, and the count is considered reassuring if it equals or exceeds a previously established baseline count. If the mother perceives fewer than 3 movements in one hour further evaluation of fetal condition must be made.[16]

Several factors influence the subjective perception of fetal movements including maternal obesity, excessive amniotic fluid, anteriorly placed placenta and ingestion of medications. In today's world where mothers are actively involved in professional work, compliance may be a problem. Daily fetal kick count may be appropriate for low-risk patients and more reliable methods may be employed in high-risk individuals. A nonreassuring fetal kick count should prompt further fetal testing.

NON-STRESS TEST

Non-stress test (NST), first introduced by Freeman and Lee, is the most widely used primary antepartum or intrapartum fetal surveillance modality for pregnancies at risk of perinatal morbidity and mortality due to its ease of application in out-patient setting, good negative predictive value (0.3/1000) and lack of contraindication.[17] NST is based on the principle that fetal heart rate accelerates with movements in a fetus with intact autonomic function. Presence of FHR accelerations is interpreted as indicator of fetal well-being. NST is performed in either a semi-Fowler position (sitting with head elevated 30 degrees) or lateral recumbent position and FHR is monitored with an external transducer and a tracing is obtained.

A reactive NST is defined as presence of at least two fetal heart accelerations of 15 bpm above baseline for 15 seconds in a 20 minutes period with or without fetal movements (Fig. 11.1). In most cases the test is completed in 10–15 minutes. If the criteria for reactivity were not met in 20 minutes the test was considered nonreactive (Fig. 11.2). Since the most common cause of a nonreactive NST is a period of fetal inactivity during quiet sleep state, the test time may be extended for additional 20 minutes, with the expectation that the fetal state will change in 40 minutes. Since healthy fetuses may not move for a period as long as 75 minutes, Brown and Patrick suggested that an extended period may increase the positive predictive value of a nonreactive NST.[18] However, Evertson et al reported nonreactive NST in 15% of his 1000 patients in spite of the extended period of observation.[19] Also this extended testing time makes the test even more time-consuming and expensive.

Variable decelerations may be observed in up to 50% of NSTs. Variable decelerations that are episodic, non-repetitive and brief (less than 30 seconds) are not associated with fetal

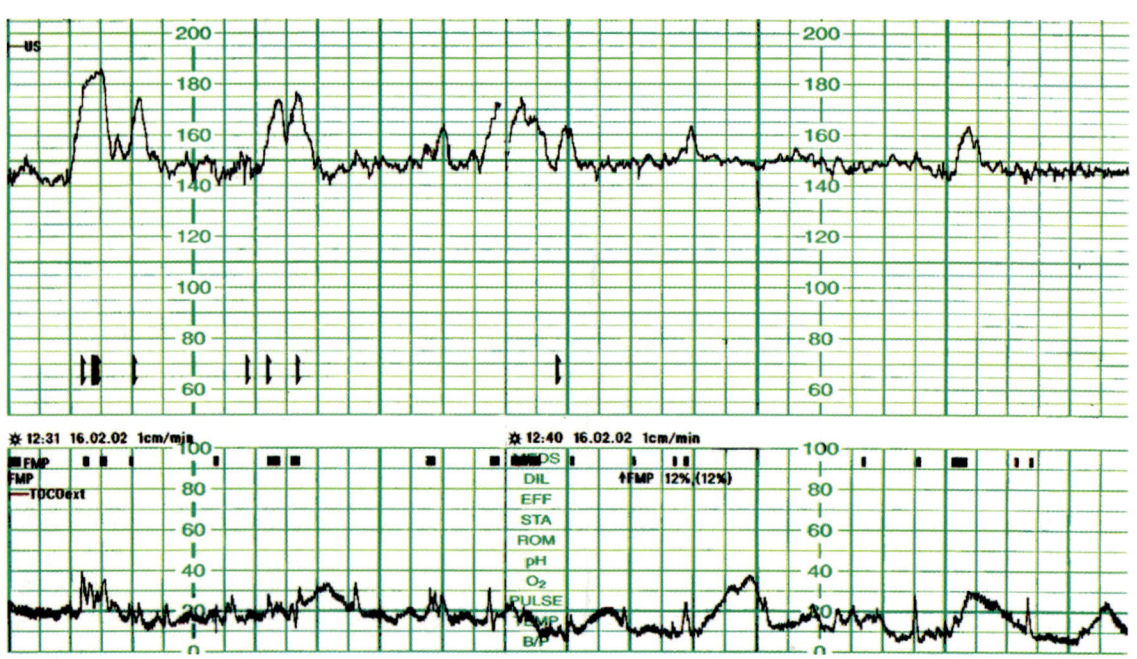

Fig. 11.1: Reactive NST

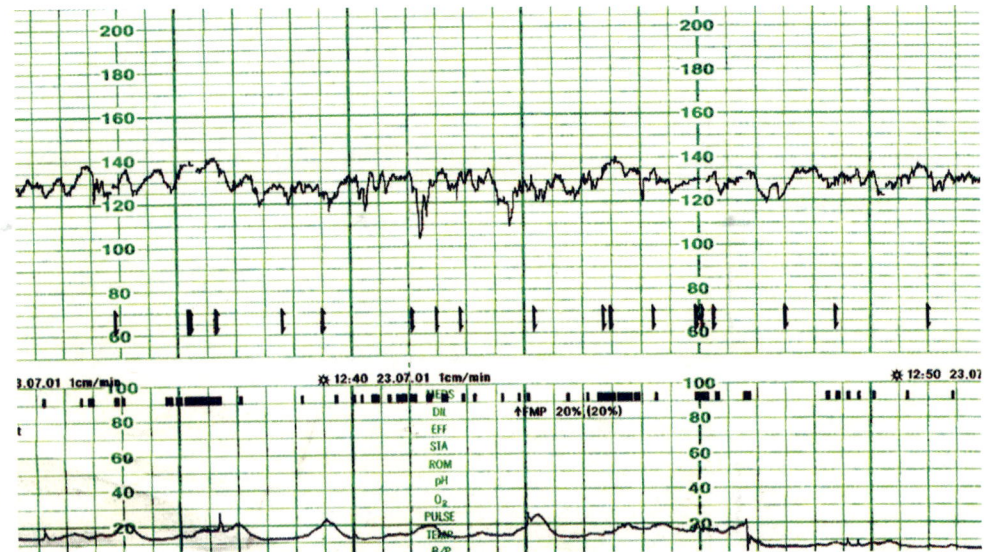

Fig. 11.2: Nonreactive NST

compromise and need no obstetric intervention. Repetitive variable decelerations (at least three in 20 minutes), even if mild, have been associated with an increased risk of cesarean delivery for a nonreassuring intrapartum FHR pattern. Fetal heart rate decelerations during an NST that persist for 1 minute or longer are also associated with increased risk of cesarean section or fetal demise *in utero*.[15]

NST for normal preterm fetuses may be more often nonreactive. Up to 50% of fetuses between 24 and 28 weeks and up to 15 % of those between 28 and 32 weeks of gestation are nonreactive. The thresholds for reactive NST in preterm fetuses <32 weeks has been revised to accelerations of 10 beats/second for 10 seconds.[15]

However, this method has its own limitations as it relies solely on the physiological changes in heart rate in response to spontaneous movement for determining the state of fetal health. And although a reactive NST is efficacious in determining that the fetus is not asphyxiated at the time of testing, it is associated with high false positive rates, i.e. absence of accelerations in an otherwise healthy fetus (50–75%).[20] This high false

positivity is primarily due to fetal sleep-wake cycle. Most fetuses with a nonreactive NST are not compromised but simply fail to show reactivity during the time period of testing due to its quiet sleep state (1F). The likelihood of a nonreactive test is higher in early third trimester (50%) as compared to late third trimester or term pregnancy (15%).[21] Thus, an unacceptable high proportion of tested fetuses will be judged to have a non-reassuring status following an NST and may be subjected to unnecessary intervention.

Due to an unacceptable high false positive rate of an NST in predicting compromise while accurately detecting fetuses that will remain well, modifications in the NST were tried. Fetal stimulation tests were introduced with the aim to improve specificity and shorten the duration of the test by stimulating the fetus to an awake state.

The BPP gives a more complete evaluation of the state of fetal health, and overcomes the problem of the high false positive rates of the NST by assessing multiple fetal parameters of well-being, giving it improved diagnostic accuracy in the detection of antepartum fetal academia.

VIBRO ACOUSTIC STIMULATION TEST

The idea that the fetus responds to extrauterine sound stimuli came from the observation that a woman felt an increase in fetal movement during the audience applause at a concert.[22] Obstetricians have used different stimuli to interrupt this quiet sleep pattern and evoke a cardio-acceleratory response. Evoked fetal response was attempted by investigators by manual fetal manipulation by vigorous shaking of uterine fundus or rocking the fetal head with no definable impact, no change in the number of nonreactive NST or the length of time needed for testing.[23] Attempts to alter fetal activity by modification of maternal glucose level also showed no absolute relationship although investigators found increased fetal activity with 100 gm glucose load during a 3-hour glucose tolerance test.[24] Another fetal interaction that was observed by a few researchers was a transient increase in fetal movement in response to amniocentesis but it has not been validated by others.

Sound or vibrations alone or in combination have been used in an attempt to arouse the fetus. External sound emitted by a speaker, an oscillator and even vibratory effect of electric toothbrushes have been tried but the most standardized and commonly used stimulus is vibroacoustic, emitted from an artificial larynx or a vibroacoustic stimulator. The first report of using vibroacoustic stimulation for fetal well-being analysis was published in 1977.[25] Vibroacoustic effects on fetal behavior are evaluated by increase in heart rate and movements and decrease in fetal breathing.

A vibroacoustic device like the Corometrics—46 sends out a vibratory sound stimulus at a frequency of approximately 75 Hz, with an intensity of around 74 dB, and has stimulation duration of 3 seconds. The electronic larynx is placed on the mother's abdomen over the fetal head and applied for period of 1–3 seconds and it is expected to induce a startle reflex in the fetus with subsequent fetal movement and fetal heart rate acceleration. This may be repeated up to three times for 3 seconds each. The distance between the stimulus and the fetal head, frequency and intensity of stimulatory impulse, its duration and the medium through which they pass determine the total amount of sound reaching the fetus. The advantage of an artificial larynx is the consistent combination of a vibratory and sound stimulus over a wide frequency range. The stimulus provokes a physiological sympathetic rage response characterized by fetal gross body movement perceived by the mother and fetal heart rate accelerations suggesting an intact non-hypoxic CNS and a healthy fetus.

Cardio-acceleratory response to vibroacoustic stimulation is gestational age dependent. Although human cochlea morphologically develops by 20 weeks, responses to acoustic stimulation are demonstrable consistently after 28 to 30 weeks. Clinicians should be aware that evoked response to acoustic stimulation may not be demonstrable prior to this gestation in a healthy fetus. As gestational age advances, the height of acceleration increases. Prior to 30 weeks FHR response was usually characterized by a single prolonged acceleration but after 30 weeks there was a prolonged increase in baseline heart rate lasting for up to 1 hour along with accelerations.

The period between the stimulation and occurrence of FHR accelerations varies but usually occurs within 10 seconds. The initial increase in FHR after stimulation is probably due to rapid withdrawal of vagal tone. A delayed increase in movements and FHR, 10 to 20 minutes after stimulation and persisting for an hour have also been observed.[26]

The state of the fetus prior to the stimulation affects the response. A negative correlation exists between pre-stimulation basal FHR and evoked response. Stimulation if performed on a fetus during the period of tachycardia may not show adequate response.[27] Jensen observed diphasic response to vibroacoustic stimulation with an initial acceleration followed by a deceleration. This was not associated with fetal pathology.[28] Habituation is a progressive decrease in response following repeated stimuli and reflects a normal central nervous system. Fetuses who fail to habituate demonstrate an increased incidence of fetal distress.

Immediate fetal movement perceive by the mother or the clinician following vibroacoustic stimulation, named fetal startle response, indicates an intact fetal brain stem in a healthy baby. This reflex involves generalized paroxysmal motion of the whole body in response to a combined sound-vibratory stimulus. Most of the startle reflex or fetal recoil is felt by the mother. Since palpable or visualized fetal movements after acoustic stimulation is almost always associated with a reactive NST (98%) this is a simple means to assess neurological status of the unborn child.[29] Fetal movement after stimulation has also correlated with a biophysical profile of >8.[30]

Vibroacoustic stimulation also alters the fetal behavioural state from 1F to 2F state or from 2F to 4F state. The time spent in state 1F decreases while 4F state is significantly more commonly seen after stimulation.[31]

SAFETY OF VIBROACOUSTIC STIMULATION

In utero sound with these devices could be louder than sound pressure levels recorded in air and may harm the tender developing auditory system. Safety of VAST has been a concern due to animal studies showing cochlear damage as a result of sound exposure especially at the time of cochlear development, disorganization of physiological state and chance of worsening fetal distress by tightening of the cord.

Most recent studies have found these devices safe for use in pregnancy. They have received US FDA approval. *In utero* sound

pressure levels were recorded when vibroacoustic stimulator was applied to the mother's abdomen and possibility of cochlear damage was assessed. Arulkumaran et al concluded that for equal sound pressures, sound intensity and sound vibrations are about 4000 times less in amniotic fluid, compared to that produced in air. Viscous and hydrodynamic features of the ear provide further protection to the cochlea. Vibroacoustic stimulation may not damage the cochlear cells because of attenuation of sound by the maternal abdominal wall and amniotic fluid.[32] Same authors further evaluated auditory acquity of 465 children at the age of 4 years who were exposed to vibroacoustic stimulation in utero and reported no hearing impairment attributable to VAST.[33]

Visser et al studied fetal behavioral states following stimulation and expressed concern about its safety as change from 1F to 4F seen with VAST is usually not seen in an un-stimulated fetus and such a change in the newborn is associated with painful stimuli.[34] The author suggests that pain perception may be present causing a disorganization of fetal behavior states. Fetal voiding reflex has also been reported following VAST. However, computerized analysis of FHR following VAST suggests that the mean baseline FHR and the overall and short term variations are affected only for the first 10 minutes if stimulation is applied once.[35]

Studies involving large numbers of antepartum high-risk patients have confirmed the validity of vibroacoustic induced fetal heart rate accelerations as a good predictor of antepartum well-being. Fetal deaths occurring within 7 days of a reactive stimulation test (VAST) has been reported as 1.9/1000 (false negative), which is comparable with a standard reactive NST.[36]

Hence vibroacoustic stimulation test (VAST) is used to alter fetal state from quiet sleep to active sleep or active state in order to reduce the time spent in performing an NST by almost 7 minutes and also reduce the likelihood of a falsely positive nonreactive NST by 40% without compromising the detection of an acidotic fetus.

BIOPHYSICAL PROFILE

Biophysical profile, originally described by Manning et al, is a method of ascertaining antenatal fetal well-being by employing five discrete biophysical variables on real time ultrasound and cardiotocography.

In adult medicine, physical examination of the patient plays a crucial role in the assessment of health. The vital signs (heart rate, respiratory rate, temperature, urine output, state of alertness and activity) are universally used in the detection of health or disease. Similarly the Apgar score for evaluation of the newborn assesses heart rate, respiratory rate, appearance, tone and responsiveness. The advent of real-time ultrasound has made it possible to perform a similar *in utero* physical examination of the fetus in order to evaluate its state of health, yielding the concept of the "fetus as a patient."

As emphasized by Manning et al, "fetal biophysical scoring rests on the sound principle that the more complete the examination of the fetus, its activities, and its environment, the more accurate may be the differentiation of fetal health from disease state." Fetal biophysical score is based on the basic principle that the greatest accuracy in differentiation of the normal from the compromised fetus is achieved when multiple fetal and environmental parameters are considered together.[37]

Technique for the Biophysical Profile

Fetal biophysical scoring is based on assessment of five discrete biophysical variables. Four of the variables (except NST), namely fetal breathing movements, gross body movements, fetal tone, and amniotic volume (fetal urine output) are monitored

simultaneously by dynamic ultrasound imaging. These variables are dependent on the integrity of the fetal CNS and are affected in situations of fetal compromise.

The first step in performing a BPP is performing a continuous fetal heart rate tracing, an NST. It is considered reactive, or reassuring, if there is a normal baseline of 110 to 160 beats per minute, and there are two or more accelerations over a period of 20 minutes. Acceleration is defined as an increase of the fetal heart rate over the baseline of at least 15 beats per minute, and lasting at least 15 seconds. Following the NST, real-time sonography is performed to evaluate the fetal biophysical activities. Amniotic fluid volume measurement is performed. Presence of a cord loop-free pocket of amniotic fluid that measures at least 2 cm in two perpendicular planes is normal. Fetal breathing movements (>1 episode of rhythmic fetal breathing movements for 30 seconds within 30 minutes) should be considered present only if they are continuous for at least 30 seconds, with breath to breath interval of less than 6 seconds. Similarly gross fetal movements are only considered present if three or more rolling movements of the trunk or limb are observed within 30 minutes. Fetal tonus is defined as > 1 episode of extension of a fetal extremity with return to flexion or opening or closing of a hand (Table 11.4).[38]

In the original BPP score described by Manning, management was based on the total BPP score according to fixed criteria. Each of the five components of the BPP is assigned a numerical value of 2 if normal or 0 if absent or abnormal. A composite value of 8 or 10 indicates that the fetal status is reassuring or normal as long as the score of 8 does not include an abnormal amniotic fluid volume. The presence of oligohydramnios (single deepest vertical pocket of fluid < 2) demands further testing or delivery, irrespective of the composite score value. A score of 6 is equivocal and requires further testing in

Table 11.4	Manning's biophysical profile scoring	
Variable	Score 2	Score 0
Fetal breathing movements	Presence of at least 30 seconds of sustained fetal breathing movements in 30 minutes	Less than 30 seconds of fetal breathing movements in 30 minutes
Fetal movements	3 or more gross body movements in 30 minutes. Simultaneous limb and trunk movements are counted as single movement	2 or less gross body movements in 30 minutes
Fetal tone	At least one episode of motion of a limb from a position of flexion to extension and rapid return to flexion	Fetus in a position of semi- or full-limb extensiom with no return to flexion with movement; absence of fetal movement is counted as absent tone
Fetal reactivity	2 or more FHR accelerations of at least 15 beats/minute and lasting at least 15 seconds and associated with fetal movements in 40 minutes	No accelaeration or less than 2 accelerations of FHR in 40 minutes
Quantitative amniotic fluid volume	A pocket of amniotic fluid that measures at least 1 cm in two perpendicular planes	Largest pocket of amniotic fluid measures less than 1 cm in two perpendicular planes
Maximal score	10	
Minimal score		0

24 hours if less than 37 weeks or consideration for delivery if > 37 weeks. A BPP score of 4 or less is suggestive of fetal compromise and an indication for delivery if salvageable weeks of gestation. If gestational age less than 32 weeks, then extended monitoring during the time of maternal steroid administration may be appropriate and individualized.[15]

The examination is completed when a normal biophysical score (of 8/8 in the Manning scoring system or 12/12 in the scoring system described by Vintzileos et al) has been achieved, or after 30 minutes of continuous real-time sonographic examination. More than 90% of tests that are normal are completed within the first 4 minutes, and the average testing time is less than 8 minutes.[39]

The false negative rate of BPP is 0.7/1000, a value better than that of the NST and similar to that of contraction stress test, i.e. antepartum death of a structurally normal fetus within one week of a normal BPP. The false positive rate of BPP is approximately 30%, better than NST or CST. The negative predictive value of BPP is similar to NST (98.5%). The positive predictive value of an abnormal test is 50.8% which is better than that of a nonreactive NST.[40] More than 97% of the pregnancies tested have normal test results.[41]

The main problem with the original BPP is the structure of the test because in the original BPP described by Manning et al, each of the five parameters were either given a score of two when normal or zero, when abnormal, with no scores in between. As a result, a fetus who had some movements and tone, but did not meet the threshold for a score of two, was awarded a score of zero, the same score as a fetus demonstrating no movement or tone whatsoever. To deal with this limitation, Vintzileos et al in 1983[42] proposed another scoring system, which gave intermediate scores, and also included placental grading as one of the biophysical variables. A normal biophysical score (of >8) is considered predictive of a nonacidotic fetus. The presence of oligohydramnios in either scoring system is considered abnormal, because reduced amniotic fluid puts the fetus at risk of cord compression, death, or adverse perinatal outcomes, regardless of the scores of the other biophysical parameters.

Whereas it seems obvious that reliance on any single test (antepartum heart rate) to detect a myriad of potential fetal diseases is inappropriate given current development, monitoring all possible variables of fetal physiology will be time consuming and prohibitive. Expanded experience also suggest that not all variables carry same power.

GRADUAL HYPOXIA CONCEPT

Centers in the human brain that control the individual biophysical activities start functioning at different times during gestation. Tone and movements which are controlled by centers in the cortical/subcortical area are the earliest biophysical activities to appear during intrauterine life, at 8–9 weeks of fetal life. Breathing movements appear next at about 21 with its center located in the ventral surface of the fourth ventricle. The last of the fetal biophysical activities to appear is the fetal heart rate reactivity which appears late in the second trimester or early third trimester, between 28 and 30 weeks and is regulated in the posterior hypothalamus and medulla.[43]

The sensitivity of each of these centres to hypoxia is different and those that become functional earlier in fetal development are more resistant to acute changes in fetal oxygenation. This 'gradual hypoxia concept' was first described by Vintzileos and colleagues which suggest that the biophysical activities that appear first during fetal life are the last to disappear when there is fetal asphyxia or intra-amniotic infection. The first biophysical activities to become compromised in the presence of fetal academia are fetal heart rate reactivity and fetal breathing movements; these are the last biophysical activities to appear in intrauterine development. As fetal acedemia progresses, fetal body movements and then fetal tone sequentially become absent.[44]

The gradual hypoxia concept questions the arbitrary assignment of equal weights to each of the five parameters of biophysical profile in predicting fetal asphyxia. This has formed the basis of several criticism of Manning's original BPP.[36] Fetal heart rate reactivity (NST), amniotic fluid volume and fetal breathing movements are the most powerful markers of perinatal outcome. Amniotic fluid volume reflects chronic changes in intrauterine oxygenation, whereas the other 4 variables are indicators of a more acute state. Impaired renal perfusion results in reduced fetal urine production and subsequently oligohydramnios.[45]

Biophysical profile is time consuming, requires ultrasound training and unless the BPP is videotaped, it cannot be reviewed. However, it has no contraindications and involves no risk for the mother or the fetus. It is both more sensitive and more specific in predicting abnormal outcome than the NST.

The Effects of Medications on the Biophysical Profile

A variety of medicines that are used in obstetrics practice have significant effects on the BPP.

- The corticosteroids are given in preterm gestations at risk of delivery, in order to promote fetal lung maturation. Reduction in fetal breathing, fetal heart rate reactivity and fetal movements are observed within 48 hours following steroid administration in at least one-third of fetuses and returns to normal by 48–96 hours.
- Magnesium sulfate, although decreases fetal heart reactivity and breathing movement, does not show any effect on other sonographically monitored BPP parameters.

• Hallak and coworkers studied the effect of terbutaline and indomethacin. They found that these tocolytics increase fetal breathing movements, but bring no change in body movements.

If one is not aware of the effects of medications on the BPP, inappropriate interpretation of the results may occur resulting in the possibility of iatrogenic and unnecessary premature delivery. Conversely, it will help prevent erroneously attributing an abnormal BPP to administration of medications that do not affect the BPP.

Modified Biophysical Profile

Because the BPP is labor intensive and requires a person trained in ultrasonic visualization of the fetus, several abbreviated modifications of the original BPP have been introduced in order to rationalize the use of human and equipment resources without impairing test performance. By limiting the number of variables included in a test, the potential for error is reduced, time is saved and interpretation is simplified.

The most powerful components of the BPP are liquor volume and NST and therefore assessment of fetal wellbeing using these two tools alone may well be as effective as formal BPP. The modified BPP, proposed by Vintzileos et al, describes a testing scheme using NST (using acoustic stimulation test, AST) with an evaluation of the amniotic fluid. Thus, this modification assesses both the most sensitive acute marker (fetal heart rate tracing) and chronic marker (amniotic fluid volume) of fetal wellbeing.[46]

Technique for Modified Biophysical Profile

Initially perform an NST and if spontaneous acceleration is not seen in 5 minutes, a single 3 seconds sound stimulation is applied with an artificial larynx to the lower abdomen (AST). Since reactive NST requires two acceleration in 10 minutes, a second stimulus may be given at 9 minutes of the first acceleration. AST is used to shorten the time required to achieve a reactive NST.

A four quadrant amniotic fluid volume is assessed by placing an ultrasound transducer perpendicular to the wall of the uterus in four abdominal quadrants and measuring the largest vertical pocket free of umbilical cord in all quadrants. A four quadrant sum of 5 cm or greater is considered normal.

The test has excellent negative and positive predictive values, is easy to interpret, has clearly defined end points, and can be performed in an average of 20 minutes. The following guidelines are useful when the MBPP is used as the primary test for fetal surveillance:[46]

• If both NST and amniotic fluid index are normal, weekly fetal surveillance with MBPP is continued.

• If both tests are abnormal (nonreactive NST, decreased amniotic fluid volume) and the pregnancy is of 36 weeks or more, the best option may be delivery. If pregnancy is less than 36 weeks, management is individualized. Doppler studies, full BPP, or delivery may be used depending on the circumstances.

• If amniotic fluid volume is decreased but the NST is reactive, a search for indicators of placental insufficiency or undiagnosed rupture of membranes need to be undertaken and management will be dependent on the final diagnosis.

• If amniotic fluid is normal and NST is nonreactive, further testing with full BPP or Doppler is indicated.

The amniotic fluid volume is always evaluated regardless of the NST results. Oligohydramnios at term or near term gestations, as well as in pregnancies complicated by intrauterine growth restriction is considered abnormal, and it is prudent to deliver women with oligohydramnios to avoid cord accident and *in utero* death. In very preterm gestation with oligohydramnios, delivery may be associated with high perinatal mortality due to prematurity. At gestational ages less than 26 weeks, each day of prolongation of gestation may be associated with increase in survival of 2–3%. Therefore, it is desirable to delay in delivery as long as safely possible. In these cases close fetal surveillance is mandatory with full BPP and fetal heart rate monitoring at least once a day. Doppler velocimetry of the uteroplacental circulation should be used in the management of preterm high risk fetuses, especially those with IUGR.[47]

By using modified BPP, fetal demise of structurally normal fetuses occurring after a reassuring MBPP assessment is a rarity. The perinatal morbidity and mortality of modified BPP compares favorably with that of the complete BPP. Overall, the MBPP has a false positive rate (60%) comparable to the NST, but higher than a full BPP. However intervention resulting from a false positive test resulted in iatrogenic prematurity in only 1.5% of cases. The false negative rate (0.8/1000) and the ease of performance of MBPP make it an excellent approach for the evaluation of high-risk patients in large numbers.[48]

This protocol has decreased the testing time, increased the number of tested individuals and, therefore, improved the efficiency of a unit. The American College of Obstetricians and Gynecologists (1999) has concluded that the MBPP is an acceptable means of antepartum fetal surveillance.[49]

FETAL GROWTH RESTRICTION

Identification of Risk Factors for Predicting FGR

All pregnant women should be assessed for risk factors for a SGA neonate at first registration, so as to identify ones that would require more fetal surveillance. Previous birth of a SGA neonate increases the risk of subsequent SGA twofold and even more after previous two SGA. Females with previous history of placental mediated maternal disease such as

prior pre-eclampsia, prior preterm or term stillbirth are at increased risk. Table 11.5 gives a summary of all risk factors and has been adapted from the RCOG guideline no. 31.[50]

Biochemical Markers for Prediction of FGR

Low Pregnancy associated protein A (PAPP-A) and placental growth factor (PLGF) are 1st trimester serum analytes that are associated with early abnormality in placental angiogenesis leading to FGR. A low level of PAPP A < 0.4 MOM is a major risk factor for FGR fetus.s Advantage of PAPP-A is its availability as part of 1st trimester integrated screen for aneuploidy.

Early 2nd trimester markers studied for their association with SGA are serum estradiol, human placental lactogen (HPL), human chorionic gonadotropin (HCG), Maternal serum alphafetoprotein (MSAFP). Increased MSAFP and HCG are most useful for their association with abnormal placentation and FGR. A single unexplained elevated value of MSAFP or HCG >2 MOM increases the risk of growth restriction in the fetus by fivefold (in the absence of aneuploidies, neural tube defects and abdominal wall defects).

Screening for FGR by Uterine Artery Doppler

Uterine artery Doppler reflects directly the uteroplacental perfusion and is an important screening tool in all high-risk pregnancies for prediction of FGR, pre-eclampsia and adverse pregnancy outcome. Abnormal uterine artery Doppler at 20–24 weeks is defined by increased pulsatility index (PI) >95th percentile or presence of diastolic notching. Abnormal flows in uterine artery are consistent with abnormal placental implantation and indicates increased risk of pre-eclampsia, placental abruption and early onset of FGR.

Meta-analysis involving 61 studies has shown that 2nd trimester uterine artery Doppler PI has a sensitivity and specificity of 18% and 95% respectively for predicting fetal

Table 11.5	Risk factors for fetal growth restriction
Maternal age	>35 years, more so if > 40 years
	Teenage pregnancy < 20 yrs
Parity	Nulliparity
BMI	Underweight BMI< 20
	Obese BMI > 30
Maternal substance abuse	Smoker (dose dependant)
	Cocaine
	Alcohol
	Caffeine in the 3rd trimester
Social deprivation	Unmarried single mothers
	Immigrants
Maternal intake of medications	Antineoplastic, antiepileptic, antithrombotic
IVF pregnancies	
Exercise	Vigorous exercise during pregnancy
Previous pregnancy outcome	SGA fetus
	Stillbirth
	Pre-eclampsia
Maternal medical history	Chronic hypertension
	Diabetes with vasculopathy
	Chronic renal disease
	Hereditary thrombophilia
	Antiphospholipid antibody syndrome
Maternal and paternal birth weight	SGA at birth
Inter-pregnancy interval	< 6 months
	>60 months
Present pregnancy complications	Threatened miscarriage
	Pregnancy induced hypertension / Pre-eclampsia
	Unexplained APH
	Placental abruption
	Poor maternal weight gain
Low PAPP-A	< 0.4 MOM
Echogenic bowel	

growth restriction. This sensitivity and specificity increases to 67% and 95% respectively for severe growth restriction (<3rd percentile). Hence in high-risk population Uterine artery Doppler at 20–24 weeks has a moderate predictive value for severe SGA.[51] Patients with abnormal uterine artery Doppler should be serially monitored from 26 to 28 weeks for interval growth and umbilical artery Doppler.[50]

In females with abnormal uterine artery Doppler at 20–24 weeks, subsequent normalization at 26–28 weeks is still associated with higher risk for SGA. The risk for SGA is significantly higher in females with persistence of abnormal uterine artery Doppler. Repeating uterine Doppler at 28 weeks is of limited value.[50]

The presence of normal uterine artery flow velocity bears a negative predictive value with a LR of 0.5 and 0.8 for development of pre-eclampsia and FGR respectively. So females with normal uterine artery Doppler do not require serial monitoring for growth and Doppler and can be triaged into low risk pregnancies (unless they develop other pregnancy related complications).

First trimester uterine artery Doppler is being investigated for prediction of early onset FGR and pre-eclampsia. This screening can be integrated with 11–13+6 weeks aneuploidy screening. The ability to predict women at risk for FGR and pre-eclampsia in early pregnancy might decrease maternal and fetal morbidity through closer surveillance and early intervention with low dose aspirin. However, 1st trimester uterine artery Doppler although has a high specificity (91–96%) and a high negative predictive value (91–99%), sensitivity remains low (12–25%) for SGA neonate.[51]

Antenatal Diagnosis of Fetal Growth Restriction and Surveillance Tests

1. Early establishment of gestational age
2. Clinical suspicion of growth restriction
3. Role of ultrasound in diagnosis, evaluation and surveillance

ULTRASOUND DIAGNOSIS OF SGA

Royal College of Obstetricians and Gynaecologists recommends Fetal Abdominal Circumference (AC) or Estimated fetal weight (EFW) < 10th percentile for gestational age as cut-off for diagnosis of SGA.[50] Most common method used for identifying poor fetal growth on ultrasound is estimation of fetal weight using multiple biometric measurement. Formulas combining BPD, HC, AC, FL dimensions have been used to optimize accuracy of fetal weight assessments. Most (Haddlock, Sheppard) of these formulae have an accuracy of +/– 10% in determining EFW. About 70% of these fetuses with EFW <10th centile are normally grown and not at risk. Using lower cut-off of 3rd centile may identify more truly growth restricted fetus while milder form may be missed.

Fetal AC is related to hepatic glycogen storage and liver size and correlates closely with the nutritional status. When fetal growth is compromised, the first parameter to be affected is the AC. The most accurate AC is the smallest directly measured circumference obtained in a perpendicular plane of upper abdomen at the level of hepatic vein between fetal respiration. AC is the single best measurement for detection of IUGR. Using 10 centiles as cut-off, AC has higher sensitivity (98% v/s 85%) but lower positive predictive value than EFW (36% v/s 51%) for sonographic diagnosis of IUGR.[52]

Biometric Ratio

Head to abdominal ratio (HC: AC) is used to improve detection of asymmetric growth restriction. It compares the most preserved organ (brain) with the most compromised organ (liver). In normally growing fetus, HC: AC ratio is >1 before 32 weeks, approximately 1 at 32–24 weeks and falls below 1 after 34 weeks of gestation. While in asymmetric growth restricted fetus HC: AC ratio remains elevated > 95th centile, ratio remains normal in symmetrically growth restricted fetus as both measurements are equally affected.

Femur to abdominal ratio (FL:AC) remains constant at all gestational age from 21 weeks till term (FL:AC ratio 22+ 2). It is again a comparison of a relatively spared organ (femur) to the most affected organ and values above 23.5 suggests IUGR. It is a gestational age independent method of diagnosis of asymmetric FGR and is useful if patient is unsure of dates.

Ponderal index is an index of neonatal size (weight/length^3) and reflects nutritional state of the neonate. The same has been extrapolated to fetal life but is not clinically used often.

Reduction in subcutaneous fat deposition at face, abdomen, arms and thighs have been evaluated for their correlation with FGR and may be used as an adjunct to other more established methods.

Customized/Population-based Growth Charts

The definition most widely used for Small for Gestational Age is EFW below the population 10th centile corrected for gestational age. A single assessment for weight may not be an accurate measure of dynamic fetal growth. Plotting fetal growth using population based growth curves over a period of time may detect more accurately truly growth restricted fetuses. Serial USG for growth at interval of not less than 2 weeks may be charted on these standardized population based growth curves. Loss or decrease in growth velocity may suggest truly growth restricted fetus. Growth curves for constitutional small fetus will continue at normal velocity but along lower centile (Figs 11.3a–c).

Customized growth charts give an individual growth chart for every patient by making adjustments for maternal height, weight, parity, ethnicity and fetal sex (GROW curves) and are considered more accurate than population based charts.

Investigation to Identify Etiology

Fetal Anatomical Survey

Once a diagnosis of FGR is made, attempt should be made to determine the etiology. A

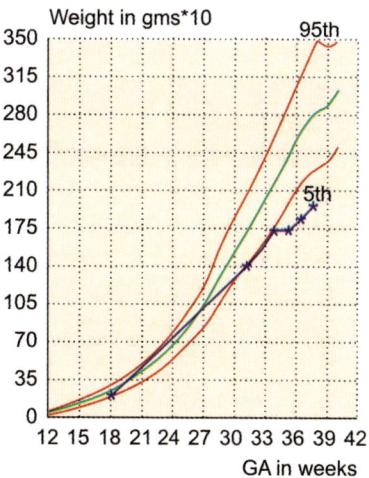

Fig. 11.3b: Growth curve for late onset asymmetric FGR

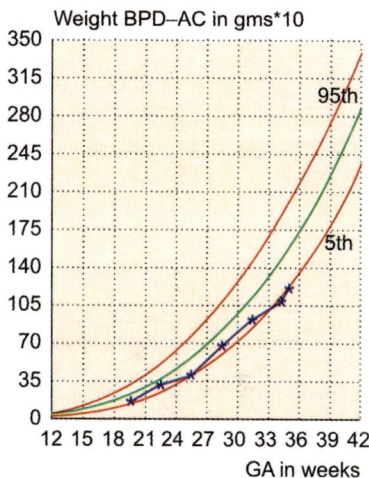

Fig. 11.3c: Growth curve for constitutional small fetus

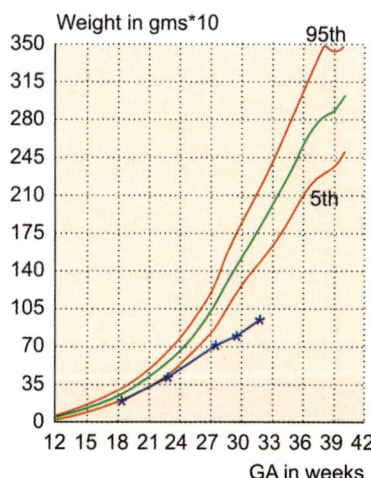

Fig. 11.3a: Growth curve for early onset symmetric FGR

detailed anatomical survey should be done by ultrasound to determine any structural anomalies, look for markers of aneuploidy, non-aneuploidy syndromes and signs for fetal infections. Identification of any of these abnormalities on ultrasound significantly impacts management and outcome.

19–20% of severe SGA are associated with chromosomal abnormalities predominantly triploidy and trisomy 13 and 18. Amniotic fluid for fetal karyotype may be offered in

- Severe SGA,
- Early onset FGR <23–26 weeks

- FGR associated with structural abnormalities or soft markers specially if amniotic fluid is normal/increased and if uterine artery Doppler are normal.

Fetal infections are responsible for 5% of SGA fetuses. Ultrasound markers for fetal infections are intracranial calcification, hepatic calcification, ventriculomegaly, hydrops, hepatosplenomegaly, ascites and echogenic bowel. In a symmetric growth restricted fetus, presence of these markers are suggestive of fetal infections, however, their absence does not exclude infections. Serological screening of mother for toxoplasmosis, rubella, CMV, syphilis and malaria should be done. If screen results positive, presence of fetal infection can be confirmed by TORCH PCR on amniotic fluid collected by amniocentesis.

DOPPLER FOR DIAGNOSIS, ETIOLOGY AND SURVEILLANCE

Once ultrasound biometry identifies a fetus as SGA (EFW <10th centile), Doppler velocimetry is extremely useful for making the critical differentiation between constitutionally small healthy (<10th centile) fetuses and the pathologically growth restricted fetuses and stratify small fetuses which require more surveillance. Doppler flow provides insight into the etiology of fetal growth restriction. Increased impedence in the umbilical artery would suggest placental insufficiency as the cause for pathological growth restriction. Doppler flow of umbilical artery along with other vessels is then used for management of the pathological growth restricted fetuses where placental insufficiency is the etiology.

Generally in obstetrics the most widely used arterial indices to quantify flow velocity with Doppler are:

- Systolic to diastolic ratio (S/D)
- Resistance index (S-D/S)
- Pulsatility index (S-D/M)

A relative decrease in end diastolic velocity causes elevation of all the indices and reflects increased or elevated downstream resistance. When end diastolic flow is lost, the S/D ratio reaches infinity and RI becomes 1. The PI offers the advantage of smallest measured error, narrower reference limit and advantage of ongoing numeric analysis even when diastolic flow is absent or reversed. S/D, RI, PI reference value varies at different gestational age and they can be plotted on centile charts.

UMBILICAL ARTERY DOPPLER (UAD)

American college of Obstetrics and Gynecology (ACOG 2013) recommends use of umbilical artery Doppler as a standard method for antepartum fetal surveillance in FGR fetus. Use of umbilical artery Doppler in SGA fetus is associated with significant reduction in perinatal mortality rates (29% reduction) as well as reduced iatrogenic intervention.[53]

Whereas uterine artery reflects trophoblastic invasion and placental development, umbilical artery Doppler reflects placental function and indicates degree of placental insufficiency. It correlates with the tertiary villous architecture and blood flow resistance. Umbilical artery Doppler resistance is elevated if at least 30% of fetal villous vasculature becomes abnormal. Absence or reversal of umbilical artery end diastolic flow suggests 60–70% damage to villous vascular architecture and an increase risk of perinatal mortality (Figs 11.4a–d).

The severity of Doppler abnormality in the umbilical artery is proportionate to risk of hypoxemia. Umbilical artery Doppler correlates well with fetal acid–base status. A normal umbilical Doppler index is consistent with absence of fetal acidemia due to uteroplacental dysfunction (high negative predictive accuracy). Reduced or absence of umbilical artery diastolic flow (elevated resistance/absent diastolic flow) indicating severe villous abnormality but the relationship

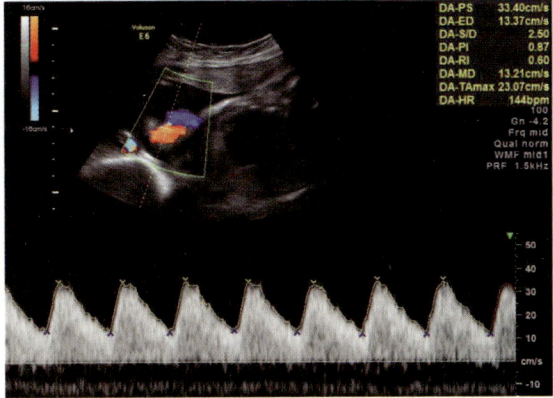

Fig. 11.4a: Normal umbilical artery Doppler flow

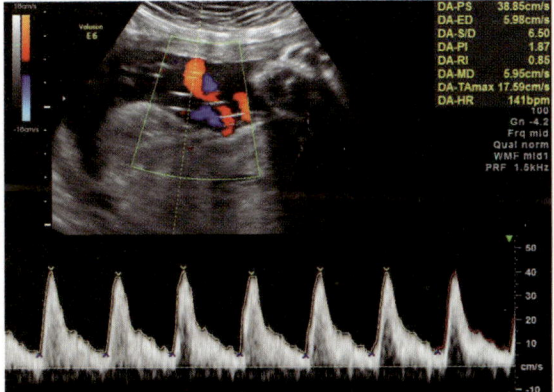

Fig. 11.4b: Elevated resistance in umbilical artery Doppler

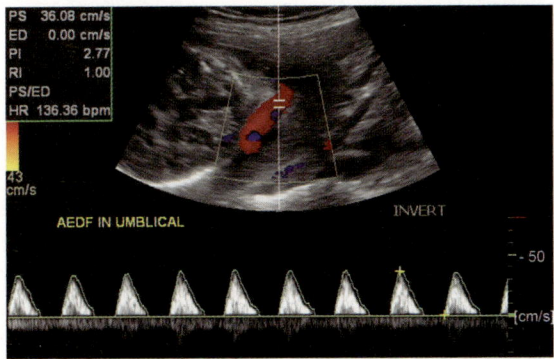

Fig. 11.4c: Absent end diastolic flow in umbilical artery

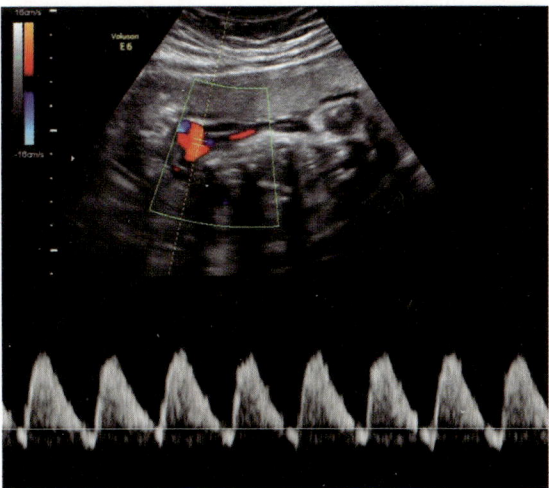

Fig. 11.4d: Reversal of diastolic flow in umbilical artery

different gestation are different. Hence it requires further assessment of other fetal vessels to determine fetal response to hypoxemia.

MIDDLE CEREBRAL ARTERY (MCA) DOPPLER FLOW

Uterine artery reflects placental development, umbilical artery indicates placental function, middle cerebral artery Doppler flows reflects fetal response to hypoxemia. The fetus adapts itself to chronic hypoxemia with preferentially diverting blood to vital organs (brain, heart and adrenals) leading to increased diastolic flow in the MCA and decrease in Doppler indices (brain sparing effect/cerebral vasodilatation/centralisation).

MCA is an ideal vessel to be sampled for assessment of fetal cerebral circulation as it is easy to sample at low insonation angle. A reduced MCA PI or MCA/umbilical artery PI ratio (cerebroplacental ratio <1) are early signs of hypoxemia in FGR (Figs 11.5a–b).

VENOUS DOPPLER

Doppler studies of venous circulation are done when a FGR fetus shows brain sparing effect. Venous Doppler start showing abnormal

between placental pathology and fetal academia is not directly proportionate as oxygen requirement of different fetuses at

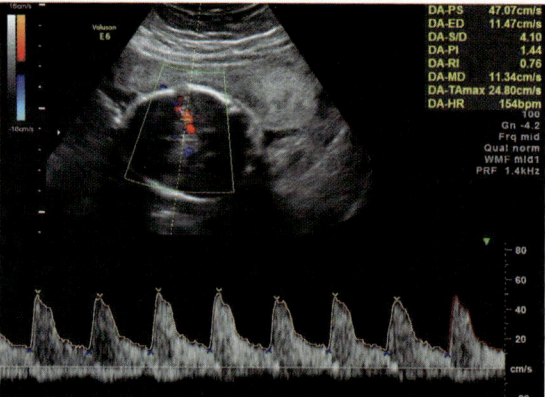

Fig. 11.5a: Normal middle cerebral artery Doppler flow

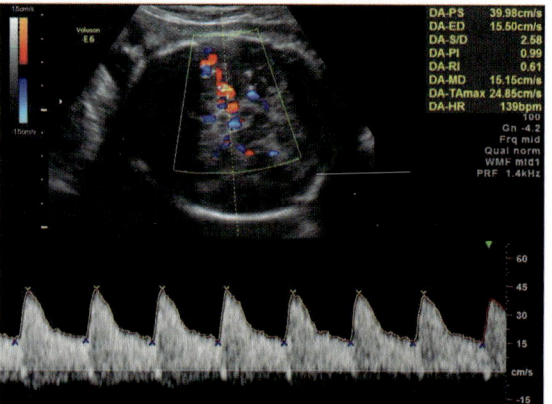

Fig. 11.5b: Brain sparing effect in middle cerebral artery

flows once umbilical and MCA Doppler becomes abnormal.

Ductus venosus is the vessel most commonly sampled and reflects cardiac function secondary to hypoxia. Ductus venosus has a characteristic triphasic flow pattern (M shape) that reflects pressure volume changes in right atria through cardiac cycle. The descent of the AV ring during ventricular systole and passive diastolic ventricular filling generates the systolic and diastolic peaks respectively (S and D wave). The sudden increase in right atrial pressure with atrial contraction in late diastole produces the second trough after the D wave (a wave). In venous system, flow reversal during atrial contraction is seen normally in

the IVC and hepatic vein. However, normal blood flow in DV is forward flow throughout the cardiac cycle.

Marked placental insufficiency can lead to impaired cardiac function. Ineffective preload, decline in cardiac output, myocardial dysfunction may present in DV with decline in forward flow in DV (deep a wave or reverse a wave). Reversed a wave in ductus venosus in a growth restricted fetus would suggest cardiac failure due to hypoxia (Figs 11.6a–b).

Umbilical vein normally has a non-pulsatile flow. Pulsation in the umbilical vein occurs at a late stage of fetal compromise.

METHODS OF FETAL SURVEILLANCE USED IN SGA FETUSES

1. *Amniotic fluid volume* is a marker of chronic uteroplacental insufficiency. It is assessed either by amniotic fluid index (AFI) which is calculated by adding vertical liquor pockets in each of the four quadrants of maternal abdomen or by calculating a single maximum vertical pocket (MVP). Recent Cochrane study suggests that AFI <5 cm is associated with more induction without improvement in fetal outcome. MVP >2 cm has been recommended for amniotic fluid volume assessment.

2. **Biophysical profile/modified biophysical profile** may be used after 30–32 weeks for fetal surveillance in FGR.

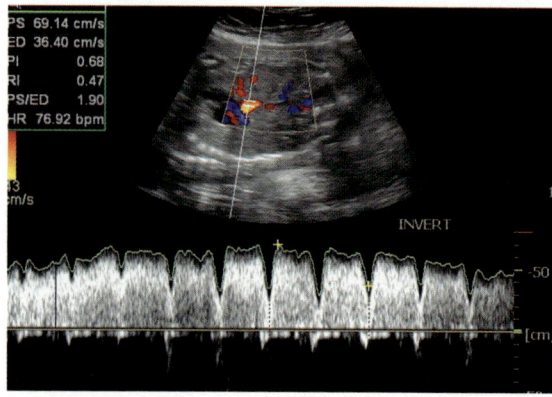

Fig. 11.6a: Normal ductus venosus forward flow

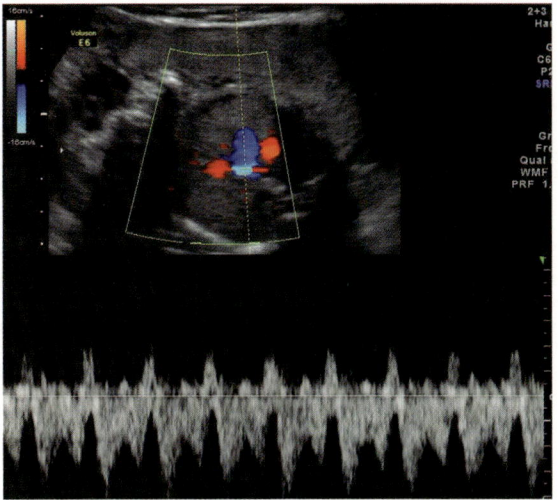

Fig. 11.6b: Reverse a wave in ductus venosus

3. *Cardiotocography (CTG/ NST)* is a marker of acute hypoxemia. Recent Cochrane review supports the use of computerized CTG instead of conventional CTG as it is more consistent and has shown reduction in perinatal mortality.
4. *Color Doppler:* With progressive placental dysfunction, Doppler flows across multiple vessels, in most growth restricted fetuses, show a typical progression which may vary slightly from fetus to fetus (Flowchart 11.1). Umbilical artery Doppler used in conjunction with standard fetal surveillance methods such as CTG and biophysical profile improves neonatal outcome in FGR.[53]

Flowchart 11.1: Progression of multivessel Doppler flows in placental insufficiency

1	• Elevated resistance in umbilical artery
2	• Brain sparing effect in middle cerebral artery
3	• Absent end diastolic flow in umbilical artery
4	• Reversal of flow in the umbilical artery
5	• Absent or reversal of flow in the ductus venosus
6	• Pulsatile umbilical vein

Constitutional Small Fetus

Fetuses who have EFW or AC <10th centile with normal umbilical artery Doppler with normal liquor can be monitored on OPD basis. Interventions in the form of bedrest, maternal oxygen administration, dietary modification, protein supplements, calcium, NO donors, plasma volume expanders, sildenafil, betamimetic drugs have all been evaluated but none have shown significant benefit. However in low socioeconomic strata where there may be a substrate deficiency due to poor or malnutrition, it may be useful to counsel mothers for appropriate isoenergetic diet.

These genetically small babies are not at significant risk of hypoxia and fetal surveillance may be done by fortnightly interval growth, Doppler and liquor assessment.

Delivery: No adequate randomized trial has been performed to determine optimal time for delivery in the constitutionally small fetuses <10th centile. In the absence of any co-morbidities, awaiting spontaneous labor is appropriate. However, delivery by induction of labor may be attempted for those progressing over their due date.

These constitutional small babies can be allowed vaginal delivery with routine intrapartum monitoring. Cesarean section may be reserved for obstetric indications. These low growth profile healthy fetuses, however, have poor reserves and may not tolerate prolonged latent or active labor.

Placental Insufficiency Mediated FGR

Fetuses which are < 10th centile with abnormal Doppler flows or severe growth restricted fetuses <3rd centile are the ones which are at increased risk for adverse outcomes and require advanced fetal surveillance.

Elevated resistance in umbilical artery (PI or RI >95th centile) can be monitored with weekly/biweekly umbilical and middle cerebral artery Doppler flows and weekly interval growth and amniotic fluid index. Consider delivery at 37 weeks if other parameters for fetal well-being are normal or earlier >34 weeks if static growth for 3–4 weeks, oligohydramnios or worsening of Doppler flows.

Absent/reversed end diastolic flow in umbilical artery (AREDF) should be monitored with daily umbilical artery and ductus venosus Doppler flows, computerized CTG and weekly interval growth and amniotic fluid volume. These growth restricted fetuses with AREDF should be delivered if pregnancy is >32 weeks even if the ductus venosus is normal or earlier if ductus venosus becomes abnormal or pulsatile umbilical vein appears or abnormal cCTG.[50]

Delivery: Fetuses with absent or reversal of flows in the umbilical artery or severe form of growth restriction cannot tolerate the brunt of labor contraction and must be delivered by cesarean section. FGR fetuses with elevated resistance in umbilical artery and normal amniotic fluid may be allowed vaginal delivery although emergency cesarean section rates are high. They are at increased risk of fetal distress in labor and must be strictly monitored with intrapartum fetal heart rate monitoring preferably continuous CTG.

As per expert consensus, a joint conference of the Eunice Kennedy Shriver National Institute of Child Health and Human Development, the Society for Maternal-Fetal Medicine and ACOG suggested delivery at 38 + 0 to 39 + 6 weeks of gestation in cases of isolated growth restriction (i.e. normal Doppler, normal liquor, weight <10th centile with no other comorbidities). However, delivery is indicated at 34 weeks to 37+6 weeks of gestation in case of FGR with additional risk factors like oligohydramnios, abnormal umbilical artery Doppler, maternal risk factors or co-morbidities.[53] Flowchart 11.2 summarizes the management of SGA fetus.

CONCLUSION

Management of Abnormal Result of An Antenatal Surveillance Test

- An abnormal antepartum fetal test result should always be considered in the context of the overall clinical picture.
- Pregnancies with decreased fetal movement should be evaluated by and NST or BPP or a modified BPP
- Normal tests exclude imminent fetal jeopardy within 7 days (high negative predictive value of >99.9) for most tests
- A nonreactive NST or an abnormal modified biophysical score is an indication for complete biophysical profile and/or Doppler
- BPP score of 4 indicates prompt delivery with or without steroids
- BPP score of 6 is equivocal and indicates delivery if >37 weeks or repeat BPP after 24 hours if <37 weeks
- In cases of fetal growth restriction use umbilical artery Doppler along with middle cerebral artery Doppler to be used as primary surveillance tool in conjunction with standard fetal surveillance tools.
- Normal umbilical artery Doppler with normal liquor in a constitutionally small fetus may be managed at 2–3 weekly umbilical artery Doppler with interval growth and liquor
- More frequent Dopplers are needed for high resistance flow in umbilical artery
- Absent or reversed end diastolic flow require admission, steroid and close surveillance if <32–34 weeks and delivery if >32–34 weeks.
- The use of deepest vertical pocket (<2 cm) as opposed to amniotic fluid index (<5 cm) to diagnose oligohydramnios is associated with reduction in unnecessary intervention without an increase in adverse perinatal outcome

Abnormal fetal surveillance is based on physiologic changes that alter fetal heart rate,

Flowchart 11.2: Management of small for gestational age fetus

```
┌─────────────────────────────────────┐
│ Pregnancy at risk for SGA            │
│ Presence of risk factors,low PAPP-A, │
│ High MSAFP, uterine artery           │
│ Doppler elevated                     │
└─────────────────────────────────────┘
                 │
                 ▼
┌─────────────────────────────────────┐
│ Clinical suspicion of SGA            │
│ Lag in SFH >3 cm, OR                 │
│ SFH <10 centile                      │
└─────────────────────────────────────┘
                 │
                 ▼
┌─────────────────────────────────────┐
│ Ultrasound confirmation of SGA       │
│ EFW <10th centile, OR                │
│ AC <10th centile, OR                 │
│ lag in interval growth on centile    │
│ charts                               │
└─────────────────────────────────────┘
                 │
                 ▼
┌─────────────────────────────┐    ┌──────────────────────┐    ┌──────────────────────────┐
│ Assess liquor, anatomical   │───▶│ Fetal anomaly/USG    │───▶│ Non-placental mediated    │
│ survey, umbilical A Doppler │    │ markers of infection/│    │ FGR-Manage os per         │
└─────────────────────────────┘    │ aneuploidy markers   │    │ fetal pathology           │
                                    └──────────────────────┘    │ Offer amniocentesis for   │
                                                                │ karyotype and/or infection│
                                                                └──────────────────────────┘
```

Normal UA Doppler Constitutional small fetus	Abnormal UA Doppler Placental mediated FGR
Monitor with EFW, liquor, UAD every 14 days	High resistence in UAD PI >95th centile — Absent or reversed end diastolic flow in UAD
Deliver between 38+0 and 39+6 weeks if no co-morbidities. Deliver >34 weeks after steroid cover if static growth, oligohydramnios, abnormal UAD or MCA PI<5th centile	Monitor UAD, MCA and liquor biweekly EFW weekly — >32 weeks give steroid and deliver even if DV normal < 32 weeks Give steroids and monitor daily UAD and DV, CTG and liquor deliver if abnormal DV/ pulsatile umbilical V/ abnormal CTG
	Deliver at 37 weeks if normal liquor, growth and no worsening of UAD Deliver >34 weeks after steroid cover, if static growth, oligohydramnios, worsening UAD, or MCA

fetal activity and Doppler blood flows. Fetal heart rate, fetal movement, and tone in particular are impacted by uteroplacental fetal blood flow alterations and are thereby sensitive to fetal hypoxemia and acidemia. While nonreassuring fetal surveillance is associated with fetal hypoxemia and acidemia based on these physiologic adjustments, these indicators can neither predict the degree or duration of the fetal acid–base disturbance nor precisely predict neonatal outcome.

An abnormal test is a warning sign, while normal tests are highly reassuring since fetal deaths within 1 week of normal test are rare. The negative predictive values (true negative test) for most of the tests are 99.8% or higher. In contrast, the positive predictive values (true positive tests) for abnormal results are quite low (10–40%).

According to the American College of Obstetricians and Gynecologists there is no 'best test' to evaluate fetal well-being. Management depends on gestational age, maternal condition, and which antenatal testing or combination thereof is abnormal.

Questions which still remain unanswered after so many years are: Does antenatal fetal testing really make a difference in perinatal outcome? Fetal surveillance increased from less than 1% of pregnancies in the early 1970s to 15% in the mid-1980s but there are contrasting opinions on the benefits of antenatal fetal testing. Another important and unanswered question is whether antepartum fetal surveillance identifies fetal asphyxia early enough to prevent brain damage. These tests also do not predict stillbirths related to acute changes in maternal fetal status such as cord prolapse or abruption placentae.

REFERENCES

1. Confidential Enquiry into Maternal and Child Health Perinatal mortality 2005: England,Wales and Northern Ireland. CEMACH, London, UK (2007).
2. Signore C, Freeman RK, Spong CY. Antenatal testing—a reevaluation: executive summary of a Eunice K: executive summary of a Eunice Kennedy Shriver National Institute of Child Health and Human Development Wornnedy Shriver National Institute of Child Health and Human Development workshop. Obstekshop. Obstet Gynecol. 2009; 113: 687–701.
3. Maurice LD, James FS Jr, Steven G, Kathryn LR: Obstetrics-Normal and Abnormal pregnancies, 5th ed. Philadelphia, Churchill Livingston 2007, p 269.
4. Andres SR: Practical Guide to High Risk Pregnancy and Delivery: A South Asian Perspective, 3rd ed.Elsevier, 2008, p 14.
5. Rayburn WF. Clinical significance of perceptible fetal motion. Am J Obstet Gynecol 1980;138: 210.
6. Patrick J, Campbell K, Carmichael L, Natale R, Richardson B. Patterns of human fetal breathing during the last 10 weeks of pregnancy. Obstet Gynecol 1980;56:24–30.
7. Nijhuis JG, Prechtl HFR, Martin CB Jr, et al: Are there behavioural states in the human fetus? Early Hum Dev 6: 177, 1982.
8. Patrick J, Campbell K, Carmichael L, et al. Patterns of gross fetal body movements over 24-hour observation intervals during the last 10 weeks of pregnancy.Am J Obstet Gynecol 142:363, 1982.
9. Pillai M James: The importance of the behavioural state in biophysical assessment of the term human fetus.Br J Obstet Gynecol 97:1130,1990.
10. Pawar A, Damania K. Identification and Antepartum surveillance of High Risk pregnancy. In: Bhide A, Arulkumaran SS, Damania K, Daftary SN eds. Arias Practical guide to High risk pregnancy and delivery, A South Asian perspective. 4th ed, Elseiver. 2015, 116–133.
11. Kontopoulos EV, Vintzelous AM. Condition specific antepartum fetal testing. AmJ Obstet Gynecol; 191: 1546–51, 2004.
12. Talaulikar VS, Arulkumaran SS. Antenatal fetal surveillance: Some for all, more for those at risk. In: Arulkumaran SS, Haththotuwa R, Tank J. Tank P eds. Antenatal and Intrapartum fetal surveillance. Hyderabad: Universities Press. 3–15, 2013.
13. National Institute of Clinical Excellence. Antenatal Care for uncomplicated pregnancies. Clinical guidelines 62, London; 2008.
14. Moore TR, Piaquadio K:A prospective evaluation of fetal movements screening to reduce the incidence of antepartum fetal death. Am J Obstet Gynecol 160: 1075, 1989.
15. ACOG Practice Bulletin Number 145: Antepartum Fetal Surveillance, July 2014. Obstet Gynecol. 2014; 124:182–92.
16. Sadovsky E, Yaffe H, Polishuk W: Fetal movement monitoring in normal and pathologic pregnancy. Int J Gynaecol Obstet 12: 75, 1974.
17. Devoe LD, Jones CR. Nonstress test: Evidence-based use in high risk pregnancy. Clin Obstet Gynecol.2002;45:986–92.
18. Brown R, Patrick J . The Nonstress test: How long is enough? Am J Obstet Gynecol.1981;141:646.
19. Evertson L, Gauthier R, Schifrin B et al. Antepartum fetal heart rate testing. Evolution of the Nonstress Test. AmJ Obstet Gynecol.1979;133:29.
20. ThackerSB, Berkelman RL. Assessing the diagnostic accuracy and efficacy of selected antepartum fetal surveillance techniques. Obstet Gyn Surv.1986; 41:121–41.
21. Lavin J, Miodovnik M,Barden T. Relationship of nonstress test reactivity and gestational age. Obstet Gynecol.1984;63:338.
22. Forbes HS, Forbes HB. Fetal sense reactions: hearing. J Comp Physiol Psychol.1927;7:353–552.

23. Druzin ML, Gratacos J, Paul RH et al. Antepartum fetal heart rate testing. The effect of manual manipulation of the fetus on the nonstress test. Am J Obstet Gynecol.1985;151:61.

24. Miller FC, Skuba H, Klapholz H. The effect of maternal blood sugar levels on fetal activity. Obstet Gynecol .1978;52:662.

25. Read JA, Miller FC. Fetal heart rate acceleration in response to acoustic stimulation as a measure of fetal wellbeing. Am J Obstet Gynecol. 1977;129:512.

26. Gagnon R, Hunse C, Carmichaell, et al. Effects of vibratory acoustic stimulation on human fetal breathing and gross fetal body movements near term. Am J Obstet Gynecol. 1986;155:1227–30).

27. Gagnon R, Hunse C, Carmichaell, et al. Human fetal responses to vibratory acoustic stimulation from twenty six weeks to term. Am J Obstet Gynecol. 1987;157:1375–8.

28. Jenson OH. Fetal heart rate response to controlled sound stimuli during the third trimester of normal pregnancy. Acta Obstet Gynecol Scand. 1984; 63:193–7.

29. Michael Y, Divon MD, Laurence D, et al. Evoked fetal startle response: A possible intrauterine neurological examination. Am J Obstet Gynecol. 1985;153:454–6.

30. Sarinoglu C, Dell J, Mercer BM, Sibai BM. Fetal startle response observed under ultrasonography: a good predictor of a reassuring biophysical profile. Obstet Gynecol. 1996;88:599–602.

31. Devoe LD, Murray C, Faircloth D et al. Vibroacoustic stimulation and fetal behaviour state in normal term human pregnancy. Am J Obstet Gynecol.1990;163:1156.

32. Arulkumaran S, Anandkumar C, Ratnam SS et al. In-utero sound levels when vibroacoustic stimulation is applied to the maternal abdomen: an assessment of the possibility of cochlear damage in the fetus. BJOG 1992;99:43–45.

33. Arulkumaran S, Skurr B, Tong H, Kek LP et al. No evidence of hearing loss due to fetal acoustic stimulation test. Obstet Gynecol.1991;78:283–5.

34. Visser GHA, Mulder HH, Wit HP et al. Vibroacoustic stimulation of the human fetus: Effect on behavioral state organization. Early Human Dev 1989;19:285–96.

35. Montan S, Arulkumaran S, Ratnam SS. Safety of fetal acoustic stimulation test (FAST).Singapore J of Obstet Gynecol, 1991;22(2):35–39.

36. Smith CV, Phelan LP, Nguyen HN et al. Continuing experience with the fetal acoustic stimulation test. J Reprod Med.1988;33:365–8.

37. Manning FA, Morrison I, Lange IR, et al: Fetal assessment based on fetal biophysical profile scoring: experience in 12,620 referred high-risk pregnancies. Am J Obstet Gynecol 151: 343, 1985.

38. Manning FA, Platt LD, Sipos L: Antepartum fetal evaluation : Development of a fetal biophysical profile score. Am J Obstet Gynecol 136: 787, 1980.

39. Manning FA: Fetal biophysical profile. Obstet Gynecol Clin North Am 26: 557, 1999.

40. Manning FA, Lange IR, Morrison I, et al: Fetal biophysical profile score and the nonstress test: A comparative trial. Obstet Gynecol 64: 326–331, 1984.

41. Manning FA, Morrison I, Harman CR, et al: Fetal assessment based on fetal biophysical profile scoring: Experience in 19,221 referred high-risk pregnancies, 2. An analysis of false- negative fetal deaths. Am J Obstet Gynecol 157: 880, 1987.

42. Vintzileos AM, Campbell WA, Ingardia CJ, et al: The fetal biophysical profile and its predictive value. Obstet Gynecol 62: 271, 1983.

43. Vintzileos AM, Feinstein SJ, Lodeiro JG, et al. Fetal Biophysical Profile and the effect of premature rupture of mebranes. Obstet Gynecol 1986;67:818.

44. Vintzileos AM, Gaffney SE, Salinger LM, et al. The relation between fetal biophysical profile and cord pH in patients undergoing cesarean section before the onset of labor. Obstet Gynecol 1987;70:196.

45. Nicolaides KH, Peters MT, Vyas S, Rabinowitz R, Rosen DJD, Campbell S Relation of rate of urine production to oxygen tension in small for gestational age fetuses. Am J Obstet Gynecol 162:387, 1990.

46. Sarmiento A R. Antepartum Care of High Risk Pregnancy. Fernando Arias, Shirish N Daftary, Amarnath G Bhide (eds.) in Fernando Arias, Shirish N Daftary, Amarnath G Bhide. In Practical guide to High Risk Pregnancy and Delivery. Elsevier publishers, 3rd edition, pg. 3–31, 2008.

47. Martin R C, Yinka O, Anthony MV. Antepartum Fetal Assessment by Ultrasonography: The Fetal Biophysical Profile. Peter W Callen (ed). In Ultrasonography in Obstetrics and Gynecology. WB Saunders Company, 5th Edition, 780–793, 2007.

48. Miller DA, Rabello YA, Paul RH: The modified biophysical profile: Antepartum testing in the 1990s.Am J Obstet Gynecol 174: 812, 1996.

49. American College of Obstetricians and Gynecologists: Antepartum fetal surveillance. Practice Bulletin No. 9, October 1999.

50. Royal College of Obstetritians and Gynecologists. The investigation and the management of the small for gestational age fetus. Greentop Guideline No. 31, London: RCOG; 2013.

51. Cnossen JS, Morris RK, Riet G, Mol BW et al . Use of uterine artery Doppler?ultrasonography to predict pre-eclampsia and intrauterine?growth restriction: a systematic review and bivariate meta-analysis. CMAJ 2008; 178:701–11.

52. Baschat AA, Galan HL, Gabbe SG. Intrauterine growth Restriction. In Gabbe SG, Niebyl JR, Galan H, et al(eds). Obstetrics: Normal and Problem pregnancies, 6th ed. New York: Elseiver, 2012, 706–741.

53. ACOG 2013. Fetal Growth Restriction. ACOG Practice Bulletin No.134. American College of Obstetricians and Gynecologists, Washington DC. Obstet Gynecol.2013; 121:1122–1133.

Concept of High Risk Pregnancy

Vidya A Thobbi

Care of pregnant women is based on one overriding objective that each pregnancy should result in a healthy mother and baby. Whilst majority of pregnancies will progress satisfactorily with minimal intervention, there will always be a need to identify high risk pregnancy groups for whom greater degree of care is required.

The concept of risk assessment has become increasingly important. "Forewarned is forearmed" is especially applicable to obstetric practice. One of the first things that an obstetrician learns during the training period is to identify high risk pregnancies while caring for the antenatal mothers. Depending on the level of care, the incidence of high risk pregnancies varies around 5–10%. With proper care, 90–95% of high risk pregnancies will have a good outcome.[9]

There is no universally accepted definition. As per NICHD: "A high risk pregnancy is one of greater risk to the mother or the fetus than an uncomplicated pregnancy".

As per WHO survey, factors indirectly linked to health services such as transportation, access to mobile telephone technology (thus communication, channels for information and social assistance), as well as education and economic status also determine the risk of pregnancy.

The process of identifying high risk pregnancy involves a good medical history, examination and investigations.

There are five major categories of risk factors[12]
1. Personal factors (Table 12.1)
2. Obstetrical history (Table 12.2)
3. Past health history (Table 12.3)
4. Family health history (Table 12.4)
5. Ongoing maternal and/or fetal problems (Table 12.5)

The following tables describe these categories in detail along with the potentially adverse effects.

Risk factors for complications during pregnancy include:[8]
- Pre-existing maternal disease
- Physical and social characteristics
- Age
- Problems in previous pregnancy
- Problems developing during pregnancy
- Problems developing during labor and delivery

A. PRE-EXISTING MATERNAL DISORDER

In the first antenatal visit, details of pre-existing disorders are revealed by previous records, thorough history taking, examination and investigations. Patients may hide diseases like epilepsy due to social stigma. In a country like ours, a woman's first contact with a health care person will be during antenatal visit and may identify diseases like rheumatic heart disease.

Table 12.1	Personal factors and high-risk pregnancy
Personal factors	*Potentially adverse effects on pregnancy*

Less than 18 years old	• Unplanned pregnancy
	• Poor antenatal care
	• Increased incidence of:
	  – Abortion
	  – Fetal growth restriction
	  – Preterm labor
	  – Pre-eclampsia

More than 35 years old	• Down syndrome
	• Increased incidence of:
	  – Pre-eclampsia
	  – Fetal growth restriction
	  – Fetoplacental dysfunction
	  – Prolonged labor
	  – Obstructed labor

Access to hospital far away	• Delivery on the way to hospital • Birth trauma • Neonatal asphyxia or hypothermia
Consanguinity	• Congenital malformations • Repeated/ habitual abortion
Smoking	• Spontaneous abortion, prematurity • PROM, low birth weight and • Abruption • Chronic fetoplacental dysfunction
Infertility treated	• Anxiety with pregnancy • Multiple pregnancy • Preterm labor • Increased incidence of ectopic pregnancy

Once identified, pre-existing disorders require a multidisciplinary approach and proper counseling regarding how the disease, its treatment and investigations affect the pregnancy. In this approach, early assignment of role of each team member is decided. Obstetricians have to be the team leader and a link between patient and subspecialist care giver.[5] Pre-existing maternal disorders:[9]
• Anemia[1]
• Asthma
• Autoimmune disorders
• Cancer
• Diabetes mellitus
• Heart diseases
• Hepatic disorders
• Hypertension
• Infectious diseases like tuberculosis
• Renal disorder
• Seizure disorder
• Thromboembolic disorder
• Genitourinary tract diseases

Risk assessment is a part of antenatal, intra-partum and postnatal care. Several pregnancy monitoring and risk assessment system are available. The most widely used is the PRAMS7, i.e. Pregnancy Assessing Monitoring System developed by CDC in 1987, introduced in 5 states and currently in 32 states in USA. Using states vital statistics as its population based sampling frame, PRAMS7[7] "follows back" a stratified sample

Table 12.2	Obstetrical history and high-risk pregnancy
Obstetrical history	*Potential adverse effects on current pregnancy*
Parity ≥ 5	• Prolonged/ obstructed labor • Uterine rupture • Postpartum hemorrhage (PPH) • Chronic feto-placental dysfunction
No spacing	Nutritional deficiencies
Previous IUFD/ neonatal death	Dependent on the cause of death • Prematurity • IUGR • Fetal malformation • Chronic fetoplacental dysfunction • Recurrence of risk factors
Previous small for gestational age	• Fetal growth restriction • Chronic fetoplacental dysfunction • Fetal demise • Recurrence of risk factors
Previous large for gestational age	• Gestational diabetes • Diabetes mellitus • Recurrence of risk factors • Dystocia • Birth trauma • Excessive weight gain • Polyhydramnios
Previous fetal malformation	• Congenital anomalies • Hereditary disorders
Previous spontaneous 2nd trimester abortion/preterm labor	• Persistence of risk factors • Preterm delivery • Incompetent cervix
Recurrent first trimester abortion	• Recurrence of risk factors
Previous hypertensive disorders during pregnancy	• Pregnancy-associated hypertension • Renal causes • Chronic fetoplacental dysfunction • Fetal growth restriction • Prematurity
Previous cesarean section delivery	• Dehiscence of previous scar • Uterine rupture
Previous retained placenta or (PPH)	Recurrence of the problem
Previous Rh isoimmunisation	• Stillbirth • Feto-maternal incompatibility
Duration of labor <4 hours	• Delivery on the way to hospital • Neonatal asphyxia • Neonatal hypothermia • PPH
Previous instrumental delivery	• Prolonged/obstructed labor • Uterine rupture • Cephalopelvic disproportion

Table 12.3	Past history and high-risk pregnancy
Past history	*Potential adverse effects on current pregnancy*
Hypertension	• Pre-eclampsia • Renal causes • Chronic fetoplacental dysfunction • Chronic hypertension
Heart disease	• Heart failure • Pulmonary edema • Respiratory distress
Tuberculosis	• Congenital infection of newborn • Teratogenicity of antituberculous drugs
Epilepsy/antiepileptic medication	• Teratogenicity of antiepileptic drugs (AEDs) • Traumatic seizures
Chronic illness	Illness may affect pregnancy or vice versa
Uterine anomalies including fibroids/ pelvic masses	• Second trimester miscarriage • Preterm labor • Fetal growth restriction • Abnormal placentation • Malpresentation • Antepartum/ postpartum hemorrhage (APH/PPH) • Abdominal pain due to fibroid degeneration • Uterine rupture
Previous myomectomy	• Uterine rupture • Abnormal placentation • APH/ PPH • Retained placenta
Previous cerclage	• Incompetent cervix • Increased incidence of abortion/preterm labor
Previous successful classical repair	• Difficult delivery • Soft tissue obstruction • Ruptured vagina
Previous successful repair of fistula	Increased incidence of recurrence
Previous blood transfusion	Fetomaternal incompatibility

Table 12.4	Family history and high-risk pregnancy
Family history	*Potential adverse effects on the current pregnancy*
Fetal abnormality	Fetal malformations
Multiple pregnancy in mother/sister	Increased incidence of multiple pregnancy
Hypertension	Pregnancy induced hypertension
Diabetes	• Gestational diabetes • Spontaneous abortion • Congenital anomalies • Macrosomic neonate

Table 12.5	Ongoing maternal and/or fetal problems and high-risk pregnancy[12]
Current situation	*Potential adverse effects on the current pregnancy*
Unknown last menstrual period (LMP)	• Post date • Failure to diagnose IUGR
Gait: "Limping"	• Cephalopelvic disproportion • Obstructed labor
Pallor	• Anemia • IUGR • Preterm labor
Jaundice	• Biliary colic • Obstructive jaundice • Acute cholecystitis • Pancreatitis
Maternal weight > 90 kg	• Gestational diabetes • Hypertension • Inadequate maternal weight gain • Macrosomic infants • Prolonged second stage • Shoulder dystocia • Primary cesarean delivery • Wound/episiotomy infection
Maternal weight <45 kg	• Fetal growth restriction • Preterm labor • Fetoplacental dysfunction
Maternal length ≤150 cm	• Cephalopelvic disproportion • Obstructed and/or prolonged labor
Marked varicosities of lower limbs	• Severe leg pain • Increased incidence of vulval/uterovesical plexus varicosities • Increased incidence of hemorrhoids
Hyperemesis gravidarum	• Hypotension • Tachycardia • Dehydration • Weight loss • Electrolyte imbalance • Jaundice • Retinal changes • Oliguria
Non-immune against tetanus	Tetanus neonatorum
Absent fetal movements	• Molar pregnancy • IUFD
Marked changes in frequency and/ or intensity of fetal movements	• Fetoplacental dysfunction • IUFD • Fetal growth restriction
Smaller uterine size than gestational age	• Inaccurate LMP • IUGR

Contd...

Table 12.5 Ongoing maternal and/or fetal problems and high-risk pregnancy[12] *(Contd...)*

Current situation	*Potential adverse effects on the current pregnancy*
Larger uterine size than gestational age	• IUFD • Oligohydramnios • Missed abortion • Inaccurate LMP • Diabetes mellitus • Multiple pregnancy • Molar pregnancy • Polyhydramnios
Vaginal bleeding in early pregnancy	• Ectopic pregnancy • Threatened abortion • Missed abortion • Molar pregnancy
Blood pressure ≥ 140/90 mmHg	• Increased incidence of eclampsia • Fetal growth restriction • Renal disease • Placental abruption • Fetoplacental dysfunction
Excess amniotic fluid	• PPH • Fetal malformations • Preterm labor • Maternal respiratory distress • Fetal macrosomia
Diminished amniotic fluid	• Fetal malformations (renal, pulmonary hypoplasia) • FGR • Stillbirth
Preterm uterine contractions	Preterm labor
Third trimester vaginal bleeding	• Placenta previa/ abruption • IUGR • IUFD • Intrapartum hemorrhage/ PPH • DIC
Sudden gush of vaginal watery fluid	• Ruptured membranes prematurely • PROM • Prematurity • Cord prolapse • Maternal infection (chorioamnionitis) • Fetal infection • IUFD • Neonatal infection
Hemoglobin <11 gm	• Anemia in pregnancy • IUGR • Prematurity • Anemia of the newborn

Contd...

Table 12.5	Ongoing maternal and/or fetal problems and high-risk pregnancy[12] (Contd...)
Current situation	Potential adverse effects on the current pregnancy
Proteinuria > +	• Urinary tract infection • Renal disease • Pre-eclampsia
Glucosuria	• Gestational diabetes
Rubella exposure	• Severe fetal damage or death if infection occurred during first 4 months of pregnancy • Congenital fetal infection (heart damage, cataract, deafness, mental retardation) • Severe disease during neonatal period (bleeding hepatosplenomegaly, myocarditis, thrombocytopenia)
Herpes	• Spontaneous abortion • Generalized flu-like symptoms • Reactivation • Neonatal herpes infection • Prematurity • FGR and IUFD
Non-engagement of fetal head at 40 weeks in primigravida malpresentation	• Cephalopelvic disproportion • Malposition (mainly occipito-posterior) • Prematurity • PROM • Uterine rupture • Prolonged/obstructed labor • Cord presentation/prolapse
Bacteriuria (> 100,000 bacteria in urine culture)	• Urinary tract infection • Pyelonephritis/pyelitis

Table 12.6	Risk scoring system for high risk pregnancy[6,11]	
Category	Risk factors	Score
A. Pre-existing		
1. Cardiovascular and renal disorders	Moderate to severe pre-eclampsia	10
	Chronic hypertension	10
	Moderate to severe renal disorders	10
	Severe heart failure class II-IV NYHA	10
	History of eclampsia	5
	History of pyelitis	
	Infection of renal pelvis	5
	Mild heart failure class I NYHA	5
	Mild pre-eclampsia	5
	Acute pyelonephritis	5
	History of cystitis	1
	Acute cystitis	1
	H/o pre-eclampsia	1

Contd...

Table 12.6	Risk scoring system for high risk pregnancy[6,11] *(Contd...)*	
Category	*Risk factors*	*Score*
2. Metabolic disorder	Obesity class III BMI>40	10
	Insulin dependent diabetes	10
	Previous endocrine ablation	10
	Thyroid disorder	5
	Obesity class II BMI 35–39.9	5
	Gestational diabetes	5
	Family history of diabetes	1
3. Obstetric history	Fetal exchange transfusion because of Rh incompatibility	10
	Stillbirth	10
	Late abortion 16–20 weeks	10
	Post-term pregnancy >42 weeks	10
	Pre-term newborn <37 weeks and <2.5 kg	10
	FGR <10th centile for estimated gestational age	10
	Abnormal fetal position	10
	Polyhydramnios	10
	Multi-fetal pregnancy	10
	Previous brachial plexus injury	10
	Neonatal death	5
	Cesarean delivery	5
	Habitual abortion >3	5
	Neonate >4.5 kg	5
	Shoulder dystocia	5
	Multiparity >5	5
	Seizure disorder or cerebral palsy	5
	Fetal malformation	1
4. Other disorders	Abnormal cervical cytology finding	10
	Sickle cell disease	10
	Thrombophilia	10
	Positive serology result of STDs	5
	Severe anemia Hb< 9 gm/dl	5
	History of Tb/PPD injection site induration >/10 mm	5
	Pulmonary disorder	5
	Mild anemia Hb 9–10 gm/dl	1
5. Anatomical abnormality	Uterine malformation	10
	Insufficient or incompetent cervix	10
	Small pelvis	5
6. Maternal characteristics	Age >35 years or <15 yrs	5
	Wt <45.5 kg or >91 kg	5
	Psychiatric disorder or intellectual disability	1
B. Antepartum		
1. Exposure to teratogens	Group B streptococcal infection	10
	Certain viral infection (e.g. rubella, CMV)	5
	Flu syndrome (severe)	5

Contd...

Table 12.6	Risk scoring system for high risk pregnancy[6,11] *(Contd...)*	
Category	*Risk factors*	*Score*
	Excessive use of drug	5
	Smoking >1 pack /day	1
	Alcohol (moderate to severe)	1
2. Pregnancy complications	Preterm labor at<37 weeks	10
	PPROM	10
	Rh sensitization (not requiring exchange transfusion)	5
	Vaginal spotting	5
C. Intrapartum		
Maternal	Moderate to severe pre-eclampsia	10
	Polyhydramnios/ oligohydramnios	10
	Uterine rupture	10
	Post-term (>42 weeks)	10
	Mild pre-eclampsia	5
	Premature rupture of membranes>12 hr	5
	Preterm labor at >/37 wk	5
	Primary dysfunction labor	5
	Secondary arrest of dilatation	5
	Labor>20	5
	Second stage>2.5 H	5
	Medical induction of labor	5
	Precipitous labor (<3 hr)	5
	Primary cesarean delivery	5
	Repeat cesarean delivery	5
	Elective induction of labor	1
	Prolonged latent phase	1
	Oxytocin augmentation	1
Placental	Placenta previa	10
	Abruption placenta	10
	Chorioamnionitis	10
Fetal	Abnormal presentation	10
	Multifetal pregnancy	10
	Fetal bradycardia >30 min	10
	Prolapsed cord	10
	Fetal weight < 2.5 kg	10
	Fetal weight > 4 kg	10
	Fetal acidosis pH < 7	10
	Fetal tachycardia > 30 min	10
	Operative delivery using vacuum extractor or forceps	5
	Breech vaginal delivery	5

No risk with the score of 0–3, low risk with the score of 4–9, high risk with the score of 10 or >10[6]

of women several months postpartum, surveying them about their own and their infant's pre-natal, birth and postpartum behavior and experiences.

B. PHYSICAL AND SOCIAL CHARACTERISTICS

1. Maternal Age

Teenage pregnancy (<19 years) is common in India. Poor nutritional status, physical and psychological immaturity puts teenage pregnancy at risk.

On the other hand, delaying child bearing more than 35 years is associated with infertility, chromosomal abnormalities, chronic diseases like hypertension and diabetes, increase chances of stillbirth, placenta preavia, abruption, etc.

2. Social Characteristics

According to WHO survey, in countries with decline in maternal mortality, viz. Bangladesh, Cambodia, it was seen that maternal risk is affected by factors directly and indirectly linked to health services. Social factors like improved transportation, access to mobile telephone technology, better education and economic status are associated with improved maternal health. Pregnancy in women, in remote rural communities with poor education, road infrastructure and connectivity should be considered high-risk pregnancy.

3. Maternal Weight

Both underweight and overweight women have increased complications in pregnancy. Underweight women should gain at least 12.5 kg during pregnancy while obese women should restrict weight gain to less than 11.5 kg.

4. Maternal Height

Short women (<145 cm) are likely to have small pelvis, dystocia and CPD.

C. OBSTETRICS HISTORY—PROBLEMS IN PREVIOUS PREGNANCY

1. *Number of pregnancies:* Primigravida are at increased risk of complications, viz. pre-eclampsia, cesarean deliveries. Grand multiparity is associated with anemia, PPH, APH, etc.
2. *Bad obstetric history[1]:* Detailed evaluation of bad obstetric history and their management improves the outcome of pregnancy. Progesterone therapy and cervical encerclage help to reduce the preterm deliveries.
3. *Prior neonate with a genetic/congenital disorder:* Genetic screening, USG and evaluation by genetic specialist is recommended.

D. PROBLEMS THAT DEVELOPED DURING PREGNANCY

These can be due to

1. Disorders a woman had before she became pregnant
2. Disorders that developed during pregnancy but are directly related to pregnancy, e.g. anemia[3]
3. Disorders that are likely to occur during pregnancy.

Pregnancy which was labeled low-risk can develop complications and become high risk. Continuous surveillance is required to detect such pregnancies.

Disorders that are not directly related to pregnancy develop during pregnancy. Some of them increase the risk for pregnant women and the fetus.

1. Disorders that cause high fever
2. Infections
3. Disorders requiring abdominal surgeries

Such patients need care from multi-disciplinary team.

Some disorders can occur due to physiological changes in pregnancy. Hyperemesis gravidarum requires inpatient care, proper fluid and nutrition therapy. Thomboembolic disorders are likely to occur in pregnancy due to raised clotting factors. Anemia (60–70%) may be pre-existing or develop due to increased demand for iron and hemodilution.[3]

Urinary tract infections are common due to physiological changes. Asymptomatic bacteriuria should be diagnosed and treated. Pyelonephritis increases the incidence of PROM, preterm labour and sepsis warranting inpatient care.

Hypertension (10–15%) in pregnancy is diagnosed by blood pressure measurement during each antenatal visit. Urine protein should be tested detect pre-eclampsia. Early identification and treatment improves the outcome of hypertensive[4] disorders in pregnancy.

Diabetic is diagnosed in 15% of pregnancy, either as overt or gestational diabetes. GDM is diagnosed by oral glucose challenge test (DIPSI test) in which the patient consumes 75 gm of glucose irrespective of earlier meals and blood glucose is tested after 2 hours. Values >140 mg/dl is suggestive of GDM. Treatment includes dietary modification, exercise and insulin. Goal of treatment is to reduce macrosomia, stillbirth and maternal complications.

Sexually transmitted disease: Routine antenatal screening for STD is done for HIV, syphilis and HbsAg. Early diagnosis and treatment reduces the risk of fetal infection. Bacterial vaginosis, gonorrhea, genital chlamydia infections increase risk of preterm labor.

Polyhydramnios and oligohydramnios: These are suspected when uterine size does not correspond to gestational age or discovered during routine USG. Since abnormality of liquor volume have poor fetal birth outcome, they are categorized as high-risk pregnancy.

Multifetal pregnancies: The incidence has increased due to ART and infertility treatments. The determination of chorionicity is important for the management of multi-fetal pregnancies.

Threatened preterm labor/PROM: It is usually due to genital infections. Management with steroids, tocolytics and antibiotics improve outcome. These cases should be managed in tertiary care where NICU facilities are available.[2]

Bleeding in pregnancy: Any bleeding in pregnancy is abnormal and should be evaluated. The most common cause for APH is placenta previa and abruptio placentae, which need specialized care.

Pregnancy with previous cesarean section: Incidence is rising due to rising incidence of cesarean section. There is a risk of placental abnormalities and rupture uterus.

Fetal growth restriction: This is diagnosed when fetal weight is less than 10th percentile. Regular monitoring, Doppler studies, good intrapartum and NICU care are necessary.

Gynecological disorders like fibroid, uterine prolapse, cervical dysplasia, ovarian tumors during pregnancy are associated with increased maternal and fetal morbidity.

Abnormalities and complication of labor and delivery: These can convert low risk pregnancy to a high risk one. A proper referral or transfer protocol should be in place in all delivery units in case any problem arises.

- *Problems present before labor:* Multifetal pregnancies, post-term pregnancies, PROM, abnormal fetal presentations, induction of labor.
- *Complication developing during labor:* Amniotic fluid embolism, shoulder dystocia, prolonged/obstructed labor, cord prolapse, abruption, uterine rupture.
- *Third stage complications:* PPH, uterine inversion, placenta accreta
- *Neonatal problems:* Fetal distress, fetal injuries, meconium aspiration
- *Postpartum complications:* Puerperal fever, puerperal sepsis, secondary PPH, peripartum cardiomyopathy, breast abscess, cerebral venous thrombosis.

In the journey through pregnancy, a woman may enter a high-risk category at any point of time.[10] Fortunately, majority of pregnancies succeed in having healthy mother and child. In unfortunate few, there may be devastating outcome even with timely intervention.

Thus, the concept of high-risk pregnancy helps in identifying the woman with special needs in pregnancy.

REFERENCES

1. Yadava B Jeve, William Davies. Evidence based management of recurrent miscarriages. J Human Reprod Sci.2014 jul-sep; 7(3):159–169 doi:10.4103/0974-1208.142475.
2. RCOG Green Top Guidelines no.1B.
3. WHO-Treatments of iron deficiency anemia in pregnancy.
4. NICE Guidelines 2010. Hypertension in pregnancy: diagnosis and management.
5. Singh M, Paul VK, Maternal and child health services in India with special focus on perinatal services. J Perinatol, 1997 Jan -Feb; 17(1):65–9.
6. Sapna Jain, Sweta Anand, Rupa Aherwar. High risk scoring for prediction of pregnancy outcome: A prospective study. Int J Reprod Contracept Obstet Gynecol, 2014 Sep; 3 (3):516–522.
7. Pregnancy risk assessment monitoring system (PRAMS)
8. Risk Factors for complications during pregnancy-Gynecology and Obstetrics-Merck Manuals Professional Edition.
9. G Dangal, Hish Risk Pregnancy, The Internet Journal of Gynecology and Obstetrics 2006. Volume 7 Number 1.
10. Leontine Alkema, Global, regional and national levels and trends in maternal mortality between 1990 and 2015, with scenario based projections to 2030: A systematic analysis by UN Maternal Mortality Estimation Inter-Agency Group.
11. Hobel J Calvin, Hyvarinen A Marcia. Prenatal and intrapartum high risk screening vol.117 Number 1 AJOG.
12. Standards of Practice for Integrated MCH/RH Services: First Edition, June 2005.

Medical Disorders and Perinatal Problems: An Overview

Siddesh Mahadev Iyer, Reena J Wani

It is often found that young women do not see a physician as a routine and often the obstetrician is the primary care provider or first point of contact with the health care system, more so in rural areas.

Medical disorders have a great impact on pregnancy and its outcome. For example, anemia is the commonest disorder we see in India and may be the underlying contributory factor in up to two-thirds of maternal deaths. Certain medical conditions like hepatitis E are benign in non-pregnant women but can be associated with even 100% maternal mortality in cases with fulminant hepatic failure.

In many larger institutes the management is done by a multidisciplinary team, with available specialists as required. However, every obstetrician must be aware of the challenges one can face when encountering such medical disorders in pregnancy, and how to deal with them. Another important issue is to realize when to refer or transfer to a higher center. Within tertiary care too, be more in favor of ICU admission in case of any deterioration as pregnant women with medical disorders can suddenly worsen and die, but may be saved if aggressive timely help is available.

We have simplified the medical disorders and the problems associated with them and given recommendations to assist management in Tables 13.1–13.3.

Table 13.1	Medical disorders		
Disorder	Antenatal	In labor/during LSCS	Postnatal
Anemia	Poor weight gain, poor immunity, preterm labor, congestive cardiac failure in cases of severe anemia at 30–32 weeks, decreased work capacity	Dysfunctional labor, congestive cardiac failure, risk of anesthesia	Post partum hemorrhage, sepsis, subinvolution of uterus, pulmonary embolus, failure of lactation, delayed wound healing
Pregnancy induced hypertension	Eclampsia, HELLP syndrome, acute renal failure, damage to vital organs, need for early termination of pregnancy, placental abruption, disseminated intravascular coagulation.	Eclampsia, need for transfusion of blood and blood products, anesthesia related complications	Eclampsia, postpartum hemorrhage, sepsis, subinvolution of uterus, pulmonary embolus, chronic hypertension, recurrence in subsequent pregnancies

Contd...

Table 13.1		Medical disorders	
Disorder	*Antenatal*	*In labor/during LSCS*	*Postnatal*
Gestational diabetes	Abortions, congenital anomalies, recurrent infections, growth restricted fetus, macrosomia, pre-eclampsia, sudden intrauterine fetal death, need for medication.	Sudden intrauterine fetal death, macrosomic baby, shoulder dystocia, birth trauma, increased risk of cesarean section, anesthesia related complications	Postpartum hemorrhage, sepsis, delayed wound healing, risk of developing diabetes mellitus
Heart disease	Recurrent infections, congestive cardiac failure, arrhythmia, preterm labor, sudden maternal collapse and death	Congestive cardiac failure, sudden maternal collapse and death.	Embolus to other vital organs, postpartum hemorrhage, sepsis, sudden maternal collapse and death
Thyroid disorders	Miscarriages, preterm labor, congestive cardiac failure, placental abruption, pulmonary embolus, thyroid storm	Thyroid storm, congestive cardiac failure, anesthesia related complications	Postpartum hemorrhage
Epilepsy	Congenital malformation, growth restricted fetus, stillbirth, status epilepticus	Status epilepticus	Postpartum hemorrhage, thromboembolism
Asthma	Preterm labor, premature rupture of membranes, growth restriction, status asthmaticus, arrhythmia, respiratory failure.	Status asthmaticus	Status asthmaticus, sepsis
Jaundice (cholestasis)	Itching all over the body, sudden intrauterine fetal demise	Meconium stained liquor, increased risk of cesarean section	Postpartum hemorrhage, sepsis
Jaundice (hepatitis)	Itching, sudden intrauterine fetal demise, fulminant hepatic failure	Meconium stained liquor, increased risk of cesarean section	Postpartum hemorrhage, sepsis
Malaria	Hypoglycemia, sudden intrauterine fetal demise, fulminant organ failure	Fetal distress	Postpartum hemorrhage, sepsis

Table 13.2	Fetal and neonatal complications
Anemia	Growth restricted fetus, fetal anemia, poor APGAR at birth, low birth weight baby
Pregnancy induced hypertension	Iatrogenic prematurity, growth restricted fetus, poor APGAR.
Gestational diabetes mellitus	Fetal anomalies, macrosomic fetus, low birth weight babies, sudden fetal demise, birth injuries, neonatal hypocalcemia, neonatal hypoglycemia,
Heart disease	Low birth weight babies, anomalies.
Epilepsy	Congenital anomaly like cleft lip, palate, mental retardation, cardiac defects, limb defects
Asthma	Growth restriction
Thyroid	Growth restricted fetus, prematurity, preterm birth, stillbirth, increased perinatal mortality
Jaundice	Low birth weight, preterm labor, intrauterine fetal death

Table 13.3	Clinical management
Disorder	*Recommendation*
Anemia	• Iron rich diet to be taken. • Oral iron therapy to be given to all patients. • Hemoglobin to be documented on antenatal visit. • Parenteral iron therapy with intravenous iron sucrose in antenatal and postnatal period can be given based on clinical situation. • Deworming advisable. • Anemia typing should be done and hemoglobinopathies should be ruled out. • Cut short second stage of labor. • Active management of third stage of labor and pph prophylaxis to be done. • Blood transfusion in severe anemia. • Iron prophylaxis to be continued 3 months postdelivery
Pregnancy induced hypertension	• Blood pressure and urine albumin to be checked on all antenatal visits. • Severe PIH to be managed indoor. • Identify risk factors for developing pre-eclampsia. • Anti-hypertensives to be started timely. • Labetolol is the drug of choice. • Strict watch on premonitory symptoms and injection magnesium sulfate to be given for eclampsia. • CBC, LFT, RFT to be sent to diagnose HELLP syndrome early. • Early delivery to be planned based on the clinical situation. • PPH prophylaxis to be given.
Gestational diabetes mellitus	• Screening for diabetes in all antenatal patients. • Oral glucose tolerance test with 75 gm glucose to be done for patients with deranged sugars. • Patients to be started on metformin/insulin as per the clinical situation. • Blood sugar levels to be monitored. • Possibility of sudden IUFD to be kept in mind. • Senior obstetrician to be present at delivery and shoulder dystocia drills to be practiced. • Neonate to be screened at birth by a neonatologist.
Heart disease in pregnancy	• Patient to be treated in conjunction with a cardiologist. • High-risk patients to be managed at tertiary care hospitals. • Monitor signs of congestive cardiac failure. • Cut short second stage of labor. • Injection frusemide to be given after delivery of the baby to prevent cardiac failure.
Thyroid	• Screening with TSH to be done in early gestation. • Hypothyroid patients to be started on tab thyroxine (TSH to be maintained at 2.5 in the first trimester)
Epilepsy	• Teratogenic drugs to be avoided. • Levetiracetam is now the drug of choice in epilepsy.
Asthma	• Prostaglandin to be used with caution.
Jaundice	• Prevent hypoglycemia, high CHO diet • Monitoring with serial LFT, INR. • Fetal distress and meconium stained liquor is common in third trimester. • Delivery to be planned after 37 completed weeks of gestation when maternal condition is stable. • Monitor coagulation profile, LFT and deliver when improving trends
Malaria	• Hydration, prevention of hypoglycemia • Prefer artemisinin-based combination therapy

SUGGESTED READING

1. American College of Obstetricians and Gynaecologists: Anaemia in pregnancy. Committee opinion No.95, July 2008.
2. American College of obstetricians and Gynaecologists: hypertension in pregnancy. Task force and work group.
3. Fifth International Workshop—Conference On Gestational Diabetes. Diabetes Care 30(Suppl 2): S251, 2007.
4. Shorvon S, Antiepileptic drug therapy during pregnancy: the neurologist's perspective. J Med Genet 2002;39:248–250.
5. Siu S, Colman JM, Cardiovascular problems and pregnancy: An approach to management, hyperlink "http://www.ccjm.org/content/71/12/977.abstract" \t "_blank", Cleveland Clinic Journal of Medicine, 2004; 71: 977–985.

PPTCT: Update and Impact of Changed NACO Guidelines on PPTCT Program

Reena J Wani

INTRODUCTION

Although heterosexual transmission is responsible for 88.2% of HIV infections in India, 5% were attributed to parent to child transmission (NACO Report 2011-12). HIV infection in young children is mostly due to mother to child transmission and prevention of MTCT will greatly reduce the infection in young generations. Nevirapine (NVP) therapy given as single dose to mother just before delivery and to newborn within 72 hours of birth was being done at most centers across India, including ours, till 2013. WHO recommends ART in all HIV positive women in pregnancy which was accepted by NACO in 2013 and implemented from 2014.[1, 2]

EFFECT OF PREGNANCY ON HIV AND HIV ON PREGNANCY

Pregnancy does not influence the course of the disease and neither has it effected the survival of the HIV infected mother.[3,4] Women and infants transmission study (WITS) did not show difference in CD4+ lymphocyte count or HIV RNA trajectory or clinical AIDS rate in one or more pregnancies after diagnosis with HIV.[5] Hormonal effect of pregnancy may increase the toxicity to anti-retroviral therapy. Chances of hepatic failure and lactic acidosis with nucleoside reverse transcriptase inhibitor therapy may cause maternal deaths with long-term therapy. Pregnant women should be warned regarding the signs and symptoms of liver failure and lactic acidosis like nausea, vomiting, fatigue, tachycardia, dyspnea, hyperventilation and abdominal pain and can mimic normal signs and symptoms of pregnancy and may be overlooked. Perinatal complications increase in HIV positive mothers with or without ARV medications.[6]

There is an association with preterm delivery and low birth weight, fetal growth restriction, stillbirth and infant death. Mothers receiving antiretroviral therapy with low CD4+ counts have an additional risk factor for adverse outcome.[7] Zidovudine monotherapy is not associated with preterm birth or low birth weight. HAART therapy specifically with protease inhibitors has increased risk of preterm delivery in some previous studies.[8] However, there has been a paradigm shift in thought and research and this is reflected in the newer guidelines discussed below.

4 PRONGS FOR PPTCT (Fig. 14.1)

PRONG 1	• Primary prevention of HIV • HIV negative, i.e. general population
PRONG 2	• Prevent unintended pregnancies • HIV positive: Family planning counselling in ICTC / ART centers
PRONG 3	• Prevention of mother to child transmission • HIV positive and pregnant
PRONG 4	• Care, support and treatment • HIV positive mother and child

Fig. 14.1

ESSENTIAL PACKAGE OF PPTCT SERVICES
(Fig. 14.2)

Routine offer of HIV counseling and testing with "opt-out" option
Ensure involvement of spouse and family members
Provide ART to HIV positive pregnant and breastfeeding women
Promote institutional deliveries of positive pregnant women
Provision of care for STI / RTI, TB and opportunistic infections
Nutritional counseling and psychosocial support to positive pregnant women
Antiretroviral prophylaxis to infants
Follow-up of HIV-exposed infants
Cotrimoxazole prophylactic therapy and early infant diagnosis

Fig. 14.2

PERINATAL TRANSMISSION AND FACTORS INVOLVED (Figs 14.3 and 14.4)

Risk of perinatal transmission of HIV to the infant is a matter of serious concern. In untreated women without breastfeeding chances of transmission is 20–30%.[6] Two-thirds of this transmission will occur at delivery and one-third during late antenatal period. Breastfeeding has an additional transmission risk of 15–20%. Maternal HIV RNA level is most important predictor for transmission. ART during pregnancy and neonatal period reduces the risk of transmission.[1,2]

Risk of HIV transmission	Transmission rate
During pregnancy	5–10%
During labor and delivery	10–15%
During breastfeeding	5–20%
Overall without breastfeeding	15–25%
Overall with breastfeeding up to 6 months	20–35%
Overall with breastfeeding for 18–24 months	30–45%

Fig. 14.3: Estimated risk of mother to child transmission in absence of any intervention[1, 2]

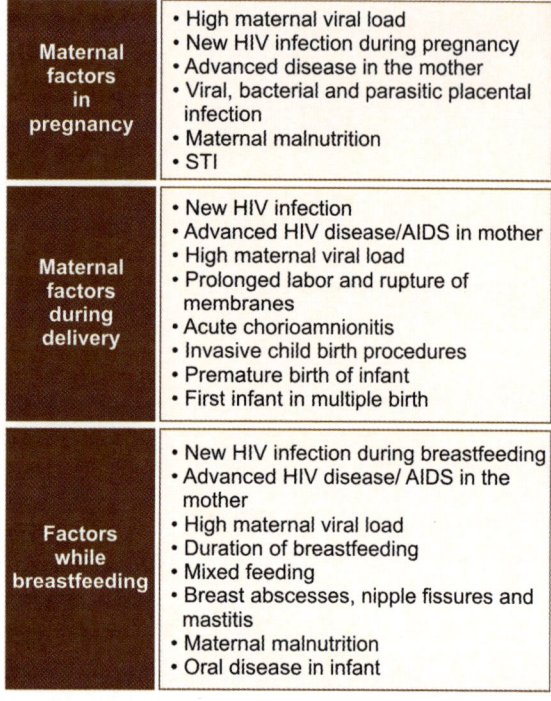

Maternal factors in pregnancy	• High maternal viral load • New HIV infection during pregnancy • Advanced disease in the mother • Viral, bacterial and parasitic placental infection • Maternal malnutrition • STI
Maternal factors during delivery	• New HIV infection • Advanced HIV disease/AIDS in mother • High maternal viral load • Prolonged labor and rupture of membranes • Acute chorioamnionitis • Invasive child birth procedures • Premature birth of infant • First infant in multiple birth
Factors while breastfeeding	• New HIV infection during breastfeeding • Advanced HIV disease/ AIDS in the mother • High maternal viral load • Duration of breastfeeding • Mixed feeding • Breast abscesses, nipple fissures and mastitis • Maternal malnutrition • Oral disease in infant

Fig. 14.4: Factors that increase transmission of HIV to the child from the mother[1, 2]

NATIONAL RECOMMENDATION ON PPTCT [1]

• All positive pregnant women including those presenting in labor and breast-feeding women with HIV should be initiated on a triple ART irrespective of CD4, for preventing mother-to-child transmission risk and should continue lifelong ART.

• The duration of NVP to infant be minimum 6 weeks but more if ART to mother was started in late pregnancy, during or after delivery and has not been on adequate period of ART as to be effective to achieve optimal viral suppression (which is at least 24 weeks), then the infant NVP should be increased to 12 weeks.

• This recommendation on extended NVP duration applies to infants of breastfeeding women only and not those on exclusive replacement feeding.

PPTCT SCENARIOS[1]

I. Pregnant women newly initiating ART
II. Pregnant women already receiving ART
III. Pregnant women having prior exposure to NNRTI for PPTCT and ART regimen
IV. Pregnant women presenting directly-in-labor.

I. Pregnant Women Newly Initiated ART (Fig. 14.5)

- TLE (Tenofovir, Lamivudine/3TC and Efavirenz) regime as fixed dose combination is now the recommendation for ART
- Start ART as soon as possible after proper preparedness counselling and continue ART throughout pregnancy, delivery, and thereafter life long
- Even if the pregnant women present very late in pregnancy (including those who present after 36 weeks of gestation) the ART should be initiated promptly
- This ART shall be initiated at ART centres only, hence all efforts need to be made, to ensure that pregnant women reach ART centers.

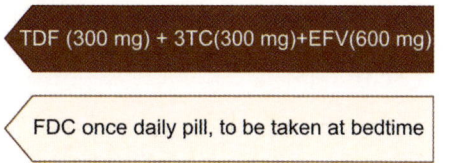

TDF (300 mg) + 3TC(300 mg)+EFV(600 mg)

FDC once daily pill, to be taken at bedtime

Fig. 14.5: TLE ART regime to be given to ALL HIV +ve pregnant women

II. Pregnant Women Already Receiving ART

- Pregnant women who are already receiving a NVP-based ART regimen should continue receiving the ART regimen
- Pregnant women who are already receiving EFV-based ART regimen: Continue
- There is no indication for abortion/ termination of pregnancy in women exposed to EFV in the first trimester of pregnancy.

III. ART Regimen for Pregnant Women having Prior Exposure to NNRTI for PPTCT (Fig. 14.6)

The prior exposure to Nevirapine in the previous pregnancy may reduce the effectiveness of Efavirenz, hence it should be substituted with Lopinavir/Ritonavir.

IV. Women Presenting Direct-in-labour: Care and Assessment (Fig. 14.7)

This situation is a special challenge and systems should be in place for dealing with it as follows:

Monitoring of Pregnant Women Receiving ART[1]

This is also an important part of the continuum of care:

- WHO clinical staging—will help in monitoring the clinical progress and potential disease progression

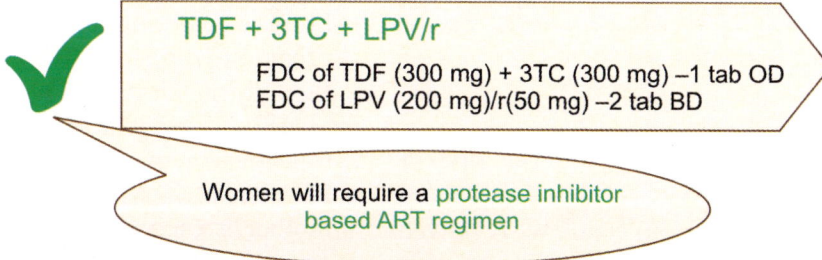

TDF + 3TC + LPV/r

FDC of TDF (300 mg) + 3TC (300 mg) –1 tab OD
FDC of LPV (200 mg)/r(50 mg) –2 tab BD

Women will require a protease inhibitor based ART regimen

Women, who need to be initiated on (LPV/r), the drugs can be proved at the ART center itself

Fig. 14.6: ART regimen to pregnant women

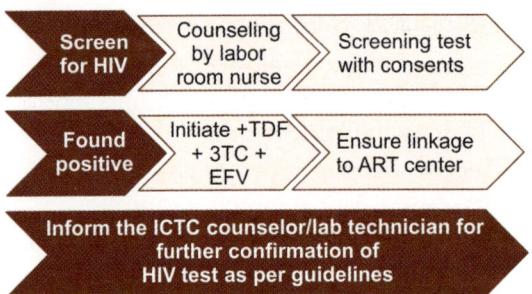

Fig. 14.7: Duration of ARV prophylaxis to HIV exposed infants

- Body weight—but can be difficult to assess in pregnancy
- Hemoglobin, functional status
- CD4 cell count—6 monthly
- *ART-related side-effects:* These may overlap with that of common pregnancy conditions, e.g. nausea and vomiting. Minor symptoms should be controlled symptomatically with medicines

Duration of ARV Prophylaxis to HIV Exposed Infants (Fig. 14.8)

Testing by DBS (dried blood spots) at appropriate intervals as per guidelines for early infant diagnosis (EID) should be done.[1]

6 Weeks (regular)	If adequate ART to mother, regardless of whether exclusively breastfed or exclusive replacement fed
12 Weeks (extended)*	If ART to mother was started in late pregnancy (<24 weeks/6 months*) and breastfeeding

* Adequate period of ART to optimal viral suppression
* Applicable to infants to breastfeeding women only and not those on exclusive replacement feeding

Fig. 14.8

Our Experience

To assess the effectiveness of the PPTCT Program and the impact of changed guidelines from 2014 on MTCT at our centre we reviewed all ANC cases presenting at our hospital over 3 years (2014 to 2016), and data of sero-positive women delivering at our institute in this period, and newborns screened for HIV by DBS as per NACO Guidelines was analysed.

Results

Out of 13,701 ANC cases registered over 3 years, totally 12,037 were screened using opt-out policy (88%). Overall 24 cases, i.e. 0.2% were detected to be HIV-positive, with a decreasing trend over the 3 years. However, of actual deliveries, 0.6% (69) were seropositive as many cases (45) had been referred to us for delivery due to sero-positive status. There was a declining trend in sero-discordance, and in cesarean section delivery in this group over the study period. Of the 67 livebirths, 82.1% neonates underwent DBS (dried blood spot) testing, and most importantly only 2 out of 55 were HIV positive (3.6% transmission rate). In fact these 2 cases were in the first year of the changed policy, and in the last 2 years (2015 and 2016), there has been ZERO positivity, i.e. 100% prevention of MTCT of HIV in our institute.

We emphasize hereby that the rate of MTCT can be reduced to zero by antenatal screening, highly active anti-retroviral therapy (ARV) during antenatal and intra-natal period and treating newborns with ARV as per revised NACO Guidelines.

SUMMARY OF THE PPTCT GUIDELINES

- Primary prevention of HIV in childbearing women
- Prevention of unwanted pregnancies
- Periodic antenatal visits of pregnant women
- Provide HIV counseling and testing to ALL pregnant women
- *Plan and monitor HIV-positive pregnancies:* "THE" most important factor for transmission of HIV is maternal viral load
- Provision of ART to all HIV infected mother, irrespective of clinical stage or CD4 count
- Provide ART lifelong for all HIV positive pregnant women as part of PPTCT

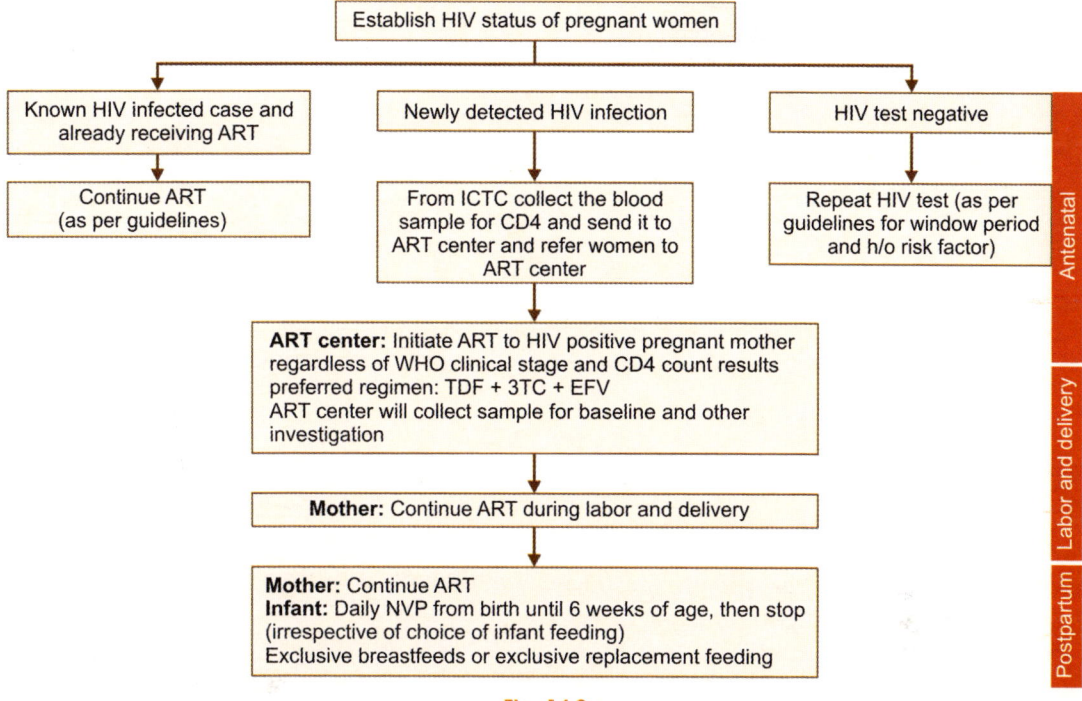

Fig. 14.9

- Practice safe delivery techniques
- Promote exclusive breastfeeding
- Provision of EID to HIV exposed infants

ACKNOWLEDGMENTS

Our PPTCT Program is run in collaboration with Departments of Microbiology, Obstetrics and Gynecology and Pediatrics with the help of MDACS. We thank all team members who are working to implement the NACO guidelines and Dean sir (HBTMC) for his support and guidance.

REFERENCES

1. Updated guideline for Prevention of Parent to child transmission (PPTCT) of HIV using Multi drug Anti retroviral regimen in India. NACO. Govt Of India, MHFW. New Delhi; Dec 2013:34–41.

2. Consolidated guidelines on HIV prevention, diagnosis, treatment and care for key populations, 2016 update WHO.

3. Watts DH. Management of human immuno-deficiency virus infection in pregnancy. N Engl J Med. 2002 Jun 13. 346(24):1879–91. [Medline].

4. Minkoff H, Hershow R, Watts DH, Frederick M, Cheng I, Tuomala R. The relationship of pregnancy to human immunodeficiency virus disease progression. Am J Obstet Gynecol. 2003 Aug; 189(2):552–9. [Medline].

5. Amy R. Sheon, Harold E. Fox, Kenneth C. Rich, Pamela Stratton, Clemente Diaz, Ruth Tuomala, Hermann Mendez, Jane Carrington, Geraldine Alexander, and Women And Infants Transmission Study Group. The Women and Infants Transmission Study (WITS) of Maternal-Infant HIV Transmission: Study Design, Methods, and Baseline Data. Journal of Women's Health. April 2009, 5(1): 69–78. doi:10.1089/jwh.1996.5.69.

6. Tuomala RE, Shapiro DE, Mofenson LM, Bryson Y, Culnane M, Hughes MD. Antiretroviral therapy during pregnancy and the risk of an adverse outcome. N Engl J Med. 2002 Jun 13. 346(24):1863–70. [Medline].

7. Watts DH. Human immunodeficiency virus in a book on high risk pregnancy—management options, Ed James Steer Weiner, Gonik, Crowther Robson, Elseveir Saunders Publication, 4th ed. 2011:479–91.

8. Combination antiretroviral therapy and duration of pregnancy. AIDS. 2000 Dec 22;14(18):2913–20.[Medline].

Oligohydramnios

Gauri Karandikar, Manjiri Khare

INTRODUCTION

Oligohydramnios is an aberrantly low amniotic fluid volume surrounding the fetus than is expected for the gestational age. The diagnosis usually becomes apparent at the anomaly scan, a fetal growth scan or in an acute presentation of decreased fetal movement. Oligohydramnios is known to be associated with a number of adverse antepartum, intrapartum and neonatal outcomes. Early onset of oligohydramnios is associated with increased perinatal mortality and morbidity.[1] Amniotic fluid is necessary for fetal lung development and also other systems like the gastrointestinal systems, muscular system and limb movements.

INCIDENCE

The reported rates of oligohydramnios vary depending on the population screened whether low risk or high risk pregnancy, the threshold used, the diagnostic criteria and the gestational age at the time of ultrasound examination.[2]

PHYSIOLOGY

The amniotic fluid in the first trimester is the transudate of plasma across the maternal surface of the uterine decidua, the placental surface and the fetal skin.[3] In the second half of pregnancy, fetal urine production and fetal lung secretions are the main contributors. Amniotic fluid volume increases from 16 to 34 weeks and decreases until term. The maintenance of amniotic fluid is a dynamic process throughout the pregnancy (Table 15.1). Pathology affecting any of these processes can affect the liquor volume.

DIAGNOSIS

Whilst symphysio-fundal height measurement less than expected for gestational age could raise the suspicion of oligohydramnios subjectively, ultrasound can be used to measure the amniotic fluid objectively.

The two commonly used ultrasound methods for measuring amniotic fluid include:

AFI (AMNIOTIC FLUID INDEX)

AFI provides a mean for quantifying liquor. AFI is measured using four cord free

Table 15.1	The estimates of daily amniotic fluid volume flows in the near term fetus (adapted from Underwood et al, 2005)	
Source of amniotic fluid		*Fluid ml/day*
Fetal urine production		800–1200
Fetal lung liquid secretion		170
Fetal swallowing		500–1000
Intramembranous flow		200–400
Oral nasal secretions		25
Transmembranous flow		10

quadrants of the uterus around the fetus. There are physiological variations in amniotic fluid volume depending on the swallowing and urination of the fetus. AFI of ≤ 5 cm as the cut off is clinically relevant for the diagnosis of oligohydramnios. In general AFI below the 5th centile for the gestation can be used to prompt further assessment.

SDP (SINGLE DEEPEST POOL/ POCKET)

A single deepest pocket (SDP) < 2 cm can also be used to objectively determine oligo-hydramnios.[4]

A randomized trial comparing AFI and SDP for prediction of adverse pregnancy outcome at term in low and high-risk pregnancies found that the use of AFI increased the frequency of diagnosis of oligohydramnios and labor induction but there was no change in perinatal outcome compared with SDP.[5]

Anhydramnios

It can be defined as the absence of measurable AFI or SDP.

Multiple Gestations

Assessment of liquor by AFI is difficult in multiple gestations and here SDP <2 cm is used for diagnosis of oligohydramnios. In particular, it is relevant for the assessment in twin to twin transfusion and selective fetal growth restriction in multiple pregnancies.

Finberg reported a possible pitfall in determining the amniotic fluid in twin pregnancies,' the amniotic wrinkle' which may give a false reassurance of adequate liquor volume when there is actually oligo-hydramnios in one of the twin.The intertwin membrane may fold on itself creating a wrinkle or an intrauterine sling in which a fetus appears to be suspended within the amniotic space of the other twin. This is clinically relevant for diagnosing 'stuck twins' and twin to twin transfusion syndrome.[6]

Use of color Doppler to highlight presence of cord in the quadrants being measured is useful in clinical practice to avoid erroneous measurements of liquor pools.

The AFI and SDP measurements could be affected by the abdominal pressure applied on the transducer by the sonographer, presence of umbilical cord or extremities in the pocket of amniotic fluid being measured, echogenic shadows.[7]

ETIOPATHOLOGY

The etiopathology of oligohydramnios can be broadly considered as:
- Prelabor rupture of the membranes (PROM)
- Uteroplacental insufficiency
- Fetal abnormalities
- Drugs
- Idiopathic

Second Trimester

Disorders related to the fetal renal/urinary system play a prominent role in the aetiology in the second trimester. The liquor abnormalities seen on ultrasound can vary from mild to severe oligohydramnios and possibly anhydramnios. Renal reserve in the fetus, as in the adult, is such that only one functioning kidney is required to maintain normal renal function and adequate urine production. Structural ultrasound findings may not necessarily correlate with the fetal renal function.

Bilateral renal dysplasia, polycystic kidney disease, bilateral multicystic dysplastic kidney disease and renal tubular dysgenesis result in interstitial renal dysfunction and failure causing oligohydramnios and possibly anhydramnios.

In bilateral renal agenesis the fluid volume may be normal even until 20 weeks due to other sources of fluid production. With this a number of dysmorphic features evolve over the next weeks, called Potter facies. In addition to Potter facies, oligohydramnios leads to limb deformities, generalised growth restriction and pulmonary hypoplasia, i.e. the oligohydramnios sequence. The perinatal mortality in this group is mainly due to pulmonary hypoplasia. Severe oligohydramnios for any aetiology in the period up to 26 weeks, the end of the cannalicular phase of fetal lung development is likely to be associated with severe pulmonary hypoplasia.

Urinary tract obstruction as seen with posterior urethral valves, ureteroceles, urogenital septum malformations and bilateral uterovesical junction or pelviureteric junction obstruction can result in renal damage due to back pressure changes.

A number of genetic conditions may also be associated with renal abnormalities, renal dysfunction and subsequent oligohydramnios.[8]

Maternal and placental factors as well as premature rupture of membranes may be considered in the probable causes.

Third Trimester

Oligohydramnios primarily detected in the third trimester may be a consequence of PROM or uteroplacental insufficiency as a result of maternal vascular disorders or pre-eclampsia which is usually associated with fetal growth restriction. Fetal anomalies may also be seen in the third trimester.

Fetal hypoxia leads to redistribution of blood flow with fetal cerebral perfusion being maintained at the expense of visceral perfusion, including renal blood flow, resulting in decreased amniotic fluid volume. These changes in blood flow as apparent in Doppler flow examination reveal reduced vascular resistance in the fetal middle cerebral artery and increased resistance to flow in the renal arteries. These changes in blood flow are mediated through different mechanisms, viz. Chemobaroreceptors, the influence of renin angiotensin system and the release of vasopressin.

Post-term pregnancy can be associated with decreased liquor volume.

Drugs, for example, indomethacin, when used for prolonged period is known to be associated with oligohydramnios due to reduction in fetal urine production. It is advisable to stop these after 34 weeks of gestation.

Many cases of third trimester oligohydramnios are idiopathic.

Management

The clinical management of oligohydramnios will be determined by underlying pathology if known and gestational age at diagnosis.

History and Clinical Examination

A careful history should include
1. History for leaking fluid per vaginum
2. History to assess for any known risk factors for oligohydramnios and fetal growth restriction

    It must include risk factors associated with fetal intrauterine growth restriction (Table 15.2).[9]
3. A history of decreased fetal movements and change in pattern.

Table 15.2	Risk factors associated with fetal growth restriction and oligohydramnios	
Fetal	Maternal
Chromosomal abnormalities	Pregnancy induced hypertension
Congenital malformations	Maternal hydration
Multiple gestations	Drugs like prostaglandin synthase inhibitors
Infection	Extremes of malnutrition
	Vascular/renal disease
	Congenital or acquired thrombophilic disorder
	Drugs
	Lifestyle (smoking)
	High altitude or significant hypoxic disorder

Clinical Examination

This should include abdominal examination for symphysiofundal height, sterile speculum examination if there is a clinical suspicion of PROM based on history. Appropriate assessment of mother's well-being should include blood pressure, weight and urine check. Cardiotocograph (CTG) if presenting with reduced fetal movements.

Sonographic Evaluation

Comprehensive sonographic evaluation to include fetal biometry, growth velocity plotted on centiles, fetal anatomy and additional evaluation on the basis of biophysical profile, umbilical artery Doppler studies and placental pathology. During ultrasound of the fetal anatomy particular attention should be given to assessing fetal kidneys and filling of fetal bladder during the scan.

Evaluation in Case of Suspected PROM

Sterile speculum examination for vaginal pooling of liquor and an arborisation or ferning pattern may be observed, when dried posterior vault fluid is examined microscopically. Cervical mucus, semen and blood may cause false positive results. Nitrazine paper turns blue as the amniotic fluid is more basic, pH 6.5–7 than normal vaginal discharge, pH 4.5.[10] Indigocarmine dye tests may be performed, only if invasive testing is clearly indicated.

If pooling of amniotic fluid is not observed, an insulin like growth factor binding protein 1 test or placental alpha microglobulin 1 is tested for in the vaginal fluid by means of a rapid dipstick test. The test helps to detect microrupture and the result is not affected by interference from blood, other body fluids, disinfectants or medicines.[11]

Biochemical Evaluation

Maternal serum alpha fetal protein (MSAFP), a placental marker can be elevated in early onset fetal growth restriction. An elevated level has been linked to oligohydramnios in the presence or absence of anomalous fetus. The combination of elevated MSAFP and oligohydramnios carries a poor prognosis.[12]

Amniocentesis

Amniocentesis may reveal abnormal fetal karyotype especially in cases with fetal anomalies. A chorion villous biopsy may be preferred in cases where the placenta is anterior and there is not enough fluid to collect for sampling.

Managing Oligohydramnios

The management will depend on the gestation at diagnosis and the underlying cause for oligohydramnios.

PREMATURE RUPTURE OF MEMBRANES (PROM)

PROM is associated with brief latency between membrane rupture and delivery, increased risk of perinatal infection, and *in utero* umbilical cord compression. Management of PROM depends on gestational age and evaluation of the relative risks of preterm birth versus intrauterine infection, placental abruption, and cord complications that could occur with expectant management.

Expectant management includes observation for signs of clinical chorioamnionitis. This would include daily temperature, monitoring for signs of clinical infection and cardiotocography to assess for fetal tachycardia. Prophylactic antibiotics are given for reducing the risk of infection. This also delays delivery and therefore allows time for the effect of antenatal steroids for fetal lung maturity.

Erythromycin or penicillin are the antibiotic of choice. If group B Streptococcus is isolated in cases of PPROM, penicillin or clindamycin in women who are allergic to penicillin. Antenatal corticosteroids should be administered in women with PPROM between 24 and 34 weeks of gestation.

Delivery should be considered at 34 weeks of gestation. Where expectant management is considered beyond this gestation, increased risk of chorioamnionitis and decreased risk of respiratory problems to the neonate should be discussed with the woman.[10]

Amnioinfusion

There is no treatment of oligohydramnios that has proven to be effective long term. Short term improvement is possible and may be considered under certain circumstances where fetal visualisation for a detailed anatomic survey is recommended.

Amnioinfusion has been used to increase amniotic fluid temporarily in certain settings viz.

To Improve Detection of Fetal Anomalies

Oligohydramnios may limit assessment of fetal anatomy in the second trimester. Transabdominal infusion of approximately 200 ml saline under ultrasound guidance can improve diagnostic precision and contribute to better pregnancy management.

A review of patients with midtrimester oligohydramnios and intact membranes revealed that the overall rate of adequate visualization of fetal structures improved from 50.98 to 76.79% (p < 0.0001). In fetuses having preinfusion-identified obstructive uropathy, there was improvement in identification of associated anomalies from 11.8 to 31.3%.[13]

Fetal MRI alone and in combination with amnioinfusion has been used in midtrimester oligohydramnios to detect fetal anomalies.

To Prevent the Sequelae of Oligohydramnios

Serial transabdominal infusions have been performed to improve fetal outcome in pregnancies with idiopathic oligohydramnios or early oligohydramnios due to premature rupture of membranes.[10]

Additional studies to confirm these findings are needed prior to clinical utilisation of the procedure in second trimester oligohydramnios.

Fetal Intervention

In majority of cases of renal abnormality, no intervention is possible and management will depend on whether the condition is expected to be lethal or whether survival is anticipated. In the latter cases, the timing of delivery will depend on the fetal growth, the severity of oligohydramnios, gestation and availability of neonatal and paediatric surgical facilities.

In some cases of lower urinary tract obstruction, fetal intervention, viz. placement of a vesico-amniotic shunt in bladder outflow obstruction may be considered. More invasive procedures including open fetal surgery, fetal cystoscopy, electrocautery of posterior urethral valves have been undertaken with a high intrauterine loss rate or neonatal morbidity, they are therefore applicable only in limited cases.[8]

Other Treatment Modalities

There are other treatment modalities that are not currently used in routine clinical practice but there is data emerging from smaller studies or used in the context of clinical trials currently to explore their efficacy.

Maternal Hydration

It has shown to demonstrate a transient positive effect on the liquor volume which is a result of the changes in the maternal plasma osmolality and sodium concentration. This results in osmotically driven maternal to fetal flux and improves uteroplacental perfusion.[14]

Maternal hydration may be a safe, well-tolerated and useful strategy to improve amniotic fluid volume especially in cases of isolated oligohydramnios. In view of many obstetric situations in which a reduced liquor volume may pose threats, particularly to the fetus, the possibility of increasing it with a simple and inexpensive method like hydration may have useful clinical applications in obstetric care in the future.[15]

dDAVP-1-deamino-[8-D-arginine] vasopressin (a selective anti-diuretic agonist)

When dDVAP has been administered to the mother, it induces maternal plasma hypo-osmolality and increased placental water flow, resulting in fetal plasma hypo-osmolality and a likely induced diuresis and hence increase in liquor volume. Use of desmopressin (dDAVP) for this indication is experimental and is approved for use under research protocols.[2]

L-Arginine

L-arginine is an amino acid with a wide range of biological functions. It serves as a precursor not only to proteins but also nitric oxide which has been identified as endothelium-derived relaxing factor. L-arginine increases uteroplacental blood flow through nitric oxide mediated dilatation of vessels, thereby increasing the supply of nutrients to the fetus aiding its growth. However, extensive long-term studies are required to demonstrate not only its efficacy but also its effect on maternal and perinatal outcome which would help in establishing its role as a potent treatment option for oligohydramnios.[16]

Fetal Membrane Sealants

A variety of tissue sealants, e.g. fibrin glue, gelatine sponge, Amniopatch have shown some success in cases of leaking from ruptured membranes.There is insufficient evidence to recommend fibrin sealants as routine treatment for second-trimester oligohydramnios caused by PPROM.[9]

Monitoring and Prognosis

The fetal/neonatal prognosis depends on the cause, time of onset, severity of oligohydramnios and the duration of oligohydramnios.

First Trimester

Reduced liquor in the first trimester is an ominous sign with the pregnancy resulting in spontaneous abortion. Patients need to be counseled regarding the poor prognosis.

Second Trimester

The management depends on the severity of oligohydramnios and the underlying aetiology.

Pregnancies with borderline or low liquor volume, in the absence of fetal malformations need to be followed up with serial sonographic assessment to monitor the course of the process which may remain stable, resolve or further decrease and/or be associated with fetal growth restriction.

Oligohydramnios in the second trimester is more likely to have worse outcome including fetal death and/or preterm labor. Many of these women choose termination because of the poor prognosis. Early onset of severe oligohydramnios or anhydramnios is associated with poor prognosis secondary to pulmonary hypoplasia. Also prolonged oligohydramnios can be associated with limb contractures. It is good practice to arrange joint counseling with the neonatal team if possible to explain the risks to the fetus and mother.

Both duration of severe oligohydramnios exposure and gestational age at premature rupture of membranes were independent significant predictors of increased neonatal risk. Severe oligohydramnios > 14 days after premature rupture of membranes at <25 weeks' gestation has a predicted neonatal mortality of >90%.[17]

Third Trimester

Adverse outcomes in the third trimester are related to umbilical cord compression, uteroplacental insufficiency and meconium aspiration.

Pregnancies with decreased AFI between 24 and 34 weeks, including borderline AFI as well as oligohydramnios, were significantly more likely to be associated with major fetal malformations. In the absence of malformations, pregnancy was found to be complicated by fetal growth restriction and preterm birth.[18]

Idiopathic isolated olighydramnios carries a better prognosis.

The duration of oligohydramnios is also an important prognostic factor, those presenting with idiopathic oligohydramnios at an earlier age are at risk of increased risk of adverse perinatal outcome as compared to those presenting later.[19]

A combined evaluation done using CTG, biophysical profile and AFI helps to improve the perinatal outcome. Doppler velocimetry may have a role in identifying pregnancies with idiopathic oligohydramnios who are at a higher risk of adverse outcome and to reduce interventions, thus avoiding iatrogenic morbidity related to prematurity.[20]

Most studies have reported good fetal outcomes in general, and no increased risk of fetal acidosis in isolated oligohydramnios when compared to pregnancies with normal amniotic fluid volumes.

A sonographic diagnosis of oligohydramnios after 40 weeks carries an increased risk of adverse perinatal outcome, even in low-risk pregnancies. The serial trend in amniotic fluid volume reduction does not seem to have prognostic significance.[21]

CONCLUSION

Pregnancies complicated by oligohydramnios have a greater risk of intrauterine growth restriction, preterm delivery, labor induction and operative intervention. The underlying cause for oligohydramnios determines the outcomes for these pregnancies. Oligohydramnios is an indicator to look for possible underlying fetal, placental and maternal problems.

REFERENCES

1. Ulkumen BA, Pala HG, Baytur YB, Koyuncu FM, Outcomes and management strategies in pregnancies with early onset oligohydramnios, Clinical and Experimental Obstet Gynecol. 2015; 42(3):355–7.
2. Ron Beloosesky, Michael G Ross, Oligohydramnios, Up-to-date, January 2016
3. Underwood MA, Gilbert WM, Sherman MP, Amniotic fluid: not just fetal urine anymore, Journal of Perinatology. 2005 May; 25(5):341–8.
4. Magann EF, Sandlin AT, Ounpraseuth ST., Amniotic fluid and the clinical relevance of the sonographically estimated amniotic fluid volume: oligohydramnios., J Ultrasound Med. 2011 Nov;30(11):1573–85.
5. Paolo Rosati, Lorenzo Guariglia, Anna Franca Cavaliere, A comparison between amniotic fluid index and the single deepest vertical pocket technique in predicting adverse outcome in prolonged pregnancy-Journal of Prenatal Medicine. 2015 Jan-Jun; 9(1-2): 12–15.
6. Finberg HJ. The amniotic wrinkle: a pitfall in evaluating amniotic fluid for twins, Journal of Ultrasound Med. 2010 Feb; 29(2):249–54.
7. Flack N, Dore C, Southwell D, Kourtis P, Sepulveda W, Fisk NM, The influence of operator transducer pressure on ultrasonographic measurements of amniotic fluid volume. Am J ObstetGynecol 1994;171: 218–22.
8. Richard Brown, Abnormalities of amniotic fluid, Progress in Obstetrics and Gynaecology, Studd,Volume 18.
9. Resnik R, Intrauterine growth restriction, Obstet Gynecol. 2002 Mar;99(3):490–6.
10. RCOG Green-top Guideline No. 44, November 2006, Preterm prelabour rupture of membranes
11. NICE guidelines, Preterm labour and birth, November 2015.
12. Frans J Los, Adriana M. Hagenaars, TitiaE. Cohen-Overbeeks And Hendrik WP Quarteros, Maternal Serum Markers In Second-Trimester Oligohydramnios, Prenatal Diagnosis, Vol. 14: 565–568 (1994).
13. Pryde PG, Hallak M, Lauria MR, Littman L, Bottoms SF, Johnson MP, Evans MI., Severe oligohydramnios with intact membranes: an indication for diagnostic amnioinfusion, Fetal Diagnosis and Therapy 2000 Jan-b;15(1):46–9.
14. Gizzo S, Noventa M, An Update on Maternal Hydration Strategies for Amniotic Fluid

Improvement in Isolated Oligohydramnios and Normohydramnios: Evidence from a Systematic Review of Literature and Meta-Analysis., PLoS One. 2015 Dec 11;10(12).

15. Patrelli TS, Gizzo S, Maternal hydration therapy improves the quantity of amniotic fluid and the pregnancy outcome in third-trimester isolated oligohydramnios: a controlled randomized institutional trial.J Ultrasound Med. 2012 Feb; 31(2):239–44.

16. Sreedharan R, et al. Effect of L-arginine on amniotic fluid index in oligohydramnios, International Journal of Reproduction, Contraception, Obstetrics and Gynecology, 2013 Mar;2(1): 80–82.

17. Kilbride HW, Yeast J, Thibeault DW, Defining limits of survival: lethal pulmonary hypoplasia after midtrimester premature rupture of membranes, Am J Obstet Gynecol. 1996 Sep; 175:675–81.

18. Petrozella LN, Dashe JS, McIntire DD, Leveno KJ, Clinical significance of borderline amniotic fluid index and oligohydramnios in preterm pregnancy, Obstet Gynecol. 2011 Feb;117:338–42.

19. Vink J, Hickey K, Ghidini A, Deering S, Mora A, Poggi S., Earlier gestational age at ultrasound evaluation predicts adverse neonatal outcomes in the preterm appropriate-for-gestational-age fetus with idiopathic oligohydramnios, Am J Perinatol. 2009 Jan;26(1):21-5.

20. Green-top Guideline No. 31, 2nd Edition, February 2013, Minor revisions-January 2014. The investigation and management of the small for gestational age fetus.

21. Locatelli A, Zagarella A, Toso L, Assi F, Ghidini A, BiffiA., Serial assessment of amniotic fluid index in uncomplicated term pregnancies: prognostic value of amniotic fluid reduction. J MaternFetal Neonatal Med. 2004 Apr;15(4):233–6.

Preterm Prelabor Rupture of Membranes

Rashmi Jalvee, Reena J Wani

Delivery before 37 weeks gestation is preterm birth. Prematurity is the most important cause of perinatal mortality. Preterm birth results in approximately 70% of neonatal deaths, 36% of infant death and 25–50% of cases of long-term neurologic impairment in children.[1]

Preterm premature rupture of membranes (PPROM) is defined as the rupture of membranes before the onset of regular uterine contractions at term. Preterm premature rupture of the membranes complicates approximately 3 percent of pregnancies and leads to one-third of preterm births.[2]

Neonatal death associated with PPROM is mainly due to:

- Prematurity
- Pulmonary hypoplasia or
- Sepsis.

It has been seen that the interval between PROM and delivery is longer when rupture of membranes occurs at an earlier gestation.

Outcomes of survivors of PPROM depend on:

- Gestational age
- Interval of latency
- Presence of infection and
- Other maternal and fetal complications.

Etiology

The etiology of PPROM is thought to be multifactorial.

Factors known to contribute to PPROM include:[3]

- Low BMI <19.8
- Inadequate weight gain during pregnancy
- Low socioeconomic status
- Cervical incompetence
- Smoking, tobacco consumption
- Polyhydramnios
- Low placental insertion
- Prior history of PROM
- Intra-amniotic infection
- Urinary tract infections
- History of subchorionic bleeding/hematoma

Diagnosis

The diagnosis of PPROM is usually made clinically by a combination of history and speculum examination. Ultrasound examination is useful to help confirm the diagnosis.

- On sterile speculum examination, presence of a pool of fluid in the vagina is highly suggestive of membrane rupture.

Tests used to confirm membrane rupture:

- Nitrazine test, which detects pH change (strip turns blue).
- Examination of the vaginal fluid under microscope for the ferning due to crystalline pattern of dried amniotic fluid due to its sodium chloride and protein content.

- Examination for lanugo hair and fetal epithelial cells stained with Nile blue.
- Per vaginal digital examination is avoided unless there is a strong suspicion of active labor as micro-organisms may be transferred from the vagina into the cervix causing prostaglandin release, intrauterine infection and preterm labor.
- Ultrasound examination showing reduced liquor in the absence of IUGR is highly suggestive of membrane rupture, however, normal liquor volume does not exclude the diagnosis.
- AmniSure is a new FDA approved immunoassay which uses amniotic fluid test strips for the detection of PAMG-1 in amniotic fluid found in vaginal discharge of pregnant women (Fig. 16.1). It has been shown to be accurate in the diagnosis of ruptured membranes with a sensitivity and specificity of 98.9% and 100%, respectively.[4]

Complications

Infection (maternal and neonatal) is the most important concern for women with PROM. Risk of infection may increase as bacteria ascend the vagina into the uterine cavity after the loss of protective barrier of amniotic membranes (Table 16.1).

Management of PPROM

Management of PPROM presents a clinical dilemma to the obstetrician.

Management will depend upon
- The gestational age at the time of the diagnosis of membrane rupture and
- The availability of neonatal care facilities.

Expectant management is known to be associated with a risk of ascending infection at all gestational ages, hence the risks must be weighed against the risks of immediate delivery. Approximately 50% of women presenting with PPROM will deliver in the first 7 days, many of them in the first 48 hours; latency of delivery in the presence of PPROM has been shown to be prolonged significantly with antibiotic treatment.

Initial assessment should include:
- History
- *General parameters:* Pulse, temperature, respiratory rate, blood pressure
- Abdominal examination to determine fetal size and presentation, presence of uterine activity and tenderness
- Speculum examination is performed to demonstrate pooling of liquor, rule out cord prolapse, to collect cervical and vaginal swabs and to assess cervical dilatation.

It is important to contemplate maternal transfer to a tertiary care center if delivery is anticipated in a setting where neonatal intensive care unit (NICU) facilities are not adequate for preterm births.

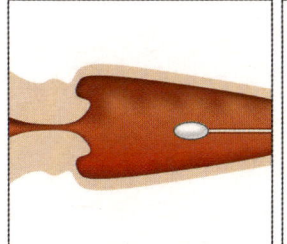

Sample of vaginal discharge is taken by sterile vaginal swab (no speculum required)

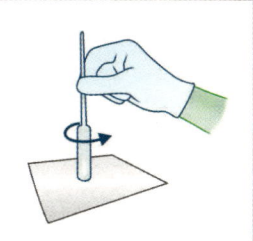

Swap is rinsed in viral with solvent and then disposed of

Test strip is inserted into vial and removed when two lines are visible, or at 10 minutes

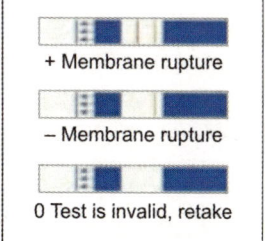

+ Membrane rupture

− Membrane rupture

0 Test is invalid, retake

Test strip is extracted from the vial and results are observed

Fig. 16.1: AmniSure

Table 16.1		Complications of PPROM	
Maternal	1. Endometritis	Repeated per vaginal examination, presence of meconium in amniotic fluid increase risk
	2. Chorioamnionitis	Characterized by: • Maternal fever (>38 degrees) • Maternal tachycardia (>100 bpm) • Increased white cell count (>15 × 10^9/L) • Uterine tenderness • Offensive smelling vaginal discharge • Fetal tachycardia (>160 bpm) The micro-organisms most commonly associated are Group B-Streptococci (GBS) and *Escherichia coli*.
	3. Abruptio placentae	Causing DIC, blood and blood product transfusion, increased operative interference, increased maternal morbidity and mortality, birth asphyxia in the baby
Fetal	1. Birth asphyxia	Due to RDS, cord compression, cord prolapse
	2. Neonatal infection	Pneumonia, meningitis, sepsis causing vomiting, diarrhea, abdominal distension, poor suckling effort, apnea, bradycardia, respiratory distress, seizures, jaundice.
	3. Limb and body deformities	Due to prolonged compression
	4. Pulmonary hypoplasia	Due to arrested alveolar development

Fetal Assessment

- Ultrasound to assess the fetal presentation, liquor volume, placental maturity, effective fetal weight
- Cardiotocography (CTG) to assess fetal condition

Delivery should be expedited in the presence of intrauterine infection, abruptio placentae or fetal compromise.

Laboratory Investigations

- Complete blood count (CBC)
- C-reactive protein (CRP)
- High vaginal swabs for microscopy and culture
- Midstream urine for microscopy and culture

Antibiotic Prophylaxis

Infections are commonly polymicrobial. Antibiotic treatment in PPROM reduces the risk of ascending infection, chorioamnionitis and delivery within 7 days. For the neonate, maternal antibiotics reduce the duration of (NICU) admission, major cerebral abnormalities and neonatal infections (Table 16.2).

Labor should be augmented under intravenous antibiotic cover. It is important to decide the optimal mode of delivery (vaginal birth versus LSCS) depending on clinical findings and the anticipated duration until birth.

It is important to rule out infection at other sites (e.g. urinary or respiratory tract) which can mimic these changes. Histological examination of placenta and membranes with evidence of acute inflammation may confirm the diagnosis after birth.

TOCOLYTICS

Prophylactic tocolysis in PPROM in the absence of uterine activity is not recommended. In the presence of uterine contractions and the absence of evidence of chorioamnionitis, RCOG guidelines (RCOG GTG No 7) recommend that tocolysis may be commenced to prolong

Table 16.2	Antibiotic treatment
Absence of chorioamnionitis[5,6]	Oral erythromycin 250 mg 4 times a day for 10 days or until delivery 48-hour course of ampicillin 2 grams intravenous (IV) 6 hourly and erythromycin 250 mg IV 6 hourly followed by 5 days of amoxicillin 250 mg per oral 8 hourly and erythromycin 333 mg per oral 8 hourly Co-amoxiclav should be avoided as it increases the risk of necrotizing enterocolitis.
Presence of chorioamnionitis[7]	Inj ampicillin (or amoxycillin) 2 g IV 6 hourly, inj gentamicin 5 mg/ kg IV and inj metronidazole 500 mg IV 12 hourly.
	If allergic to penicillin, inj clindamycin 600 mg IV every 8 hours and Inj gentamicin 5 mg/kg IV daily until delivery.
	Following delivery, once patient is afebrile and can tolerate orals, shift to Tab amoxycillin 500 mg and Tab metronidazole 400 mg 8 hourly for 5 days.
	If allergic to penicillin, Tab metronidazole 400 mg 12 hourly for 5 days and Tab azithromycin 1 g orally single dose, repeated after 7 days.

pregnancy for 48 hours until corticosteroid cover is established. The suggested dose of nifedipine is an initial oral dose of 20 mg followed by 10–20 mg three to four times daily, adjusted as per uterine activity for up to 48 hours.[8]

Corticosteroids

Corticosteroids are effective in preventing adverse perinatal outcome like respiratory distress syndrome, necrotitis enterocolitis and improves the likelihood of neonatal survival. Antenatal corticosteroids should be administered in women with PPROM between 24+0 and 34+6 weeks at a dose of Inj betamethasone 12 mg intra-muscular and repeated 24 hours later.

If inj betamethasone is not available, inj Dexamethasone 6 mg intramuscular every 12 hours can be given for 4 doses.[9]

For women in active preterm labour, even a single dose of steroids has been shown to improve neonatal survival. It is also important to ensure giving the first dose of steroids when maternal transfer to a higher center is considered for want of advanced neonatal care.

Seven days after the first course, if the woman is still at risk of preterm birth, a repeat single dose of inj betamethasone 12 mg can be given.

Magnesium Sulfate

Studies have shown that fetal exposure to magnesium sulfate given before preterm birth (24 weeks to 32 weeks) has a neuroprotective role.[10] Given within a minimum of four hours before birth, magnesium sulfate reduces the risk of cerebral palsy and protects gross motor function in infants born preterm.

Amnioinfusion

Amnioinfusion during labor is not recommended in women with PPROM (RCOG GTG No 44).[11] There is insufficient evidence

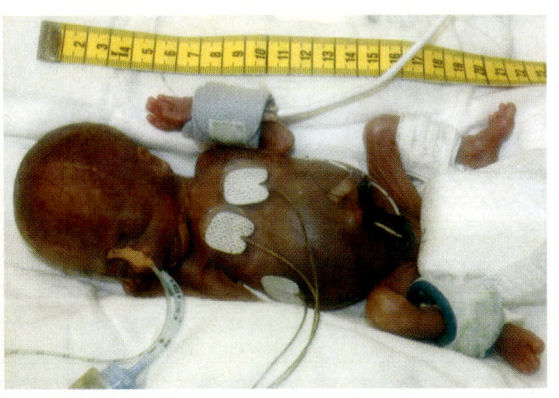

Fig. 16.2: Preterm neonate

to recommend amnioinfusion in extremely preterm PPROM as a method to prevent pulmonary hypoplasia.

To Summarize, the Management of PPROM will Depend on the Period of Gestation

PPROM <23 Weeks Gestation

Outcomes for these preterm neonates depend largely on access to NICU care. Parental decisions must be considered while deciding management. In most developing countries including India, birth before 23 weeks is still considered unviable for survival.

PPROM 23–34 weeks of Gestation

The neonatal survival improves with increasing gestational age. It is seen that neonatal survival is better after 28 weeks, access to NICU care and when antenatal corticosteroids have been administered. Delivery should be considered at 34 weeks of gestation.
- Antibiotic prophylaxis
- Birth should be allowed in the presence of active labor, presence of chorioamnionitis, fetal compromise or in the presence of antepartum hemorrhage
- Consider cesarean section if birth is not imminent.
- Expectant management is appropriate in the absence of the above. Management should include:
  - Daily clinical assessment for temperature, pulse and per vaginal loss
  - Assessment of uterine activity (abdominal pain or tenderness)
- *Fetal assessment:*
  - FHS monitoring
  - CTG twice weekly
- Investigations will depend on the available resources. Investigations should be repeated as per the agreed protocols of individual hospitals.
  - CBC, CRP twice weekly
  - High vaginal swab at weekly intervals; results may guide use of antibiotics

PPROM at >34 Weeks Gestation

The decision to deliver or manage expectantly in cases of PPROM after 34 weeks requires an assessment of the risks related to risk of infection in pregnancies managed expectantly compared with the gestational age-related risks of prematurity in pregnancies delivered earlier. Labour should be allowed to progress.

To conclude, PPROM is a major complication of pregnancies and an important cause of perinatal morbidity and mortality. Currently, there is no effective way of preventing spontaneous rupture of fetal membranes. However, it is important that women be well informed regarding maternal, fetal and neonatal complications regardless of controversies of its management.

REFERENCES

1. Management of Preterm Labor Practice Bulletin 127 [Practice Bulletin No. 127, 2012]. The American College of Obstetricians and Gynecologists.
2. Medina TM, Hill DA. Preterm Premature Rupture of Membranes: Diagnosis and Management. Am Fam Physician. 2006; 73:659–64. Romero R, Athayde N, Maymon E, Pacora P, Bahado-Singh R. Premature rupture of the membranes. In: Reece A, Hobbins J, eds. Medicine of the fetus and the mother. Philadelphia: Lippincott Raven; 1999:1581–625.
3. Chakraborty B, Mandal T, Chakraborty S. Outcome of Prelabour Rupture of Membranes in a Tertiary Care Center in West Bengal. Indian Journal of Clinical Practice. 2013:24(7):657–662. APEC Guidelines: Premature Rupture of Membranes. Available from URL: https://medicaid.alabama.gov/
4. Cousins LM, Smok DP, Lovett SM, Poelter DM. AmniSure placental alpha microglobulin-1 rapid immunoassay versus standard diagnostic methods for detection of rupture of membranes. Am J Perinatol 2005; 22:317–20.
5. Kenyon S, Boulvain M, Neilson JP. Antibiotics for preterm rupture of membranes. Cochrane Database of Systematic Reviews 2013, Issue 12. Art. No.: CD001058. DOI: 10.1002/14651858. CD001058.pub3. (Level I).
6. Mercer BM, Miodovnik M, Thurnau GR, Goldenberg RL et al. Antibiotic therapy for

reduction of infant morbidity after preterm premature rupture of the membranes; a randomized controlled trial. National Institute of Child Health and Human Development Maternal-Fetal Medicine Units Network. JAMA 1997; 278:989-95.

7. Hopkins L, Smaill F. Antibiotic regimens for management of intra-amniotic infection. Cochrane Database Syst Rev. 2002:CD003254.

8. Royal College of Obstetricians and Gynaecologists (RCOG). Green-top Guideline No.1b: Tocolytic drugs for women in preterm labour. Available from URL: https://www.rcog.org.uk/

9. Royal College of Obstetricians and Gynaecologists (RCOG). Green-top Guideline No.7: Antenatal corticosteroids to reduce neonatal morbidity. Available from URL: https://www.rcog.org.uk/

10. The Antenatal Magnesium Sulfate for Neuro-protection Guideline Development Panel. Antenatal magnesium sulfate prior to preterm birth for neuroprotection of the fetus, infant and child: National clinical practice guidelines. Adelaide: The University of Adelaide, 2010. ISBN Print: 978-0-86396-720-7.

11. Royal College of Obstetricians and Gynaecologists (RCOG). Green-top Guideline No.44: Preterm Prelabour Rupture of Menbranes. November 2006. Available from URL: https://www.rcog.org.uk/

FGR—Monitoring and Management

Geetha Balsarkar

Fetal growth restriction (FGR) is defined by the American College of Obstetricians and Gynecologists (ACOG) as 'the failure of a fetus to achieve its individual potential. It is associated with high risk for perinatal morbidity and mortality where the risk increases with severity of condition'.

SGA (Small for gestational age) fetus is defined as fetal abdominal circumference (AC) or estimated fetal weight (EFW) < 10th centile. Almost 50–70 percent of small for gestation age fetuses are constitutionally small but healthy. Such a fetus is found to be growing along the lower percentile (10th or 5th), but maintaining its growth curve. This is often called "low profile fetus". Approximately 20 percent of SGA fetuses are classified as having 'true' FGR, and another 5–10 percent are associated with chromosomal anomalies, structural anomalies, or chronic intrauterine infection.

RULES OF FETAL GROWTH

- Every fetus has its own growth rate.
- Every fetus maintains its growth rate as long as it is growing normally.
- Younger fetus grow faster than older fetuses.

Hence early and accurate dating with appropriate biometry is important in making the diagnosis of FGR. Fetal monitoring is then based on ultrasound estimation of biometry, Doppler flow and liquor assessment.

A protocol based approach starting with detailed history is essential.

History

a. Menstrual history (regularity/duration of cycle).
b. Mode of conception (natural/IUI/IVF). If ART—date of embryo transfer.
c. Date of pregnancy confirmation.
d. Medical history and previous obstetric history.

Examination

a. Parental height and weight.
b. Symphysio-fundal height at every visit.

Prior scan records: This helps to ascertain the gestational age with accuracy (transvaginal scans are more reliable for dating in early pregnancy).

Monitoring the Growth Restricted Fetus

Clinical methods such as identifying subnormal uterine size, followed by abdominal palpation and direct measurement of the symphyseal fundal height detects only 30% of IUGR fetuses. Ultrasound is the benchmark of actual pregnancy dating and diagnosis of FGR. Abdominal circumference should be considered

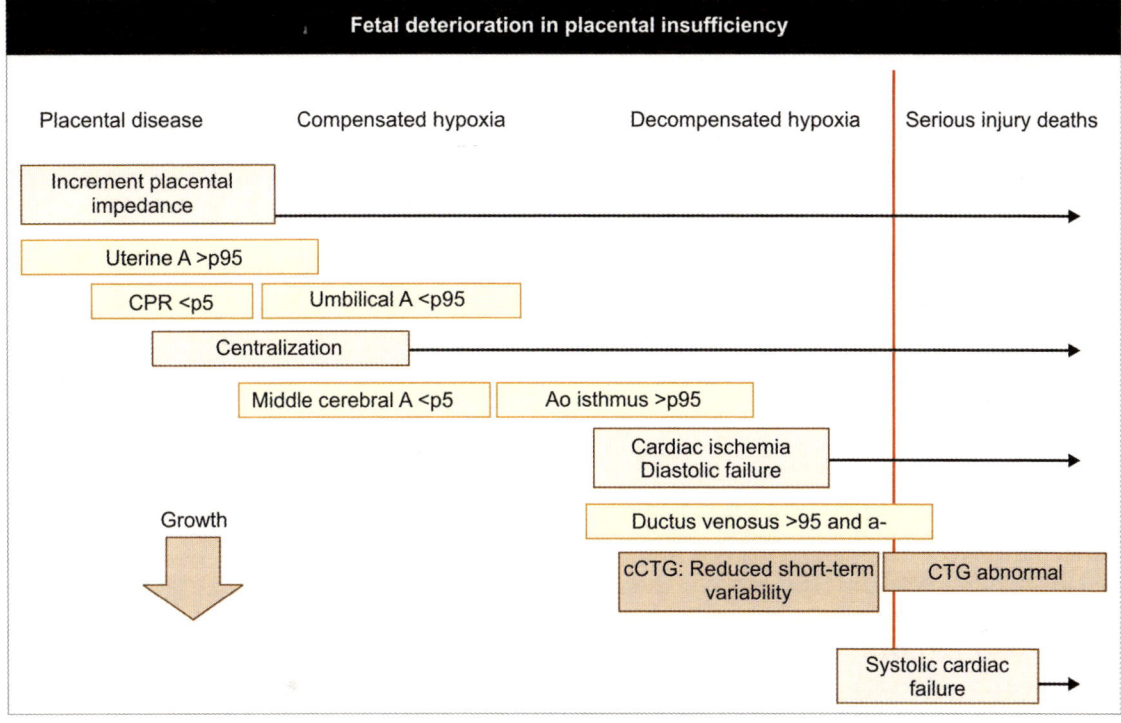

Fig. 17.1: Fetal deterioration

as the best single measurement to screen for FGR as it has good correlation with fetal weight. Although cut off value for FGR is at the 10th centile, adverse outcomes are mainly confined to fetuses below 5th or 3rd centiles.

Placental dysfunction starts with abnormal tertiary villous vessels and ends with characteristic fetal multi-vessel cardio-vascular manifestations. These effects can be documented with Doppler ultrasound examination by

1. Fetal umbilical arteries for the placenta.
2. Middle cerebral artery (MCA) for preferential brain perfusion.
3. Ductus venosus for the cardiac effects of placental dysfunction.

As growth restriction progresses, Doppler abnormalities in these vascular territories also deteriorate, suggesting a sequential pattern of disease progression. The presence of maternal disease such as pre-eclampsia or autoimmune diseases may cause a sudden and unpredictable worsening of this sequence.

1. Increased placental blood flow resistance indicates placental dysfunction as the underlying cause.
2. Increased diastolic blood flow in the cerebral circulation indicates redistribution.
3. As more placental vasculature is damaged, there is loss of diastolic flow in umbilical artery progressing to reversal.
4. Progressive elevation in venous Doppler indices indicates alterations in cardiac function that precede deterioration of biophysical parameters in the fetus.

The first clinically relevant step is the distinction of 'true' fetal growth restriction (FGR), associated with signs of abnormal fetoplacental function and poorer perinatal outcome, from constitutional small-for-gestational age, with a near-normal perinatal outcome. Classifying a fetus as having true fetal growth restriction is clinically relevant, as these fetuses are associated with poorer perinatal outcome while constitutionally small fetuses have near normal perinatal outcome. Closer monitoring and appropriate delivery timing can be instituted which would then help reducing the incidence of perinatal insult.

According to recent recommendations, such a distinction should not be based solely on umbilical artery Doppler, since this index detects only early-onset severe forms. FGR should be diagnosed in the presence of any of the factors associated with a poorer perinatal outcome, including Doppler cerebroplacental ratio, uterine artery Doppler, a growth centile below the 5th centile.

FGR presents under two different pheno-types when the onset is early or late in gestation. In general, but not always, there is a correspondence between early-onset and the most severe forms of FGR. The cut-off to define early- versus late-onset FGR has commonly been set in an arbitrary fashion at about 32–34 weeks at diagnosis or 37 weeks at delivery. Table 17.1 depicts the main differences between both clinical forms.

Early-onset Fetal Growth Restriction

Early-onset FGR represents 20–30% of all FGRs. Early FGR presents in association with early pre-eclampsia (PE) in up to 50%. Early-onset FGR is highly associated with severe placental insufficiency and with chronic fetal hypoxia. This explains that UA Doppler is abnormal in a high proportion of cases. Early severe FGR is associated with severe injury and/or fetal death before term in many cases. Management is challenging and aims at achieving the best balance between the risks of leaving the fetus *in utero* versus the complications of prematurity.

Late-onset Fetal Growth Restriction

Late-onset FGR represents 70–80% of FGR. Its association with late pre-eclampsia (PE) is low, roughly 10%. In late-onset FGR the degree of placental disease is mild, thus UA Doppler is normal in virtually all cases. Despite normal UA PI Doppler, there is a high association with abnormal CPR values. In addition, advanced brain vasodilation suggesting chronic hypoxia, as reflected by an MCA PI <p5, may occur in 25% of late FGR. Advanced signs of fetal deterioration with changes in the DV are virtually never observed.

There is a risk of acute fetal deterioration before labor, as suggested by the high contribution to late-pregnancy mortality, and a high association with intrapartum fetal distress and neonatal acidosis. Thus, the cascade of sequential fetal deterioration described above does not occur in late FGR.

Once the diagnosis is established, late FGR does not represent a management challenge, as seen in early FGR. However, still undiagnosed late FGR contributes to a large share of late pregnancy stillbirths.

Table 17.1	The main differences between early- and late-onset form
Early-onset FGR (1–2%)	*Late-onset FGR (3–5%)*
Problem: Management	Problem: Diagnosis
Placental disease: Severe (UA Doppler abnormal, high association with pre-eclampsia)	Placental disease: Mild (UA Doppler normal, low association with pre-eclampsia)
Hypoxia ++: Systemic cardiovascular adaptation	Hypoxia +/–: Central cardiovascular adaptation
Immature fetus = higher tolerance to hypoxia = natural history	Mature fetus = lower tolerance to hypoxia = no (or very short) natural history
High mortality and morbidity; lower prevalence	Lower mortality (but common cause of late stillbirth); poor long-term outcome; affects large fraction of pregnancies

Monitoring in Fetal Growth Restriction

Uterine Artey (UtA) Doppler (Fig. 17.2)

Impaired trophoblastic invasion of the maternal spiral arteries results in high uterine artery PI and persistence of the uterine artery notch. In high-risk pregnancies uterine artery Doppler helps to identify patients at risk for pre-eclampsia and FGR.

Umbilical Artery (UA) Doppler (Fig. 17.3)

UA Doppler provides both diagnostic and prognostic information for the management of FGR. The progression of UA Doppler patterns to absent or reverse end-diastolic flow correlates with the risks of injury or death and adverse perinatal outcome. Absent or reversed end-diastolic flow occurs when 60–70 percent of the placental villous tree is damaged. Use of UA Doppler in high-risk pregnancies (most of them SGA fetuses) improves perinatal outcomes, with a 29% reduction in perinatal deaths.

Middle Cerebral Artery (MCA) Doppler (Fig. 17.4)

MCA is valuable for the identification and prediction of adverse outcome among late-onset FGR, independently of the UA Doppler, which is often normal in these fetuses. MCA informs about the existence of brain vasodilation, a marker of hypoxia. Fetuses with abnormal MCA PI had a sixfold risk of emergency cesarean section for fetal distress when compared with SGA fetuses with normal MCA PI, which is particularly relevant because labor induction at term is the current standard of care of late-onset FGR.

Cerebroplacental Ratio (CPR)

The CPR is emerging an important diagnostic index. The CPR improves remarkably the sensitivity of UA and MCA alone, because increased placental impedance (UA) is often combined with reduced cerebral resistance (MCA). In late onset growth restriction CPR may be abnormal before MCA PI becomes

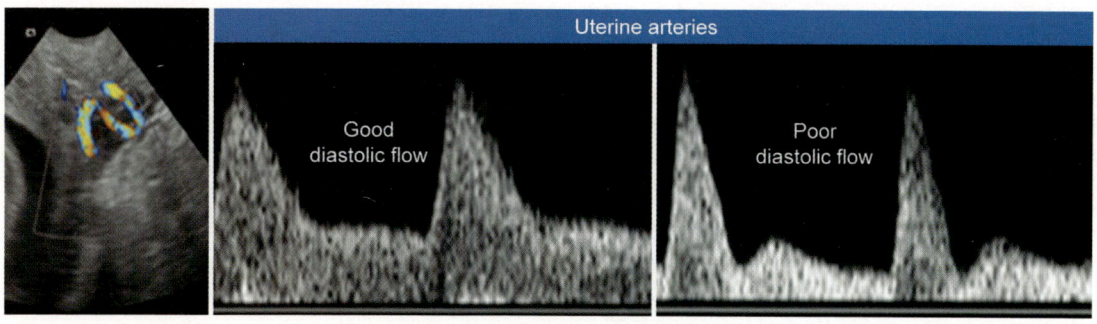

Fig. 17.2: Uterine artey (UtA) Doppler

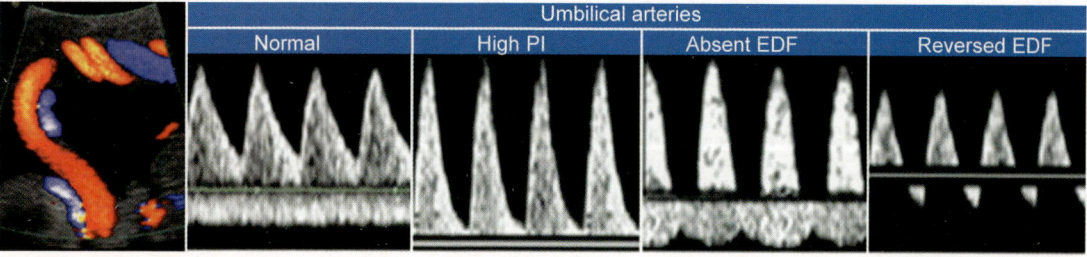

Fig. 17.3: Umbilical artery (UA) Doppler

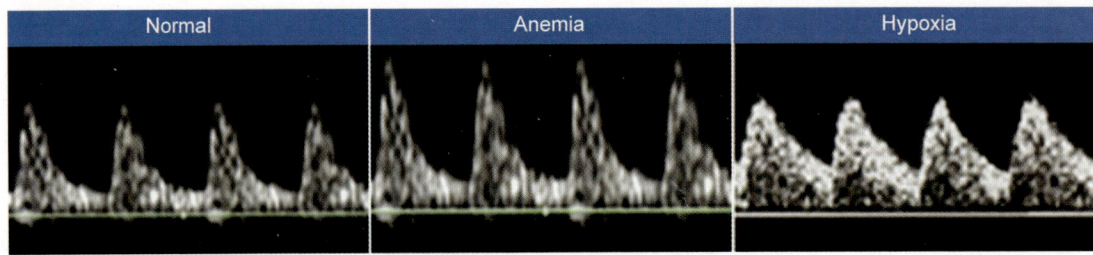

Fig. 17.4: Middle cerebral artery Doppler

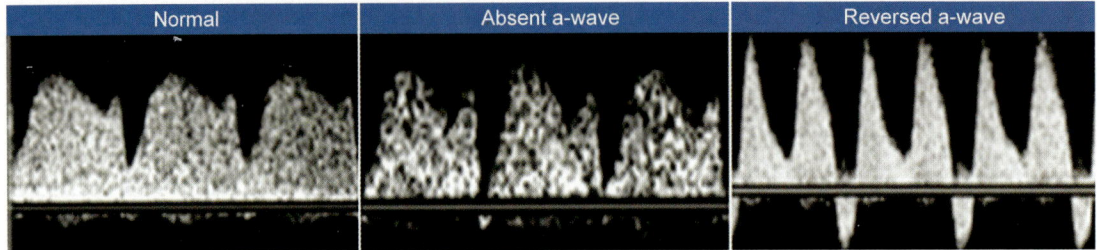

Fig. 17.5: Ductus venosus Doppler

abnormal when there is still some forward flow in the umbilical artery. This deterioration in the CP ratio is associated with poorer neonatal neurodevelopmental outcomes.

Ductus Venosus (DV) Doppler (Fig. 17.5)

DV is the strongest single Doppler parameter to predict the short-term risk of fetal death in early-onset FGR. DV flow waveforms become abnormal only in advanced stages of fetal compromise. Absent or reversed velocities during atrial contraction are associated with perinatal mortality independently of the gestational age at delivery. Thus, this sign is normally considered sufficient to recommend delivery at any gestational age, after of steroids administration.

Aortic Isthmus (AOI) Doppler

This vessel reflects the balance between the impedance of the brain and systemic vascular systems. Reverse AOI flow is a sign of advanced deterioration and a further step in the sequence starting with the UA and MCA Dopplers. The aortic isthmus (AOI) Doppler is associated with increased fetal mortality and neurological morbidity in early-onset FGR.

Biophysical Profile (BPP)

BPP is calculated by combining ultrasound assessment of fetal tone, respiratory and body movements, with amniotic fluid index and a conventional CTG. Studies show an association between abnormal BPP and perinatal mortality and cerebral palsy. However, a high false-positive rate (50%) limits the clinical usefulness of the BPP. Evidence suggests that use of BPP in high-risk pregnancies does not reduce perinatal death. The use of BPP increases C section rates without improving perinatal outcome. Consequently, whenever Doppler expertise and/or cCTG are available, the incorporation of BPP in management protocols of FGR is questionable.

Management of FGR

The aim in clinical management of FGR should be

1. To distinguish FGR from SGA
2. To monitor and identify the optimal time to deliver, balancing the *in utero* risks to the risks of prematurity and neonatal loss.

In the first step, once a small fetus (i.e. EFW <10th centile) has been identified, the Doppler study, i.e. umbilical artery PI, middle cerebral artery PI, ductus venosus and the cerebro-placental ratio (CPR) should be measured in order to classify FGR versus SGA. Once confirmed, growth assessment can be planned on two to four weekly basis depending on the gestational age.

In cases of FGR fetuses, changes in the UA, DV and AoI Doppler, and cCTG where available, are used to define stages of deterioration again depending on the gestational age.

Small-for-Gestational Age

Excluding infectious and genetic causes, the perinatal results are good.

Doppler and growth assessment is recommended fortnightly (Table 17.2). Labor induction should be recommended at 40 weeks. Fortnightly monitoring is recommended.

Stage I—Fetal Growth Restriction (Severe Smallness or Mild Placental Insufficiency)

Either UtA, UA or MCA Doppler, or the CPR are abnormal. In the absence of other abnormalities, evidence suggests a low risk of fetal deterioration before term. Labor induction beyond 37 weeks is acceptable, but the risk of intrapartum fetal distress is increased. Weekly monitoring recommended.

Stage II—Fetal Growth Restriction (Severe Placental Insufficiency)

This stage is defined by UA absent-end diastolic velocity (AEDV) or reverse AoI.

Delivery should be recommended after 34 weeks. The risk of emergency cesarean section at labor induction exceeds 50%, and, therefore, elective cesarean section is a reasonable option. Monitoring twice a week is recommended.

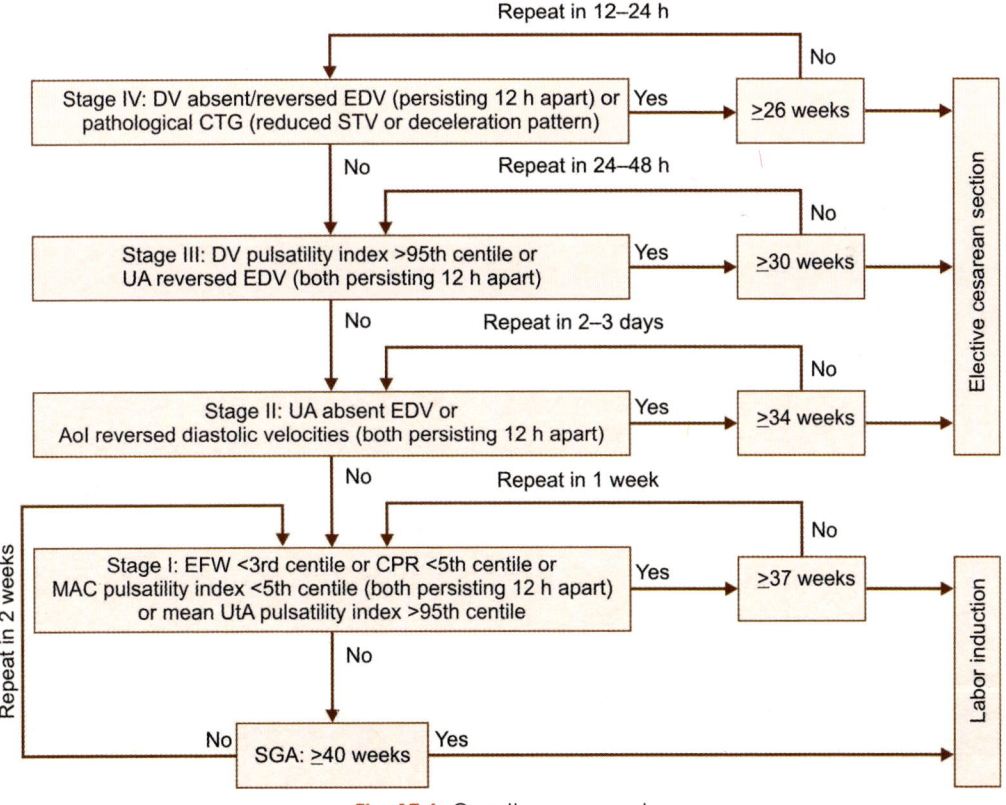

Fig. 17.6: Growth assessment

Table 17.2		Stage based classification and management of FGR		
Stage	*Pathophysiological correlate*	*Criteria (any of)*	*Monitoring**	*GA/mode of delivery*
I	Severe smallness or mild placental insufficiency	EFW <3rd centile CPR <p5 UA PI >p95 MCA PI <p5 UtA PI >p95	Weekly	37 weeks LI
II	Severe placental insufficiency	UA AEDV	Biweekly	34 weeks CS
III	Low-suspicion fetal acidosis	UA REDV	1–2 days	30 weeks CS
IV	High-suspicion fetal acidosis	DV reverse a flow cCTG <3 ms FHR decelerations	12 h	26 weeks** CS

All Doppler signs described above should be confirmed at least twice, ideally at least 12 h apart. GA: Gestational age; LI : Labor induction; CS: Cesarean section. *Recommended intervals in the absence of severe preeclampsia. If FGR is accompanied by this complication, strict fetal monitoring is warranted regardless of the stage. **Lower GA threshold recommended according to current literature figures reporting at least 50% intact survival. Threshold could be tailored according to parents wishes or adjusted according to local statistics of intact survival.

Stage III—Fetal Growth Restriction (Advanced Fetal Deterioration Low-suspicion Signs of Fetal Acidosis)

The stage is defined by reverse absent-end diastolic velocity (REDV) or DV PI >95th centile. There is an association with a higher risk of stillbirth and poorer neurological outcome. However, since signs suggesting a very high risk of stillbirth within days are not present yet, it seems reasonable to delay elective delivery to reduce as possible the effects of severe prematurity. We suggest delivery should be recommended by cesarean section after 30 weeks.

Monitoring every 24–48 h is recommended.

Stage IV—Fetal Growth Restriction (High Suspicion of Fetal Acidosis and High Risk of Fetal Death)

There are spontaneous FHR decelerations, reduced STV (<3 ms) in the cCTG, or reverse atrial flow in the DV Doppler. Spontaneous FHR deceleration is an ominous sign. cCTG and DV are associated with very high risks of stillbirth within the next 3–7 days and disability. Deliver after 26 weeks by cesarean section at a tertiary care center under steroid treatment for lung maturation. Intact survival exceeds 50% only after 26–28 weeks, and before this threshold parents should be counseled by multidisciplinary teams. Monitoring every 12–24 h until delivery is recommended.

REFERENCES

1. Manual on Intrauterine Growth Restriction by Mediscan

2. Turan OM, et al. Progression of Doppler abnormalities in intrauterine growth restriction. Ultrasound Obstet Gynecol 2008; 32:160–167.

3. Baschat AA, et al: Predictors of neonatal outcome in early-onset placental dysfunction. Obstet Gynecol 2007; 109: 253–261.

4. Cruz-Martinez R, et al. Changes in myocardial performance index and aortic isthmus and ductus venosus Doppler in term, small- for-gestational age fetuses with normal umbilical artery pulsatility index. Ultrasound Obstet Gynecol 2011; 38: 400–405.

5. Alfirevic Z, Stampalija T, Gyte GM. Fetal and umbilical Doppler ultrasound in high-risk pregnancies. Cochrane Database Syst Rev 2010:CD007529.

6. Oros D, et al. Longitudinal changes in uterine, umbilical and fetal cerebral Doppler indices in late-onset small-for-gestational age fetuses. Ultrasound Obstet Gynecol 2011; 37:191–195.

7. Hershkovitz R, et al. Fetal cerebral blood flow redistribution in late gestation: identification of compromise in small fetuses with normal umbilical artery Doppler. Ultrasound Obstet Gynecol 2000; 15: 209–212.

8. Cruz-Martinez R, et al. Fetal brain Doppler to predict cesarean delivery for non-reassuring fetal status in term small-for-gestational age fetuses. Obstet Gynecol 2011; 117: 618–626.

9. Baschat AA, Gembruch U. The cerebroplacental Doppler ratio revisited. Ultrasound Obstet Gynecol 2003; 21: 124–127.

10. Schwarze A, et al. Qualitative venous Doppler flow waveform analysis in preterm intrauterine growth-restricted fetuses with ARED flow in the umbilical artery—correlation with short-term outcome. Ultrasound Obstet Gynecol 2005; 25: 573–579.

11. Fouron JC, et al. The relationship between an aortic isthmus blood flow velocity index and the postnatal neurodevelopmental status of fetuses with placental circulatory insufficiency. Am J Obstet Gynecol 2005; 192:497–503.

12. Alfirevic Z, Neilson JP. Biophysical profile for fetal assessment in high risk pregnancies. Cochrane Database Syst Rev 2000:CD000038.

13. Figueras, et al. Fetal Diagnosis and Therapy; January 2014: Update on Diagnosis and Classification of Fetal Growth restriction and Proposal of a Stage based management protocol.

14. Figueroa-Diesel, et al. Doppler changes in the main fetal brain arteries at different stages of hemodynamic adaptation in severe growth restriction. Ultrasound in Obstetrics and Gynecology, 2007.

Small for Gestational Age—Diagnosis and Delivery

Rashmi Jalvee, Reena J Wani

Being born small for gestational age (SGA) is a risk factor for growth and development disorders, and chronic diseases later in life. The term 'small for gestational age' (SGA) *in utero* refers to estimated fetal weight less 10th percentile on ultrasound.[1,2] This diagnosis does not necessarily describe pathologic growth abnormalities, and may simply imply a fetus at the lower end of the normal range.

India has a high incidence of SGA babies. The incidence of SGA in India is about 30% babies in contrast to 5–7% in developed countries.[3]

SGA fetuses are a heterogeneous group comprising:[2]

• Constitutionally small fetuses OR
• Fetuses that have failed to achieve their growth potential (fetal growth restriction, FGR) (Fig. 18.1)

The challenge is to identify the subset of pregnancies affected with pathological growth restriction that are at higher risk of intra-uterine and postnatal complications so that interventions can be taken to reduce morbidity and mortality.

Around 50–70% of SGA foetuses are constitutionally small[2] and the lower the centile for defining SGA, the higher the possibility of FGR. On the other hand, a fetus with growth restriction may not be SGA.

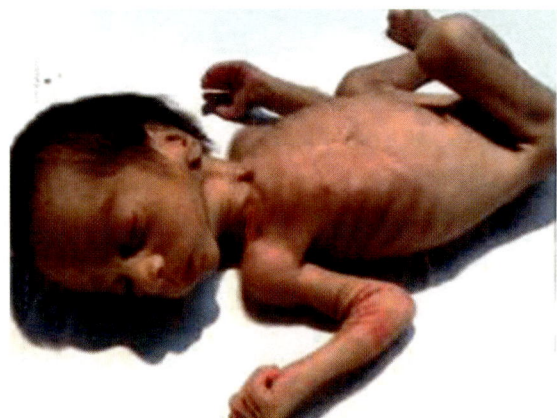

Fig. 18.1: Growth restricted neonate

The morbidity in the intrauterine period and after delivery will depend upon the underlying cause. This may be

• Placenta-mediated growth restriction
• Fetal structural anomalies
• Chromosomal abnormalities
• Infective abnormalities or
• Secondary to maternal factors.

Risk Factors for SGA

At the first antenatal visit, all women should be assessed for risk factors for a SGA fetus to identify those who require increased surveillance. The risk factors include:[4,5]

• *Maternal risk factors:* Age >35 years, nulliparity, BMI <20 or >30, smoking, daily vigorous exercise

- *Previous obstetric history:* Previous stillbirth, previous SGA, previous pre-eclampsia or pregnancy interval < 6 or >60 months
- *Maternal medical history:* Chronic hypertension, autoimmune diseases, chronic renal disease, diabetes with vascular complications, maternal SGA.
- *Paternal medical history:* Paternal SGA
- *Current pregnancy complications:* Low maternal weight gain, threatened miscarriage, pre-eclampsia, unexplained APH.
- *Fetal:* Aneuploidy, congenital infection, structural abnormalities.

Second trimester aneuploidy markers have limited predictive accuracy for diagnosis of a SGA neonate. A low level (<0.415 MoM) of the first trimester marker PAPP-A should be considered a major risk factor for delivery of a SGA neonate.[6]

Serial ultrasound measurement of fetal size and assessment of well-being with umbilical artery Doppler should be offered in cases of fetal echogenic bowel.

In women with risk factors, uterine artery Doppler at 20–24 weeks of pregnancy has a moderate predictive value for a severely SGA neonate.

Diagnosis of SGA[4,7] (Table 18.1)

Effective screening and diagnosis of SGA requires precise dating of pregnancy which involves review of the mother's menstrual history, either a first trimester or early second trimester dating ultrasound or relevant assisted reproductive technology information.

Following tests are used for screening, diagnosis and surveillance of the fetus:

Other investigations indicated in SGA fetuses include:[4]

- Detailed fetal anomaly scan and uterine artery Doppler if severe SGA identified at the 18–20 weeks scan.
- Karyotyping in severely SGA fetuses with structural anomalies detected before 23 weeks of gestation, especially if uterine artery Doppler is normal.
- Consider serological testing for congenital cytomegalovirus (CMV) and toxoplasmosis infection in severe SGA.
- Testing for syphilis and malaria should be considered in high-risk populations.

Management

Medical Management

- Antiplatelet agents: In women at high risk of pre-eclampsia or those with high-risk factors for SGA, starting low dose aspirin (60–150 mg per day) at or before 16 weeks of pregnancy may be effective in preventing SGA birth.[8]
- Quit smoking, tobacco usage
- Continue ongoing monitoring for pre-eclampsia

Bedrest, dietary modifications, oxygen therapy or progesterone supplements have not shown any consistent effect on SGA fetuses and are not recommended.

Role of Sildenafil citrate in FGR: Dosage 25 mg TID orally from the day of diagnosis of IUGR until delivery has been used with suggested benefits ranging from vasodilatation in myometrium to prolongation of pregnancy with possible long-term benefits.[9] More work is needed in this area.

Fetal Surveillance and Timing of Delivery[4,7]

Except delivery, there is no effective intervention to alter the course of FGR. Optimally timed delivery is thus a critical issue to balance the risks of prematurity against those of continued intrauterine stay with possible intrauterine death.

In women with abnormal uterine artery Doppler at 20–24 weeks, serial ultrasound measurement of fetal size and umbilical artery Doppler should be started at 26–28 weeks of pregnancy.

Table 18.1	Diagnosis of SGA	
History		*Maternal, fetal and placental risk factors*
Physical examination	Abdominal palpation Symphysio fundal height (SFH)	Limited diagnostic accuracy, subjective • Measured by first identifying the variable point, the fundus to the fixed point, the pubic symphysis, with cm value hidden from the examiner. • Low sensitivity, high false positive rates, significant intra- and inter-observer variation • Single SFH below the 10th centile or serial measurements with slow or static growth should be referred for further investigation • Customised fundal height chart (based on height, weight, parity, ethnic group) improves accuracy to predict SGA fetus.
Ultrasound biometry		• SGA defined as AC or EFW < 10th centile • Serial measurements of AC and EFW (growth velocities) 2 weekly superior to single measurement for prediction of FGR • Assessment of amniotic fluid volume (AFI or single deepest pocket)
Doppler velocimetry	Uterine artery Doppler at 20–24 weeks (PI > 95th centile ± diastolic notching)	• May identify pregnancies at risk of FGR in women with risk factors • Subsequent normalisation still associated with increased risk of a SGA
	Umbilical artery (UA) Doppler	• Primary assessment tool • Absent or reversed end-diastolic (ARED) flow in the presence of FGR is an ominous finding that requires intervention and delivery.
	Middle cerebral artery (MCA) Doppler	• Reduced MCA PI or MCA PI/umbilical artery PI (cerebroplacental ratio) early sign of fetal hypoxia in term SGA • In preterm SGA, MCA Doppler has limited accuracy to predict acidemia.
	Ductus venosus (DV) and umbilical vein (UV) Doppler	• DV Doppler has moderate predictive value for acidemia and adverse outcome • A retrograde a-wave and pulsatile flow in the umbilical vein (UV) signifies the onset of overt fetal cardiac compromise
Cardiotocography (CTG)		• CTG, although sensitive, has a 50% false positive rate for the prediction of adverse outcome • Should not be used in isolation to monitor fetuses with SGA
Biophysical profile (BPP)		• Abnormal score predictor of significant fetal acidemia • If assessed, delivery is indicated at scores ≤4/10 and close monitoring or delivery should be considered at scores ≤6/10

In fetuses with SGA, sonographic monitoring with Doppler plays a crucial role to improve perinatal outcomes (Table 18.2). No single individual test is available to predict outcome in IUGR; hence a combination of tests is recommended.

Umbilical artery (UA) Doppler is used as the primary surveillance tool.

Antenatal corticosteroids for fetal lung maturation should be given between 24+0 and 34+6 weeks but may be given up until 39 weeks, especially if delivery is by elective cesarean section at a dose of inj Betamethasone 12 mg intramuscular and repeated 24 hours later. If inj betamethasone is not available, inj dexamethasone 6 mg intramuscular every 12 hours can be given for 4 doses.[10]

Magnesium sulfate for fetal neuroprotection should be administered when delivery is anticipated prior to 32 weeks gestation.[11] This is a simple but effective intervention which can improve the outcome.

Mode of Delivery[4,7]

Decisions regarding the optimal timing of delivery, the mode of delivery and the place of delivery should be made on an individual basis and should involve senior experienced obstetrician or a fetal medicine specialist especially in the case of severe FGR.

Key clinical points to keep in mind are:

- UA AREDV, delivery by cesarean section is recommended.
- Normal UA Doppler or with abnormal UA PI but end-diastolic velocities present, induction of labor can be offered and continuous FHR monitoring from the onset of uterine contractions.
- Early admission if in spontaneous labor with a SGA fetus, continuous FHR monitoring:
  - Review the antenatal records—a constitutionally small baby if well can withstand the stress of labor
  - Assessment of the fetus at the onset of labor
  - Presentation/lie—cephalic can be considered for vaginal trial
  - Bishop's score—determines time required for induction
  - Previous obstetric history
  - Choice of mother/couple after counseling
  - Neonatology setup
- *Rates of emergency cesarean section are high:* Often a decision to deliver is in an

Table 18.2	Role of Doppler
UA Doppler normal	• Repeat CTG, growth scan, AFI and UA Doppler every 2 weeks • Delivery should be offered by 37 weeks
UA PI abnormal > 2SD but with end diastolic velocities	• Repeat AC and EFW weekly • AFI, UA and MCA Doppler, CTG twice weekly • Recommend delivery by 37 weeks • Consider delivery > 34 weeks if static growth > 3 weeks • If MCA Doppler abnormal, delivery should be recommended by 37 weeks
UA AREDV detected <32 weeks of gestation	• Administer steroids • Repeat AC and EFW weekly • UA, MCA, DV Doppler, CTG daily • Delivery recommended when DV Doppler becomes abnormal or UV pulsations appear after completion of steroids. • Even if venous Doppler is normal, delivery is recommended by 32 weeks of gestation and should be considered between 30 and 32 weeks of gestation.

emergency situation so a cesarean section seems to be appropriate as FGR babies may not withstand the stress of induction of labor/labor contractions.

- Delivery should be conducted in a tertiary care center with optimal neonatal expertise and facilities available.
- A senior pediatrician trained and competent in resuscitation of the newborn should be present at delivery.

Prognosis related to SGA can principally be determined by:[1]
- The cause of SGA
- Period of gestation
- Timing and duration of growth restriction
- Severity and symmetry of growth restriction
- Presence and degree of perinatal asphyxia and its complications
- Postnatal course

Complications associated with SGA:
- Intrauterine fetal death
- Perinatal asphyxia
- Hematological and metabolic abnormalities
- Impaired thermoregulation
- Respiratory distress syndrome
- Intraventricular hemorrhage
- Necrotising enterocolitis
- Retinopathy of prematurity
- Neonatal death

Further, adverse intrauterine environment increases disease risk in adulthood, viz. metabolic syndrome, hypertension, insulin resistance, type 2 diabetes mellitus, coronary heart disease and stroke (Barker hypothesis—fetal origin of adult disease)

Risk of recurrence and preventative strategies:
- Recurrence risk of FGR in a subsequent pregnancy is around 25%[12]
- Lifestyle factors such as smoking cessation should be addressed
- Underlying causes (placental histology, maternal comorbidities) should be reviewed
- Consideration should be given to start low dose Aspirin daily prior to 16 weeks

KEY POINTS

Diagnosis and Delivery of SGA:
- Diagnose early
- Differentiate constitutionally small from true FGR
- Diagnose possible cause and treat if possible
- Do not wait for absent/reversed EDF to deliver—may be too late!
- Do give antenatal corticosteroids in anticipation of intervention
- Do give magnesium sulfate for neuroprotection as indicated
- Decide optimum time and mode of delivery as a team
- Deliver in the best way, best equipped place with neonatologist

REFERENCES

1. Sinha S, Miall L, Jardine L. Essential neonatal medicine. 5th ed. Chichester, West Sussex: John Wiley & Sons; 2012.
2. Lausman A, Kingdom J, Gagnon R, Basso M, Bos H, Crane J, et al. Intrauterine growth restriction: screening, diagnosis, and management. J Obstet Gynaecol Can. 2013; 35(8):741–57.
3. Sangita Yadav S, Rustogi D. Review Article: Small for Gestational Age: Growth and Puberty Issues Indian Pediatrics, 2015;52: 135–140.
4. Royal College of Obstetricians and Gynaecologists. Investigation and management of the small for gestational age fetus. Guideline No. 31 London: RCOG; 2013.
5. Campbell MK, Cartier S, Xie B, Kouniakis G, Huang W, Han V. Determinants of small for gestational age birth at term. Paediatric And Perinatal Epidemiology. 2012; 26(6):525–533.
6. Spencer K, Cowans NJ, Avgidou K, Molina F, Nicolaides KH. First-Trimester Biochemical Markers of Aneuploidy and the Prediction of Small-for-Gestational Age Fetuses. Obstetrical & Gynaecolgical Survey 2009; 64:370–2.
7. SOGC Clinical Practice Guideline No. 295, Intrauterine Growth Restriction: Screening, Diagnosis, and Management, J Obstet Gynaecol Can 2013;35(8):741–748.

8. Bujold E, Roberge S, Lacasse Y, Bureau M, Audibert F, Marcoux S. Prevention of preeclampsia and intrauterine growth restriction with aspirin started in early pregnancy—a meta-analysis. Obstet Gynecol 2010; 116:402–14.

9. Ganzevoort W, Alfirevic Z. Strider: Sildenafil Therapy In Dismal prognosis Early-onset intrauterine growth Restriction--a protocol for a systematic review with individual participant data and aggregate data meta-analysis and trial sequential analysis. Syst Rev. 2014 Mar 11; 3:23. doi: 10.1186/2046-4053-3-23.

10. Royal College of Obstetricians and Gynaecologists (RCOG). Green-top Guideline No.7: Antenatal corticosteroids to reduce neonatal morbidity. Available from URL: https://www.rcog.org.uk/

11. The Antenatal Magnesium Sulphate for Neuroprotection Guideline Development Panel. Antenatal magnesium sulphate prior to preterm birth for neuroprotection of the fetus, infant and child: National clinical practice guidelines. Adelaide: The University of Adelaide, 2010. ISBN Print: 978-0-86396-720-7.

12. Voskamp BJ, Kazemier BM, Ravelli AC, Schaaf J, Mol BW, Pajkrt E. Recurrence of small-for gestational-age pregnancy: analysis of first and subsequent singleton pregnancies in The Netherlands. Am J Obstet Gynecol 2013; 208(5): 374 e1–6.

Intrapartum Monitoring: Cardiotocography and Beyond

Anahita Chauhan, Dipti Shende

INTRODUCTION AND PHYSIOLOGICAL BASIS OF CARDIOTOCOGRAPHY (CTG)

For over a hundred years, Pinard's fetoscope was used for fetal heart sound monitoring specially bradycardia for fetal distress. Even though the fetus is efficient at extracting oxygen from the maternal compartment, a complex interaction of antepartum complications, suboptimal uterine perfusion, placental dysfunction, and intrapartum events may be associated with adverse outcome. Deficiency of oxygen to the fetus during labor may progress from hypoxemia to hypoxia to asphyxia. Hypoxemia is the initial phase of oxygen deficiency in which organ and cell function is intact. It affects the arterial blood flow leading to decrease in oxygen saturation. To conserve energy, there is a decrease in fetal movements and breathing. Due to this compensatory mechanism, the fetus can balance energy for days and weeks. Hypoxia occurs due to further decrease in oxygen saturation affecting the peripheral blood flow, causing redistribution of blood to heart and brain and release of epinephrine, norepinephrine and anerobic metabolism. Fetal damage is less likely as oxygenation and glucose is maintained to the central organs. In asphyxia, oxygen supply to central organs is also diminished; in such situations delivery of the baby should be imminent.

Electronic fetal monitoring (EFM) technology was introduced in the 1960s to identify events that might result in hypoxic ischemic encephalopathy, cerebral palsy, or fetal death, and it focused on fetal heart acceleration, deceleration and variability. Cardiotocography (CTG) is done using one transducer on the maternal abdomen where the fetal heart sound is best heard or an electrode on the fetal scalp, and another transducer at the uterine fundus to record contractions. These variables are plotted graphically so that variations in fetal heart rate (FHR) can be viewed and interpreted. FHR tracing is performed for 20 minutes to account for fetal sleep cycle and uterine activity. The speed of paper may be 1 or 3 cm/ minute. FHR is plotted from 30 to 240 beats per minute (bpm). Uterine activity is measured in range of 0–100 relative units using tocometer and 0–100 mm of mercury when applying intrauterine pressure. Date and time on CTG machine should be confirmed and all CTGs must be labeled with the mother's name, hospital number and most importantly, the interpretation. Continuous electronic fetal monitoring has been shown to reduce the incidence of neonatal seizures, but there has been no beneficial effect in decreasing cerebral palsy or neonatal mortality. Unfortunately, CTG led to higher operative delivery rates without significant lowering perinatal mortality, compared to intermittent FHR auscultation alone, per a recent meta-analysis.[1] Intra-observer variability may play a major role in its interpretation, hence for universal interpretation of EFM, National Institute of Child Health and Human Development (NICHHD) convened a workshop in 2008 to revise the accepted definitions of EFM. These were adopted by ACOG and Maternal Fetal Medicine Foundation, and are given in detail later.

Indications for CTG

Antepartum

- Abnormal Doppler study
- Multiple pregnancy
- Intrauterine growth restriction
- Rh isoimmunization
- Oligohydraminos
- Antepartum hemorrhage
- Medical conditions like hypertensive disorder of pregnancy, cardiac disease, vascular disease, renal disease
- Diabetes
- Hyperthyroidism

Intrapartum

- Meconium stained amniotic fluid
- Induced or augmented labor
- Preterm or post-term pregnancy
- Previous cesarean section
- Regional analgesia
- Maternal pyrexia
- FHR <110 bpm or >160 bpm
- Epidural analgesia

Components and Terminology of CTG

Four components define the CTG and standard definitions should be uniformly applied to their interpretation: The baseline rate, variability, accelerations and decelerations. Previously used terms like "beat-to-beat variability" should be replaced by the term baseline variability.

Baseline Rate

The normal baseline fetal heart rate is 110–160 bpm. There must be at least two minutes of identifiable baseline segments in any 10 minute window. Fetal bradycardia is a baseline rate of <110 bpm and tachycardia is > 160 bpm.

Bradycardia can occur due to:
- Cord compression
- Acute fetal hypoxia
- Post maturity
- Congenital heart abnormality

Tachycardia can occur due to:
- Excessive uterine stimulation
- Maternal fever, anxiety
- Fetal infection
- Chronic hypoxia
- Prematurity < 32 weeks of gestation
- Dehydration
- Fetal tachyarrhythmia

Fetal Heart Rate Variability

FHR are irregular in amplitude and frequency; variability is the normal beat to beat changes in FHR, usually 5–15 bpm. Variability can be classified as marked if >25 bpm (Fig. 19.1), moderate if between 6 and 25 bpm (Fig. 19.2), minimal if <5 bpm (Fig. 19.3) and absent if undetectable (Fig. 19.4). Fetal hypoxia can lead to any of these. Decreased variability is seen in normal fetal sleep-wake pattern, prematurity, drugs like opioids. Sinusoidal pattern is smooth sine wave of amplitude of 10 bpm with three to five cycles/minutes. This is associated with fetal anemia and hydrops.

Accelerations

Acceleration is defined as transient increase of FHR >15 bpm from the baseline for >15 seconds. Before 32 weeks of gestation, in view of immature autonomic nervous system, accelerations are considered when there is a peak >10 bpm and a duration of >10 seconds.

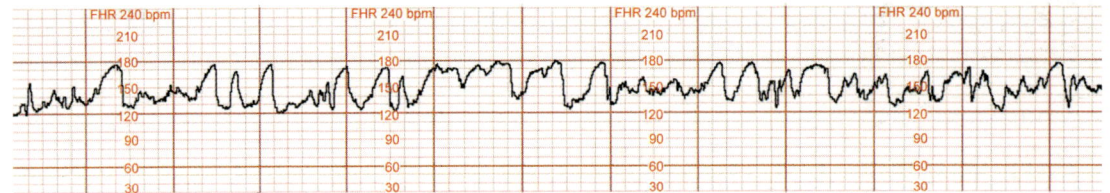

Fig. 19.1: Marked variability

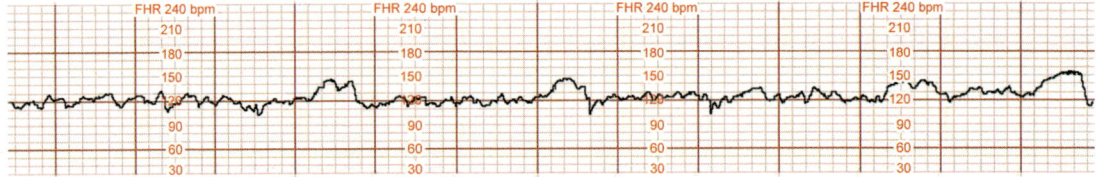

Fig. 19.2: Moderate variability

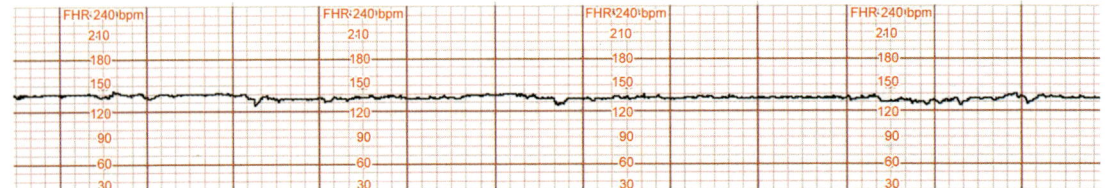

Fig. 19.3: Minimal variability

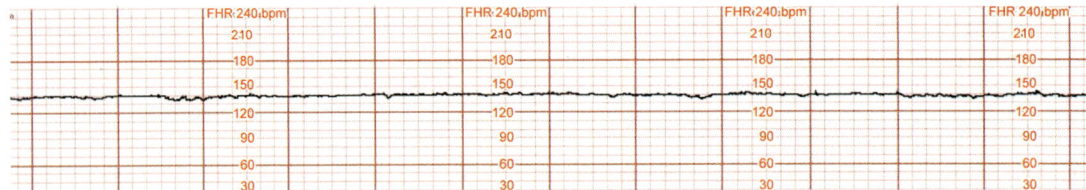

Fig. 19.4: Absent variability

Acceleration indicates normal oxygenation to the fetus and reactive means 2 accelerations in 20 minutes.

Decelerations

Deceleration is a transient decrease in FHR for 30 seconds from the onset of the deceleration to the FHR nadir. Deceleration suggests development of hypoxia and majority are related to change in fetal environment. Decelerations are classified as early, late, variable and prolonged.

Early deceleration occurs with onset of contraction and FHR recovers by end of contraction. It is mostly caused by fetal head compression and often relieved by changing maternal posture. It does not lead to poor fetal outcome (Fig. 19.5).

Late deceleration begins more than 15 seconds after the peak of the uterine contraction. It progresses or worsens with contractions and is due to decreased placental blood flow leading to hypoxia. Most common reasons are hypoxia, placental abruption,

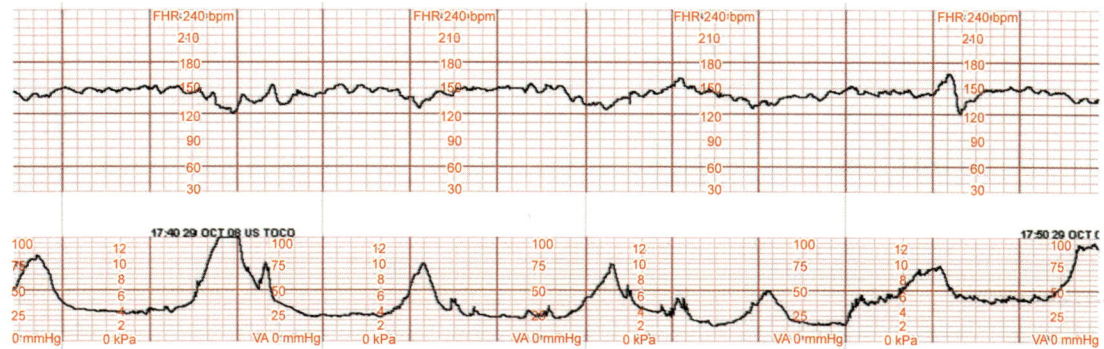

Fig. 19.5: Early deceleration

cord prolapse, excessive uterine contractions, maternal hypotension, hypovolemia (Fig. 19.6).

Variable deceleration may be variable in both time and size. Decrease in FHR of at least 15 bpm with duration of at least 15 seconds–2 minutes. It occurs due to compression of umbilical cord and reflects fetal hypoxia. The fetus can handle variable deceleration for a long period if the duration is less than 60 seconds; risk of hypoxia increases with duration in excess of this. Variable decelerations may be complicated or uncomplicated depending on other features of the CTG like fetal tachycardia, absent baseline variability, slow return to baseline FHR after the end of the contraction, presence of post-deceleration smooth overshoots (Fig. 19.7).

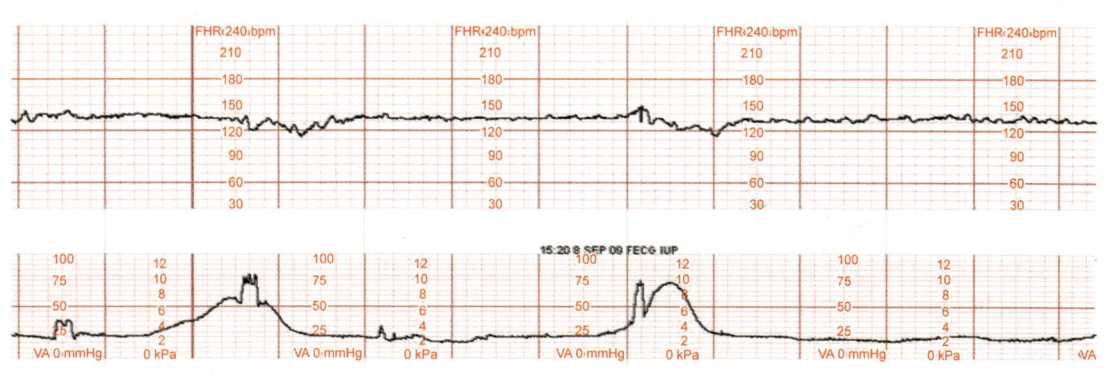

Fig. 19.6: Late deceleration

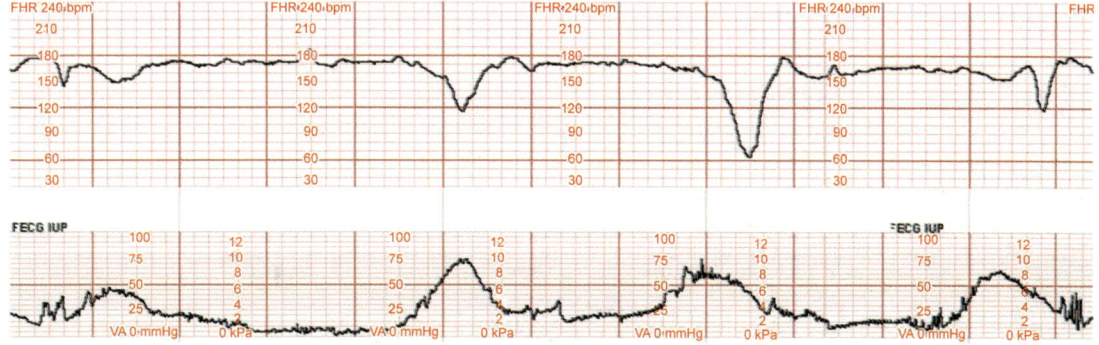

Fig. 19.7: Variable deceleration

Prolonged deceleration is a deceleration lasting for 2 minutes with a fall in FHR >30 bpm. It is usually caused by maternal hypotension, umbilical cord compression, and uterine hypertonia. Recurrent decelerations occur with >50% of uterine contractions in any 20-minute window. Intermittent deceleration occurs with <50% of uterine contraction in any 20-minute window. Presence of acceleration and moderate FHR variability predicts the absence of fetal metabolic academia. Absence of FHR variability does not reliably predict fetal acidemia (Fig. 19.8).

Uterine contraction assessment is important for FHR monitoring. Normal frequency is 2–3 contractions in every 10 minutes in initial phase of first stage of labor, which increases to 4–5 contractions in later phase of first stage. Uterine tachysystole is defined as >5 contractions in 10 min over 30 minutes; this can lead to fetal hypoxia.

Interpretation of CTG and Classification Systems

One of the main problems with CTG is the interpretation of the trace. Many scientific bodies have devised scoring systems and mnemonics to aid in understanding and decision making; the main ones, NICHD, NICE guidelines and "ALSO" (Advanced Life Support in Obstetrics) are given below.

Three Tier FHR Interpretation and Intervention System (NICHD 2008)[2]

In this system, the FHR are categorized as I (normal), II (indeterminable) and III (abnormal).

Category I FHR tracings: Normal

Predictive of normal fetal acid–base status; may be followed in a routine manner, no specific action is required. Category I FHR tracings include all the following:
- *Baseline rate:* 110 –160 bpm
- *Baseline FHR variability:* Moderate
- *Late or variable decelerations:* Absent
- *Early decelerations:* Present or absent
- *Accelerations:* Present or absent

Category II FHR tracings: Indeterminable

Not predictive of abnormal fetal acid–base status, yet do not have adequate evidence to classify as Category I or III. Continued surveillance and reevaluation is required.

Category II FHR tracings include all FHR tracings not categorized as Category I or Category III, and may represent an appreciable fraction of those encountered in clinical care. Examples of Category II FHR tracings include any of the following:

Baseline rate
- Bradycardia not accompanied by absent baseline variability
- Tachycardia

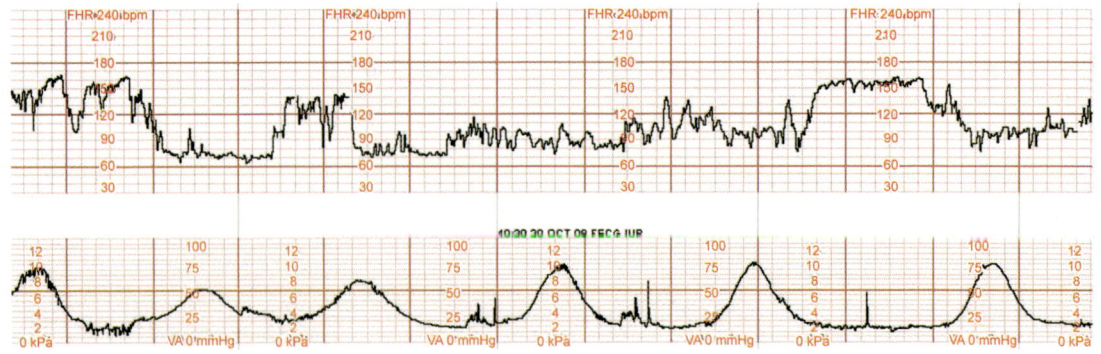

Fig. 19.8: Prolonged deceleration

Baseline FHR variability
- Minimal baseline variability
- Absent baseline variability not accompanied by recurrent decelerations
- Marked baseline variability

Accelerations: Absence of induced accelerations after fetal stimulation

Periodic or episodic decelerations
- Recurrent variable decelerations accompanied by minimal or moderate baseline variability
- Prolonged deceleration >2 minutes but <10 minutes
- Recurrent late decelerations with moderate baseline variability
- Variable decelerations with other characteristics, such as slow return to baseline, "overshoots", or "shoulders".

Category III FHR Tracings: Abnormal

Category III tracings are predictive of abnormal fetal acid–base status at the time of observation. This pattern requires prompt evaluation and efforts to resolve the abnormal FHR pattern, such as oxygen administration, change in maternal position, discontinuation of labor stimulation, and treatment of maternal hypotension.

Category III FHR tracings include:
Absent baseline FHR variability and any of the following:
- Recurrent late decelerations
- Recurrent variable decelerations
- Bradycardia

Sinusoidal pattern: This classification system makes use of a "stoplight" where category I traces are green, category II are yellow and category III are red; management is based on category (given later).

NICE GUIDELINES 2014[3]

National Institute for Health and Care Excellence (NICE) guidelines, UK in 2014 have adopted a similar system for reviewing the CTG trace, where assessment and document all 4 features (baseline fetal heart rate, baseline variability, presence or absence of decelerations, presence of accelerations) is done and traces are categorized as normal, suspicious and pathological, based on these features (Table 19.1).

Table 19.1	NICE definitions			
Category	*Definition*			
Normal	All four features (given below) are classified as reassuring			
Suspicious	One feature classified as non-reassuring and the remaining features classified as reassuring			
Pathological	Two or more features classified as non-reassuring or one or more classified as abnormal			
Feature	*Baseline (bpm)*	*Variability (bpm)*	*Accelerations*	*Decelerations*
Normal/ reassuring	100–160	5 or more	Present	None or early
Non-reassuring	161–180	< 5 for 30–90 min		Typical variable decelerations with over 50% of contractions, for over 90 min. Single prolonged deceleration for up to 3 min
Abnormal	< 100, > 180	< 5 for 90 min; sinusoidal pattern for 10 min or more		Non-reassuring variable decelerations still observed > 30 min of conservative treatment, or atypical variable decelerations with over 50% of contractions or late decelerations, both for over 30 min, or single prolonged deceleration for more than 3 min

ADVANCED LIFE SUPPORT IN OBSTETRICS (ALSO) CLASSIFICATION[4]

Advanced Life Support in Obstetrics (ALSO) classification for decision making, which incorporates maternal and fetal risk factors in mnemonic "DR C BRAVADO", is shown in Table 19.2.

DECISION MAKING IN CTG AND ALGORITHMS FOR CARE[4, 5]

Interventions for abnormal CTG, irrespective of the classification system used, include intrauterine resuscitation. The "POISON" mnemonic is easy to remember (Table 19.3):

Table 19.2	ALSO classification
DR	Determine risk (high, medium or low)
C	Assess contractions
BRA	Baseline rate (brady, normal or tachycardia)
V	Variability
A	Accelerations
D	Decelerations
O	Overall assessment and written plan

Table 19.3	Algorithm for care based on classification systems	
Category of trace	Management	
Category I—normal	• Continue care • Reevaluate and reassure	
Category II—indeterminate	"POISON"	
Category III—abnormal	• "POISON" • Prepare operation theatre and neonatal ICU • Explain assessments and interventions to the patient	

Table 19.4	Scalp lactate		
Lactate value (mmol/l)	Description	Management	
<4.2	Normal	Continue labor	
4.2–4.8	Preacidemia	Repeat FBS between 20 and 30 minutes later	
>4.8	Acidemia	Consider delivery	

- P   Position change
- O   Oxygen
- I   IV bolus
- S   Sterile vaginal examination (for abruption, cord prolapse)
- O   Off medicines (discontinue oxytocin)
- N   Notify provider

Additionally, the following measures are also recommended:

- Temperature, pulse and blood pressure monitoring
- Fetal scalp pH
- Amnioinfusion for recurrent, moderate variable decelerations
- Immediate delivery by instrumentation or cesarean section after accurate diagnosis of hypoxia.

FETAL pH

Fetal blood sampling (FBS) to estimate fetal pH was introduced in 1967 and is used by obstetricians to decrease the false positives and cesarean section rate due to CTG. While obtaining blood from the fetal scalp, the cervix should be at least 2 cm dilated. Fetal blood is collected with help of an amnioscope. The blood is collected from peripheral tissue, making it difficult to interpret due to accumulation of carbon dioxide. Drawbacks of fetal scalp pH are risk of contamination with amniotic fluid or maternal blood; contact with air leads to decrease in carbon dioxide and error in measuring metabolic acidosis. In an observational study, it was found that babies with pH of 7 or less have neonatal morbidity. Obtaining fetal scalp blood is difficult and takes approximately 20 minutes; failure rate is 20%.[6] FBS has no role in bradycardia or emergency situations like rupture uterus, cord prolapse and placental abruption; delivery has to be expedited regardless of analysis in these cases.

FETAL SCALP LACTATE

Lactate is commonly used to detect septicemia in adults and neonates. It is increased due to tissue hypoxia and tissue perfusion due to anerobic metabolism during labor. Lactate level serves as an earlier marker than fetal pH as it is increased in subcutaneous tissue. This happens due to buffering capacity, removing the hydrogen ions and maintaining pH. Due to prolonged hypoxia, production of hydrogen ions increases and pH falls.

Measuring fetal lactate is a bedside method using 5 microlitres of blood from fetal scalp, using an electrochemical micro volume device.[7] Different "lactate" devices are available which measure different blood compartments like plasma, hemolysed blood or whole blood with intact electrolytes. Advantages of lactate over pH are that sampling is earlier, it yields result more often, quick bedside investigation with better predictive value for neonatal condition. Since 2014, NICE recommends use of fetal scalp lactate provided trained staff and relevant equipment are available.

Fetal scalp lactate is superior in predicting hypoxic ischemic encephalopathy with a sensitivity of 67% and specificity of 93% as compared to 49% and 93% respectively for pH.[8]

Suggested clinical guidelines for scalp blood determination and management is as follows:

A Cochrane meta-analysis has compared lactate and pH during labor and found no differences in neonatal outcome or operative interventions but a significantly higher success rate with lactate compared with pH.[9]

FETAL ECG ANALYSIS[10]

EFM helps more in the diagnosis of a healthy fetus; critics are still doubtful about its true value in the diagnosis of intrapartum asphyxia and degree of hypoxic stress. The ST analysis (STAN) of fetal ECG is a useful adjunct to CTG; it analyzes fetal heart muscle function during labor. Due to the use of STAN, interventions and birth asphyxia are reduced.

Like the adult ECG, fetal ECG P wave represents atrial contraction, QRS complex represents ventricular contraction and T wave represents ventricular repolarization. RR stands for time between two consecutive heart beats. Rosen demonstrated T wave elevation in guinea pigs due to hypoxia.[11] This occurs due to increase in myocardial glycogenolysis and release of potassium. The increase of local catecholamine concentration with unrelieved hypoxemia leads to elevation in T wave height. The increase of T wave relative to the amplitude of the QRS complex has been identified when the fetus transitions from aerobic to anaerobic metabolism and when increased beta adrenergic release occurs. Biphasic or ST depression occurs due to chronic hypoxia, infection, immature fetus or malformation of heart. In severe asphyxia, ST waveform returns towards normal and T wave amplitude increases due to hypoxia which is directly proportional to the amount of glycogen the fetus utilizes to maintain myocardial energy. Hence the STAN fetal monitoring system includes EFM that displays standard fetal heart and uterine activity pattern along with automated STAN.

Baseline T/QRS rise: Increase in T/QRS ratio for more than 10 minutes is defined as baseline T/QRS rise. If an ST event occurs due to T/QRS rise of > 0.05, it is considered as significant. It occurs in hypoxia with anaerobic metabolism. If the electrode is placed on cervix or a dead fetus, maternal ECG is recorded. This is identified by absent P wave and wider QRS and coincides with maternal pulse. If electrode is applied to fetal membrane or vagina, then due to poor signal quality ST analysis is not detected. FDA has approved this in 2005.

Guidelines for using STAN:
- More than 36 weeks of gestational age
- Ruptured membranes
- No contraindication for scalp electrode
- First stage of labor
- Normal ECG waveform

Contradictions for STAN include active maternal herpes simplex, HIV and hepatitis. The STAN system collects fetal ECG data in a stable intrapartum environment so that accurate T: QRS baseline can be set for future comparisons. This helps in start of STAN in the first stage of labor. The main parameters for STAN are baseline T: QRS and ST segment waveforms. The T: QRS ratio is derived from 30 consecutive cardiac cycles. The baseline T: QRS is determined over 20 T: QRS ratios and then tracked for changes over time. Alerts appear as flags on the top of the monitor display and are noted in the event log. They occur when:

- There is an episodic rise in T: QRS (greater than 0.10 for less than 10 min)
- There is a baseline T: QRS rise (more than 0.05 for more than 10 min)
- There are recurrent biphasic ST segments

Like classification of CTG, STAN and CTG together are especially useful for the indeterminate or non-reassuring CTG, and are interpreted as below.

MOBILE CTG (M-CTG) USING ANDROID PLATFORM

Newer technology has overcome limitations associated with CTG machines by developing wireless CTG. FHR recording device with bluetooth module and data can be connected to an android phone as a data source. Data will be sent via Wi-Fi or GPRS to the clinician for more detailed assessment. Signal processing array includes interpolation and artifact removal, baseline assessment, acceleration/deceleration detection and features and evaluation of long-term variability as interquartile range. Finally, it includes visualization of the results. This proposed version of the android application will be useful for remote monitoring within the hospital. The main concern is Wi-Fi connectivity for transmission of data; this can be overcome by using separate Wi-Fi hotspot and smartphone data system. Backups need to be created in data repositories in case of transmission failure. Cost is a major limitation for its introduction in routine care; research in indigenous software and transmission systems for low-resource settings are ongoing.

CONCLUSIONS

Standard EFM is good at detecting the very healthy and very sick fetus. However, inter and intra-observer variations in interpretation still exist; training, practice and documentation are the only ways to eliminate errors. To this end, there are numerous online tutorials and teaching aids available for training for residents. The high false positive rates of CTG are associated with increased rate of unnecessary interventions. Adjuvant methods (FSB, lactate or STAN) help to grade fetuses between the extremes, and reduce false positives, and should be especially offered to patients with Category II or III traces.

Table 19.5	Correlation of STAN and CTG		
ST changes	Reassuring FHR	Non-reassuring FHR	Abnormal FHR
No ST change		Expectant management	
Episodic T/ QRS rise (>0.10, duration < 10 min)	Routine management and observation	Delivery within 90 min in 2nd stage	Resuscitation and delivery, regardless of ST changes
		Delivery should occur within 30 in 1st stage	
Baseline T/QRS rise (>0.05, duration < 10 min)		Delivery should occur as soon as possible in 2nd stage	
Biphasic ST	Closer observation		

REFERENCES

1. Thacker SB, Stroup D, Chang M. Continuous electronic heart rate monitoring for fetal assessment during labor. Cochrane Database Syst Rev. 2006; (3).

2. The 2008 National Institute of Child Health and Human Development (NICHD) Workshop Report on Electronic Fetal Monitoring. Update on definitions, Interpretation, and Research Guidelines. Obstet Gynecol 2008; 112: 661–6.

3. Intrapartum care: NICE guideline CG190 (December 2014).

4. Bailey RA. Intrapartum Fetal Monitoring. Am Fam Physician. 2009; 80(12):1388–1396,1398.

5. American College of Obstetricians and Gynecologists (ACOG). Intrapartum fetal heart rate monitoring: nomenclature, interpretation and general management principles. ACOG 2009 Practice Bulletin No 106.

6. Tuffnell D, Haw WL, Wilkinson K. How long does a fetal scalp blood sample take? BJOG 2006 Mar; 113(3):332–4.

7. Shimojo N, Naka K, Uenoyama H, Hamamoto K, Yoshioka K, Okuda K. Electrochemical assay system with single-use electrode strip for measuring lactate in whole blood. Clin Chem. 1993 Nov; 39(11 Pt 1):2312–4.

8. Kruger K, Hallberg B, Blennow M, Kublickas M, Westgren M. Predictive value of fetal scalp blood lactate concentration and pH as markers of neurologic disability. Am J Obstet Gynecol. 1999 Nov;181(5 Pt 1):1072–8.

9. East CE, Leader LR, Sheehan P, Henshall NE, Colditz PB. Intrapartum fetal scalp lactate sampling for fetal assessment in the presence of a non- reassuring fetal heart rate trace. Cochrane Database Syst Rev. 2015;5.

10. Amer-Wahlin I, Arulkumaran S, Hagberg H, Marsal K, Visser G. Fetal electrocardiogram: ST waveform analysis in intrapartum surveillance. BJOG 2007; 114:1191–1193.

11. Rosén, KG, Dagbjartsson A, Henriksson BA, et al. The relationship between circulating catecholamines and ST waveform in the fetal lamb electrocardiogram during hypoxia. Am J Obstet & Gynecol. 1984; 149: 190–195.

Birth Injuries: Prediction and Prevention

Shailesh Kore, Divya Kadam, Chaitra Thunga

INTRODUCTION

Birth injury as the name suggests is injury/ injuries sustained by the newborn during the birthing process as the baby transits through the birthing canal. It is often defined as an impairment of the neonate's body function or structure due to an adverse event that occurred at birth. Even though the incidence of birth injuries has decreased in modern day obstetrics with improvement in obstetrical care, it still remains a preventable cause of perinatal morbidity. The reported incidence of birth injuries is varied in western literature and is about 1–2 percent in singleton vaginal deliveries of fetuses in a cephalic position.[1] The incidence in our country where most deliveries are unsupervised may be higher. Unfortunately, data on this subject is scarce, thus the real statistics are thus unavailable. One recent study published from India showed incidence to be around 6–8 per 1000 live births.[2] Considering the neonatal mortality and morbidity associated with birth injuries, the prediction and prevention of such injuries is important and should be included as crucial part of our obstetric training.

Fortunately, most birth traumas are self-limiting and have a favorable outcome. Nearly one-half are potentially preventable with recognition and anticipation of obstetric risk factors.

RISK FACTORS

The risk of birth injuries may increased due to the factors related to fetus (e.g. fetal size and presentation), the mother (e.g. maternal size and the presence of pelvic anomalies), or the use of obstetrical instrumentation during delivery:

- *Macrosomia:* The incidence of birth injuries is directly proportional to fetal weight. As compared with normosomic neonates, the incidence of birth injury was twofold more in infants weighing 4000 to 4499 gm, three times greater in those with births weights between 4500 and 4999 gm, and 4.5 times greater in those with a birth weight greater than 5000 gm.[3]
- Cephalopelvic disproportion either because of small maternal pelvis due to short maternal stature, contracted pelvis , pelvic anomalies or large fetal presenting part compared to maternal pelvis. This leads to dystocia and difficult labor, thus increasing the risk of injuries during birthing process.
- *Oligohydramnios:* This decreases cushioning effect on baby. Also oligohydramnios is often associated with prolonged labor, increased chances of instrumental/ operative delivery, which often increases the risk of injuries.
- *Prolonged labor:* Prolonged mechanical pressure on fetal head, difficult delivery, instrumental/operative delivery can lead to

injuries to the baby's head. Even cesarean section for prolonged labor or deep transverse arrest with deeply impacted head in pelvis makes delivery of baby difficult and thus making it vulnerable to shoulder dislocation or humerus/clavicular fracture.

- *Abnormal presentations:* Vaginal breech delivery or breech extraction can lead to hip/shoulder dislocation or fracture of long bones. Sudden compression/decompression of head in breech deliver can cause intracranial bleed. Difficult head delivery in breech can rarely cause atlanto-occipital joint dislocation or injury to spinal cord, causing permanent neurological impairment or even death of the baby.
- Instrumental deliveries, especially forceps (midcavity) or vacuum particularly when applied at higher station, with difficulty, repeated failure or when pre-requisites are not fulfilled will lead to injuries, both maternal and fetal, sometimes often serious.[4]
- Versions and extractions for breech and transverse lie are obvious risk factors.
- Very low-birth-weight infant or extreme prematurity: small head and incompletely formed skull-precipitate delivery can cause 'champagne cork popping', risking intracranial hemorrhage.
- Fetal anomalies like meningocele, omphalocele can rupture during vaginal or even abdominal delivery.
- Delivery during risk hours has also been found to be an independent risk factor in an earlier study. It is found that delivery between 2:00 am and 8:00 am carried a higher risk of birth injury than at other hours may be because of less vigilance and fatigue of hospital staffs.[5]

Type of Birth Injuries (Table 20.1)

A. Extracranial Injuries (Fig. 20.1)

Caput succedaneum: This is a sero-sanguinous, subcutaneous, extraperiosteal fluid collection with poorly defined margins that spread over suture lines. It is very common after prolonged labour caused by the pressure of the scalp on the presenting part against the dilated/undilated cervix. It is seen immediately after delivery. It does not lead to any complications and resolves within a few days.

Cephalhematoma: Cephalhematoma is a localised effusion of blood beneath the periosteum of skull of a newborn, due to disruption of blood vessels. It is most commonly seen in the parietal region and sometimes the occipital region. It is usually circumscribed, soft, fluctuant and incompressible. Spread is restricted by suture lines and hence cephalhematomas are limited to one cranial bone as opposed to caput which often overlies sutures. It usually appears on 2nd or 3rd day of delivery and spontaneously resolves in a few weeks without any treatment. Rarely, large hematomas causing blood loss leads to anemia, hypotension and/or infection requiring treatment. Lysis of hematoma may sometimes cause even hyperbilirubinemia in the neonate. Rarely it may be associated with an underlying skull fracture and may warrant CT or MRI scanning for final diagnosis.

Subgaleal hematoma: This is bleeding between the periosteum and scalp galea aponeurosis mostly from rupture of emissary veins. It is usually associated with instrumental delivery but has been reported to occur spontaneously and in neonatal coagulopathy.[6] It usually appears within 12–72 hours of birth as a soft, fluctuant mass within the scalp especially over the back of the head with/without displacement of ear lobes and periorbital edema. Its spread is not restricted by suture lines. It can spread slowly and be unnoticed and present as hypovolemic shock, neurological dysfunction, seizures, etc. It is associated with high morbidity and mortality. Management of subgaleal hematoma is mostly supportive but requires NICU care and monitoring.

Table 20.1	Types of birth injuries
Birth injury	*Examples*
Extracranial	Caput succedaneum
	Cephalhematoma
	Subgaleal hematoma
Intracranial	Subarachnoid hemorrhage
	Subdural hemorrhage
	Intracerebellar hemorrhage
	Intraventricular hemorrhage
Neurological	Brachial plexus injury-Erb's/Klumpke's palsy
	Facial nerve palsy
	Phrenic nerve palsy
	Laryngeal nerve palsy
Spinal cord	Nonlethal neurapraxia to permanent weakness/paralysis
Fractures and dislocations	Clavicle fracture
	Long bones—humerus/femur
	Depressed skull fractures
Visceral injuries	Intra-abdominal organs—liver/spleen/kidneys/ lungs
Soft tissues	Face/buttocks/scrotal injuries
	Cuts and abrasions
	Subcutaneous fat necrosis
Muscles	Sternocleidomastoid hematoma—torticollis

B. *Intracranial Hemorrhage*

Intracranial hemorrhages is most likely to occur following difficult instrumental delivery, most common being difficult forceps with wrong application of blades. Nowadays delivery by vacuum extraction is replacing the use of obstetric forceps but it is also associated with increased risk for neonatal intracranial hemorrhages.[7] Excessive compression and moulding of head in prolonged labor, sudden compression-decompression of fetal head following precipitate labor, delivery of

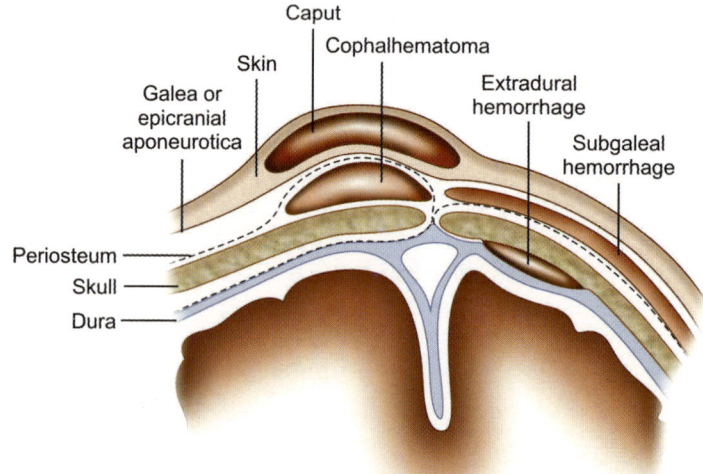

Fig. 20.1: Extracranial injuries

after-coming head of breech are other likely causes. Hemorrhage can be subdural or extradural. Outcome may range from mild cerebral and respiratory depression to stillbirth depending on site and amount of bleed. Baby may present with signs of cerebral depression like shallow respiration, depressed reflexes, apathy and refusal to suck. During recovery, signs of cerebral irritations like restlessness, inco-ordinate movements, convulsions and features of raised intra-cranial tension may be seen. Though, some infants may show full and rapid recovery, infants with large hemorrhage may show residual signs.

C. Neurological Injuries

These include cranial or other nerve injuries and rarely spinal cord injuries.

Brachial plexus injury: This occurs in about 2 per 1,000 births.[8] The injury is usually due to excessive lateral traction when the anterior shoulder is trapped in shoulder dystocia, breech extraction or injudicious hyper-abduction of the neck in cephalic presentations. Associated injuries include fracture of clavicle or humerus, subluxation of cervical spine, etc.

Erb's palsy: This is the commonest brachial plexus injury. There is damage to the C5, C6 segments of the brachial plexus causing adduction and internal rotation of arm with pronation of forearm. The grasp reflex is normally maintained but Moro's reflex is asymmetric and sometimes biceps and radial reflexes are lost. The position of the hand is said to be a waiters tip deformity.

Klumpke's paralysis: This is rare and due to damage of the nerves of segmental origin C7, C8, T1 in the brachial plexus. Here there is loss of grasp reflex but biceps reflex is present. Ipsilateral Horner's syndrome may be seen if there is damage to sympathetic fibres of T1 with symptoms of ptosis, myosis and anhydrosis.

X-rays are often required to exclude associated fractures, but sometimes MRI scan, electromyography, nerve conduction studies may be needed for nerve function assessment. These conditions usually improve over a few months but long-term deficits may persist. To prevent contractures passive range of motion exercises and physiotherapy are needed. Neurosurgical exploration and microsurgical repair with nerve grafts should be considered if there is no improvement after 3–4 months.

Other nerve injuries

Facial nerve injury: The most common cause is use of forceps with excessive traction and rotational force but rarely facial nerve injury may occur *in utero* due to compression against maternal sacral promontory. The application of forceps for either traction or rotation might result in neuropraxia of one or both facial nerves, particularly if applied incorrectly.[9] Most common site of injury is its exit from stylomastoid foramen, mostly due to pressure of wrongly applied blades. Central damage to the facial nerve causes an asymmetrical face on crying, with smoothness of the affected side and drooping of the side of the mouth, whereas peripheral damage causes paralysis to the eye, forehead or mouth only. Most cases soon start to recover but full recovery may take 2–3 months. EMG as well as auditory brainstem response to rule out auditory nerve involvement is sometimes warranted. The eye must be protected with the use of artificial tears, ointment and/or eye-patch. If there is no improvement after 7–10 days, investigations may be required.

Phrenic nerve injury: Phrenic nerve damage is rare but has significant mortality. It is mostly associated with concurrent brachial plexus injury. Symptoms include respiratory distress and elevated hemidiaphragm is seen on X-ray. Management in supportive but may need NICU care. The neonate may require positive pressure ventilation and rarely surgical repair of diaphragm.

Laryngeal nerve injury: Symptoms include stridor, dysphagia, airway obstruction and aspiration. Laryngoscopy to exclude other causes should be done. Unilateral paresis usually recovers in 4–6 weeks but may take up to a year. Bilateral paresis has a guarded prognosis and usually needs tracheotomy.

Spinal cord injury: Spinal cord injuries though rare, are associated with severe morbidity or even stillbirth or early neonatal deaths after birth, due to an inability to breathe. The most common cause is breech delivery with difficult extraction of the after coming head or traction on neck in shoulder dystocia.[10] Such traction may result in fracture of odontoid process or fracture/dislocation of cervical vertebra where instant death can occur due to compression of medulla or damage to vertebral arteries causing ischemia. The non-lethal lesion may lead to temporary neuropraxia, or the child may develop spasticity over the years. Diagnosis can be confirmed by MRI or CT myelography. Management is mostly supportive.

D. Fractures and Dislocations

Clavicle: Fractured clavicle is the commonest fracture and is mostly caused by shoulder dystocia.[11] It is usually associated with no spontaneous movement of upper limb on the affected side. Diagnosis is made by palpation which may show crepitus and uneven contour of the bone which can be confirmed with an X-ray. One must also look for associated brachial plexus injury. Most clavicular fractures are greenstick and heal within 10–14 days with the arm immobilised.

Arm and leg bones: Fracture of long bones may occur rarely in difficult deliveries, most common being delivery of extended leg/arm in breech delivery or delivery of upper arm in deeply impacted head by modified Patwardhan's manoeuvre in cesarean section. Though cesarean section was postulated to reduce such pattern of injuries, many contrary reports have shown reverse incidence.[12] Most of them are mid-shaft fractures with absence of normal movement of the limb, and the swelling becomes apparent later. Confirmation can be done with an X-ray. Treatment is immobilization of the limb and recovery is usually quick and uneventful.

Depressed skull fractures: These are uncommon and cause a palpable step-off deformity which must be differentiated from cephalhematoma where CT scan may be useful for the diagnosis. Most common cause of such injury is difficult mid-cavity forceps in an un-rotated head. It may sometimes cause neurological deficit due to brain compression/damage. Surgical elevation may be needed.

E. Others

Visceral injuries: Injury followed to intra-abdominal organs like liver, kidneys, spleen and lungs can occur during breech delivery, causing internal hemorrhage, which can turn fatal. In such cases, neonate presents with pallor and a distended abdomen, possibly bluish in color with features of shock. Diagnose can be made by paracentesis or sonography. Surgical intervention with repair of viscera is warranted.

F. Soft Tissue Injuries

Facial/Buttock/Scrotal injuries: These are more common in face/brow and breech presentation. Repeated and forceful vaginal examinations, instrumental deliveries, manipulation during vaginal delivery can cause such injuries. Commonest sites are periorbital, ocular or facial tissue injury in face/brow presentation or scrotal injury during breech delivery. There can be subcutaneous edema which mostly resolves spontaneously within a few days. Also facial/gluteal/scrotal injuries can occur during artificial rupture of membranes with Kocher's clamp or while giving episiotomy. Dislocation of the cartilaginous part of nasal septum may occur and can be seen as a deviation at birth. Treatment is by manipulation.

Cuts and abrasions: These may occur during cutting the last layer of the uterus during cesarean section or rarely while giving episiotomy. Scalp abrasions may occur during instrumentation too. Cuts usually heal by dressing but sometimes need suturing.

Subcutaneous fat necrosis: SFN though rare, is one of the delayed complications of difficult deliveries. It manifests as hard, subcutaneous nodules with overlying reddish-purple discoloration. No treatment is required as it spontaneously resolves.

Other soft tissue/muscle injuries: Rarely soft tissue hematomas may develop due to birth trauma and may lead to hyperbilirubinemia if extensive. Damage to muscle fibres and blood vessels of sternomastoid is known entity following difficult breech delivery or shoulder dystocia, which leads to hematoma formation. Contracture may lead to torticollis.

Prediction and Prevention of Birth Injuries

Preventive health care in general includes interventions done at various levels for reducing risks or threats to our well-being. It outlines primary, secondary and tertiary levels of preventive activities done to address a particular health issue. So, for preventing birth injuries, interventions are needed right from peri-conceptional period to antenatal healthcare and most importantly managing difficult labor in a tertiary health centre.

Evaluation of Risk Factors

A detailed history of prior antenatal or intra-partum complications is important like:
- Prior GDM
- Prior h/o macrosomia
- Prior h/o difficult labor/shoulder dystocia/ instrumentation
- Prior h/o preterm birth
  Also a detailed history and investigation of current health status:
- Obesity

- Maternal height
- Pre-existing or gestational diabetes mellitus, etc.

Antenatal Care

In our country where preconceptional counseling is still not available to all, the antenatal period is our best chance of screening for risk factors for difficult deliveries.
- Excessive maternal weight gain
- Second trimester deranged glucose tolerance test
- Abnormal pelvis—associated with polio, spine deformities
- Cephalopelvic disproportion—clinical pelvimetry done at least once by a senior obstetrician
- Postdatism
- Routine ultrasound to rule out congenital anomalies, macrosomia, abnormal presentation, oligohydramnios, etc.
- Macrosomia as we have established has direct correlation in developing labor complications and thus birth injuries. Macrosomia should be expected in mothers with excessive weight gain or pre-existing obesity, GDM, postmaturity or prior history. Clinical fetal weight evaluation by fundal height palpation can be done. Ultrasound estimation of fetal weight at term is important.

Intrapartum Management

- Proper evaluation and selection of women for vaginal delivery in early in labor is important. This includes evaluation of risk factors and estimated fetal weight/ presentation and maternal pelvis.
- Patients with obvious risk would benefit from an elective/selective cesarean section.
- High-risk patients should be referred and managed at higher center with neonatal intensive care set up.

- Avoid prolonged labor. Use Partogram.
- While anticipating a difficult delivery, call for help in conducting delivery or instrumentation.
- Complicated deliveries and instrumental deliveries to be handled by skilled midwives and experienced obstetricians.
- Not to apply too much traction or external force while delivering the baby.
- Training of staff and resident doctors in assessing high risk situations and actual handling of difficult cases doing mock drills and simulation training on mannequins is extremely important in keeping them prepared to manage cases with least risk of creating complications.
- An experienced neonatologist should be on standby in difficult deliveries as early diagnosis and prompt management of the injured newborns is required.
- Maintain documentation. Explain the relatives regarding the high risk and its associated complications that could be anticipated.

SUMMARY AND CONCLUSION

Birth injuries are important cause of perinatal and neonatal morbidity and occasional mortality. There are definite risk factors, antenatal and intranatal, which increases the risk of birth injuries. Predicting the risk and timely intervention by skilled personnel is important in avoiding such injuries. Although one cannot completely eliminate complications like birth injuries, with identification of the risk factors and appropriate management, incidence of such complications can be greatly decreased.

REFERENCES

1. Demissie K, Rhoads GG, Smulian JC, et al. Operative vaginal delivery and neonatal adverse outcomes; population based retrospective analysis. BMJ 2004;329:24.
2. Warke C, Malik S, Chokhandre M, Saboo A. Birth Injuries-A Review of Incidence, Perinatal Risk Factors and Outcome. Bombay Hosp J. 2012; 54 (2): 202–8.
3. Boulet SL, Alexander GR et al, macrosomic births in the United States: determinants, outcomes and proposed grades of risk. Am J ObstetGynecol 2003; 188:1372.
4. Patel RR, Murphy DJ. Forceps delivery in modern obstetric practice. BMJ. 2004 May 29. 328(7451): 1302–5
5. Linder N, Linder I, Fridman E, Kouadio F, Lubin D, Merlob P, et al. Birth trauma - Risk factors and short-term neonatal outcome. J MaternFetal Neonatal Med 2013;26:1491–5.
6. Chang HY, Peng CC et al; Neonatal subgaleal haemorrhage: clinical presentation, treatment and predictors of poor prognosis. Peaditr Int. 2007 dec;49(6):903–7.
7. Ekeus C, Hogberg U, Norman M. Vacuum assisted birth and risk for cerebral complications in term newborn infants: A population-based cohort study. BMC Pregnancy Childbirth 2014; 14:36.
8. Malessy MJ, Pondaag W; Obstetric brachial plexus injuries. NeurosurgClin N Am. 2009 Jan;20(1):1–14.
9. Laing JH, Harrison DH, Jones BM, Laing GJ. Is permanent congenital facial palsy caused by birth trauma? Arch Dis Child 1996 Jan; 74(1): 56–58.
10. Stem WE, Rand RW. Birth Injuries to the spinal cord. A report of two cases and review of the literature. Am J Obstet Gynecol 1959; 78: 498–512.
11. Hsu TY, Hung FC, Lu YJ, Ou CY, Roan CJ, Kung FT, Changchien CC, Chang SY. Neonatal clavicular fracture: clinical analysis of incidence, predisposing factors, diagnosis, and outcome. Am J Perinatol. 2002 Jan;19(1):17–21.
12. Toker A, Perry ZH, Cohen E, Krymko H. Cesarean section and the risk of fractured femur. IMAJ 2009; 11:416–418.

Congenital Malformation Diagnosed *in utero*—The Way Forward

Faram Irani, Ness F Irani

While doing an anomaly scan a lot of questions arise, whether it is normal, is it a normal variant or is it abnormal. In this chapter we have tried to dispel some of those doubts to the best of our ability.

PLACENTA

Placenta previa: If placenta reaches the internal os or overlies it.

Low lying placenta: Placental edge is within 2 cm of the internal os and does not overlie it.

Normal: Placental edge is 2 cm or more from the internal os.

CORD INSERTION

Marginal: Cord insertion less than 2 cm from the placental edge.

Velamentous: Cord insertion at a short distance away from the placental edge.

Vasa previa: Fetal vessels coursing the membranes (either velamentous or connecting two placental lobes) within 2 cm of the internal os.

BIOMETRY

- *Head measurements:* Biparietal diameter, occipitofrontal diameter and head circumference are measured in an axial plane where the CSP, both the thalami and the falx are seen.

- *Abdominal circumference:* An axial image at the level of the stomach, the umbilical vein joining the portal vein and a single rib is taken and the AP and transverse diameters are measured or AC is measured using an elliptical trace.

- *Femur length:* The femur should be as perpendicular to the ultrasound beam as possible and it is measured placing callipers at ends of the ossified diaphysis (femoral epiphysis which are the linear projection from the ends of diaphysis should not be included).

BRAIN

Routinely 3 axial planes are taken

- *Transventricular plane* (Fig. 21.1): Structures identified are the frontal horns and the

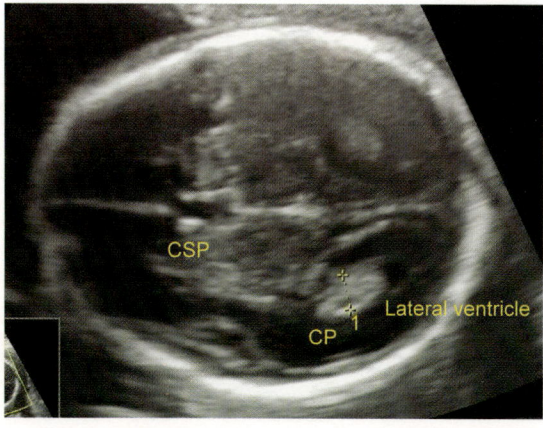

Fig. 21.1: Transventricular plane

posterior horns of the lateral ventricle, the cavum septum pellucidum (always visualised when the BPD is more than 44 mm), the choroid plexus within the lateral ventricles.

- *Transcerebellar plane:* CSP, thalami, cerebellum and cisterna magna are identified (Fig. 21.2).
- *Transthalamic plane:* It is the plane where the HC is measured (Fig. 21.3).

ABNORMALITIES SEEN IN THE BRAIN WITH THE ABOVE 3 VIEWS

- *Ventriculomegaly:* Measurement of the posterior horn is done, if it is more than 10,

it is considered as ventriculomegaly. 10–15 mm is mild and more than 15 mm is severe ventriculomegaly.

- *Anencephaly:* Absence of cranial vault.
- *Open spina bifida:* Obliteration of the cistern magna, abnormal shape of the cerebellum (banana sign), scalloping of the frontal bones (lemon sign) and ventriculomegaly.
- *Cephalocele:* The protrusion of intracranial contents through a bony defect in the skull.
- *Holoprosencephaly:*

Alobar: The interhemispheric fissure and the falx cerebri are absent, single ventricle and thalami are fused in the midline.

Semilobar: The 2 cerebral hemispheres are seperated posteriorly partially, but there is still a single ventricular cavity.

Lobar: The interhemispheric fissure is well developed, but the csp is absent.

Agenesis of corpus callosum: There is absence of csp, lateral separation of the frontal horns, elevation of the 3rd ventricle and dilatation of the posterior horn (colpocephaly)

Posterior fossa abnormalities: Dandy-Walker malformation: Splaying of the cerebellar hemispheres, communication of the 4th ventricle with the posterior fossa and ventriculomegaly may be present.

Megacisterna magna: Cisterna magna measuring more than 10 mm in the cerebellar plane.

Vermian dysgenesis: Key hole appearance of the 4th ventricle in the axial plane.

Blakepouch cyst: Cyst with keyhole appearance in axial plane but midsagittal view showing normal size and anatomy of the vermis and normal rotation.

Intracranial hemorrhage: There are various grades, initially it is hyperechoic and gradually the blood clot retracts and contains hypoechoic components. Ventriculomegaly may be present.

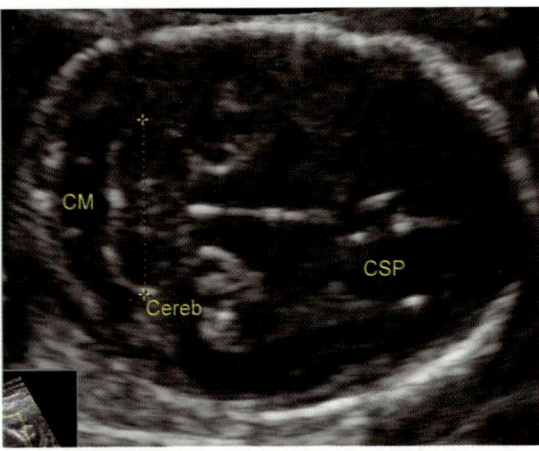

Fig. 21.2: Transcerebellar view

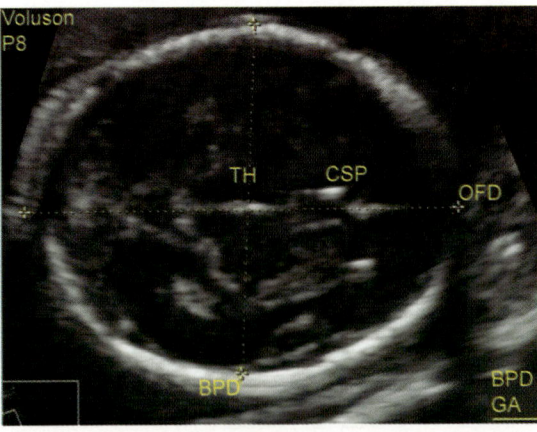

Fig. 21.3: Transthalamic view

Microcephaly: A small head with abnormal brain development considered if hc is less than 2 sd or more below the mean for gestational age.

Choroid plexus cysts: Round sonolucent structures present within the choroid plexus of the lateral ventricles.

FETAL FACE

- *Nose and lips* (Fig. 21.4): Standard view of imaging is the angled coronal nose mouth view. If a cleft lip is present, further views (axial and coronal) of the maxilla are necessary to see for cleft palate.
- *Profile view*(Fig. 21.5): Its a mid-sagittal view where the nasal bone, maxilla and the mandible are visualised. Nasal bone length is measured, midfacial anomalies and small mandible are the anomalies which can be seen in this view.
- *Mandible measurements:* On axial view of the mandible at the level of the temporo-mandibular joint, a symmetric triangle is formed by the mandibular arms. The transverse and AP diameters are measured (inner to inner point).
- *Eyes:* The standard view to measure is the axial orbital view (Fig. 21.6). Binocular, interocular and orbital diameters can be measured and there are normograms for the same to diagnose hypotelorism and hypertelorism and micropthalmia. Lens can be seen too.

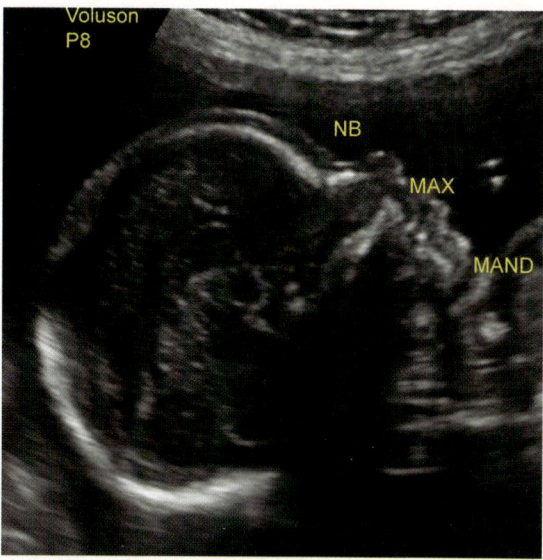

Fig. 21.5: Profile view

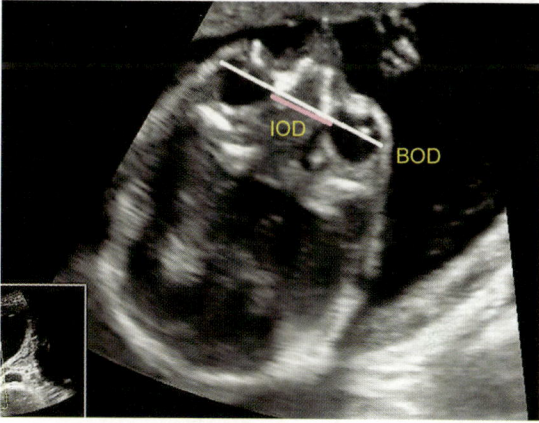

Fig. 21.6: Orbits

- *Ears:* Ear length can be measured in sagittal or coronal planes and normograms are there for the same.

FETAL SKELETAL SYSTEM

- When measuring the femur, if it falls in the 5% or below, all long bones must be measured.
- If bones are short, what segments are involved. Proximal shortening (humerus, femur) is rhizomelia. Middle segment shortening (radius, ulna, tibia, fibula) is

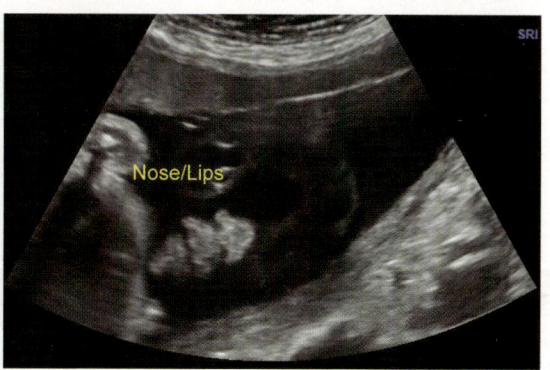

Fig. 21.4: Nose and lips

mesomelia. All segments shortening is micromelia.

- Parents height should be measured so as to not mistaken constitutional short stature.
- Various measurements and ratios should be taken to assist in the diagnosis and determine lethality.
  1. Degree of limb shortening <4 SD is always lethal.
  2. Thoracic circumference(TC), below 5% is suggestive of pulmonary hypoplasia.
  3. Femur/foot length, 1/1 ratio is normal, <0.8 is suggestive of skeletal dysplasia.
  4. FL/AC ratio less than 0.16 is suggestive of lethal skeletal dysplasia.
  5. Cardiac circumference/thoracic circumference more than 0.6 suggestive of narrow thorax.
- To look for bone morphology, whether they are fractured, underossified or angulated.
- Is the skull unusually shaped.
- Is there fusion of digits(syndactyly) or extra digits (polydactyly).
- Are there other structural anomalies, like facial clefts, genitourinary anomalies and cardiac defects which may give clues regarding diagnosis.

FETAL HEART

- Abdominal situs
- 4-chamber view (Fig. 21.7)

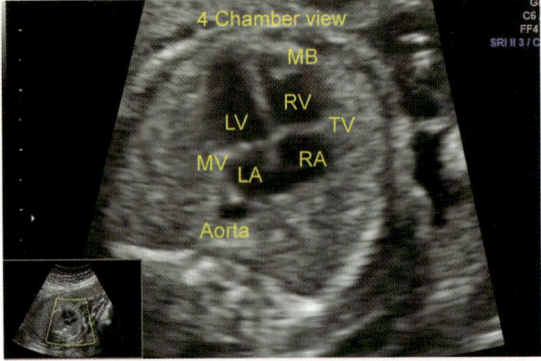

Fig. 21.7: Four-chamber view

- Left ventricular outflow tract (Fig. 21.8)
- Right ventricular outflow tract (Fig. 21.9)
- 3-vessel view and the 3-vessel trachea view

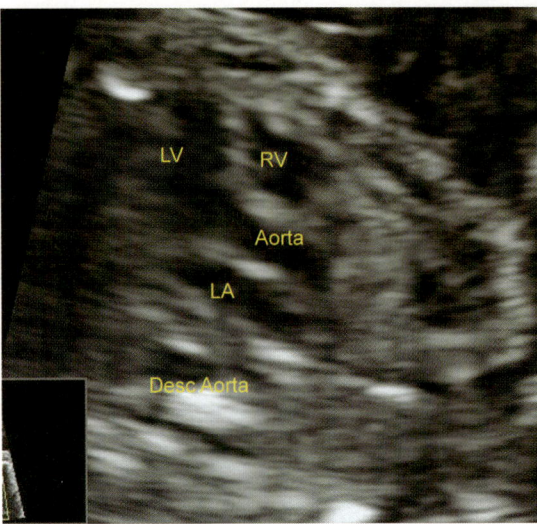

Fig. 21.8: Left ventricular outflow tract

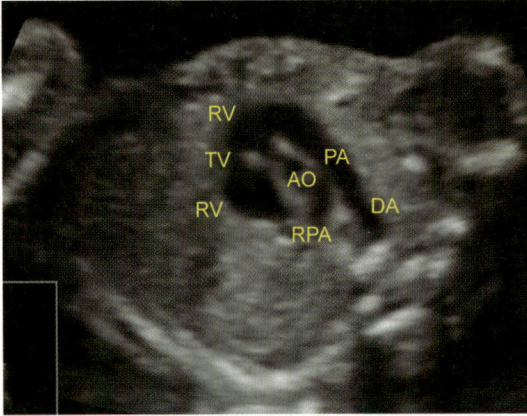

Fig. 21.9: Right ventricular outflow tract

Abdominal Situs

Obtain a transverse view of the fetal abdomen, where the stomach is to the left side, the descending aorta is posterior and to the left and the IVC is anterior and to the right. With the intrahepatic portion of the umbilical vein connects with the left portal vein, with an

L-shape to the right. Three types of visceral situs exists:

a. *Situs solitus:* Normal arrangement of organs and vessels.

b. *Situs inversus:* Mirror arrangement of vessels and organs to situs solitus.

c. *Situs ambiguous:* Heterotaxy, which is right or left isomerism.

4-Chamber View (Fig. 21.7)

a. *Cardiac axis:* The heart lies in the left chest with axis approximately at 45 degrees (range 30–60), if the axis is abnormal one must determine why.

b. *Heart size:* The cardiothoracic circumference is normally 0.55 (+/− 0.05). An increase in the same is due to cardiac causes such as ebsteins, fetal arrhythmias and dilated cardiomyopathy or extracardiac causes such as AV malformations, TTTS and severe fetal anemia.

c. *4-chamber symmetry:*

1. The 2 atria are generally equal in size, a small left atrium is seen in TAPV (total anomalous pulmonary venous drainage) and a large right atrium is seen in ebsteins and tricuspid dysplasia.

2. The 2 ventricles are usually equal in size (though in advancing gestation the right appears larger than the left). A small left ventricle can occur due to coarctation of aorta, hypoplastic left heart syndrome and TAPV, while a small right ventricle can occur in tricuspid atresia and pulmonary atresia.

d. *Atrioventricular valves:* The tricuspid valve is apically displaced compared to the mitral valve. If the mitral valve is apically displaced and the moderator band is seen in the "left" ventricle, then corrected transposition of great arteries is to be suspected. If there is a single common AV valve draining into two ventricles, then AVSD is suspected.

3-Vessel Trachea View (Fig. 21.10)

- *One great vessel is seen:* If it is normal size, then it is the aorta in transposition of great arteries, if the size of the vessel is enlarged, then it is the aorta in pulmonary atresia, the pulmonary artery in hypoplastic left heart syndrome or the common arterial trunk.
- *Dilated aortic arch:* In aortic valve stenosis. *Dilated pulmonary artery:* TOF with absent pulmonary valve syndrome and isolated pulmonary valve stenosis. *Narrow or absent aortic arch:* Coarctation of aorta or hypoplastic left heart syndrome. *Narrow or absent pulmonary artery:* Pulmonary stenosis or atresia. *4 vessels:* Presence of a left superior vena cava, where the vessel is located to the left of the pulmonary artery.

GASTROINTESTINAL SYSTEM

- Abdominal circumference is measured and a small AC is associated with fetal growth restriction (Fig. 21.11).
- Abdominal wall defects are ruled out in a majority of cases if normal cord insertion is seen.
- Extrusion of abdominal contents without a covering of a membrane is gastroschisis.

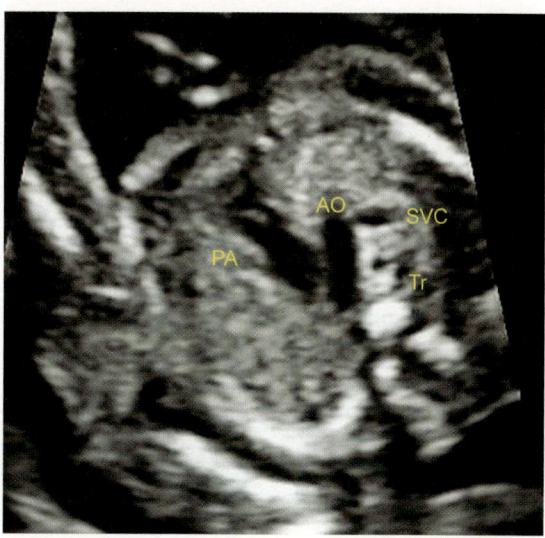

Fig. 21.10: Three-vessel trachea view

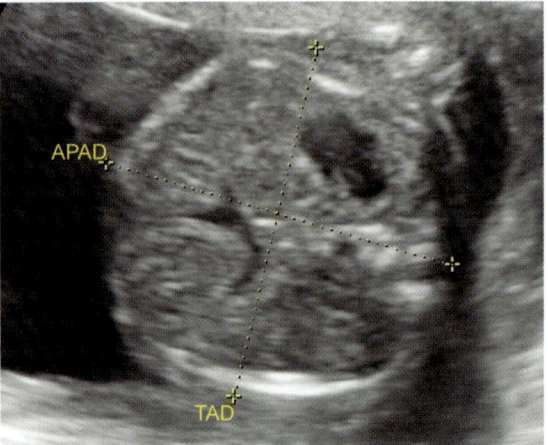

Fig. 21.11: Abdominal circumference

Most commonly small bowel is extruded, but liver, stomach and large bowel may be present. Chromosomal abnormalities are seen in gastroschisis, but are more common in omphalocele. Liver extrusion is uncommon but carries worse prognosis.

• When the extrusion of the bowel is at the base of the umbilical cord it is termed omphalocele. Small bowel only containing omphaloceles are at higher risk for chromosomal abnormalities.

• Suprapubic mass with absent bladder may be seen in bladder or cloacal extrophy.
• Unusual abdomino or thoracoschisis is seen in body stalk anomaly.
• The stomach whether it is normal, absent or small. If it is absent (after 2 or more examination), it is seen in cases of esophageal atresia. If it is small it could be due to tracheoesophageal fistula. Both the above would have some sort of poly-hydramnios especially in the 3rd trimester.
• Double bubble sign is seen commonly in duodenal atresia.
• Echogenic bowel is only considered if the bowel is as echogenic as the bone. It is seen in IUGR, infections, cystic fibrosis, aneuploidies, precursor of a bowel atresia and if there is some bleeding, due to the ingestion of blood.
• Cystic masses in the abdomen can be ovarian cysts, dilated bowel loops, meconium pseudocysts and a persistant cloaca.

GENITOURINARY SYSTEM

• When both kidneys are absent there will be anhydramnios 2nd trimester onwards

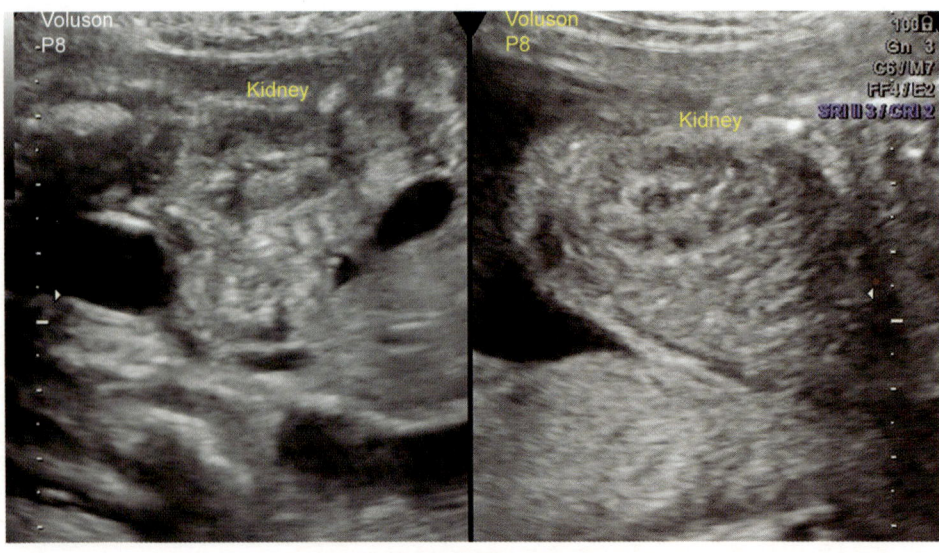

Fig. 21.12: Kidneys sagittal view

(liquor is formed majorly by the fetal urine only after 16 weeks). Also bladder will not be seen and bilateral "lying down" adrenal glands will be seen.

- Renal fossa must be carefully examined to look for both kidneys (Fig. 21.12). If only one kidney is seen then one must look for unilateral agenesis, crossed fused ectopia or pelvic kidney.

- Renal pelvis must be seen, whether it is prominent and more than 4 mm at time of anomaly scan. Also one must check for renal calyceal dilatation.

- One must also see for ureters whether they are dilated (reflux, obstruction or megaureter)

- Hypoechoic renal cysts, they do not connect with each other or the renal pelvis, these are seen in multicystic dysplastic kidneys.

- Increased echogenicity may be a normal variant or seen in polycystic kidney disease or aneuploidies.

- Increased renal size is seen in certain syndromes, compensatory hypertrophy,

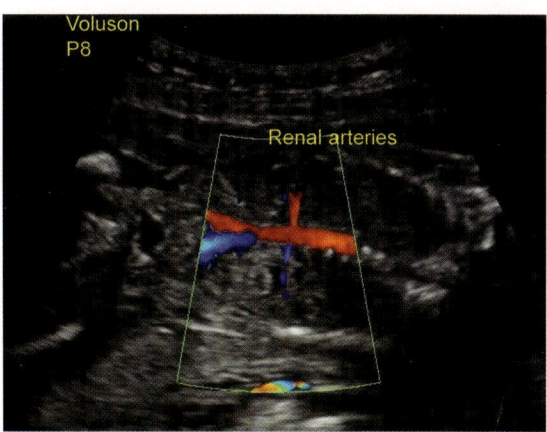

Fig. 21.13: Renal arteries

tumors or duplication of renal collecting system.

- Absent bladder is seen due to bilateral renal anomalies, leading to absence of urine production. It is also seen in donor twin of TTTS and bladder extrophy.

- If bladder is distended it is seen in lower urinary tract obstruction.

- Any suprarenal mass could be a neuroblastoma, adrenal hemorrhage or extralobar sequestration.

Baby in NICU—
Confusion in Communication

Manohar Motwani, Nanak Bhagat

A healthy baby who cries, breathes well and breastfeeds is the expected outcome of every pregnancy/delivery. The parents, family and even the doctors are anticipating this happy event. However, when things do not go well— the baby does not cry, has problems in breathing or feeding, and needs to be shifted to the NICU, problems begin.

Problems the baby in NICU presents are twofold: Medical issues which are relatively easy to tackle, and communicating with the parents and the relations which we, as doctors, find difficult.

Though communication is an essential and integral part of our day-to-day life, we are not trained to 'communicate' with patients well as a part of our syllabus in medical school.

Communication (from Latin commúnicáre, meaning "to share") is the act of conveying intended meaning to another entity through the use of mutually understood signs and semiotic rules. The basic steps of communication are the forming of communicative intent, message composition, message encoding, transmission of signal, reception of signal, message decoding and finally interpretation of the message by the recipient.[1] Communication could be verbal or non-verbal, through signals or gestures.

In effect, communication involves the Sender who composes a message which is transmitted verbally or non-verbally to the

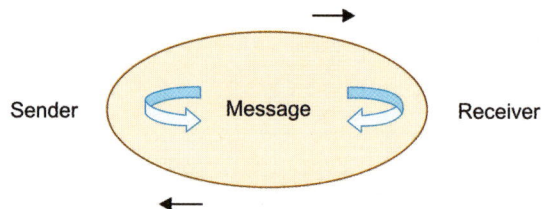

Fig. 22.1: The communication loop

Receiver who accepts the message, understands it and then acknowledges it, thus completing the communication loop.

Today, in a field like medicine, in the litigation-philic society where medico-legal problems occur at the drop of a hat, effective communication becomes all the more essential. Since the situation in the NICU is highly volatile, effective communication becomes imperative and the key issue. Treating the sick newborn is probably the easiest of the thing that we doctors in the NICU do.

The problem in NICU is complex because of the circumstances: The event was not anticipated; the parents and family cannot comprehend why things went wrong and are unable to come to terms with the event. The situation is unacceptable to them: Parents and relations are highly stressed, tempers run high and the relations are ready to pin the blame onto the doctor—usually the obstetrician—the single point of contact for the care provided till that moment. The situation is worsened because of the additional burden of the unanticipated cost of the treatment, uncertain prognosis and the doubt of perceived negligence. All of this may add up, and reduce the efficiency and standard of care provided.

The only way out of the situation is effective communication, not just between the doctors and the baby's relatives but also amongst the doctors and the staff of NICU.

In this chapter we have made an attempt to rationalize this confusion.

SET-UP IN THE NICU

Parties involved in treating the baby in the NICU are:

- *Obstetrician:* The primary caregiver who transfers the baby
- *Neonatologist/pediatrician:* Primary physician who looks after the baby, and
- *Staff:* Who actually carry out the orders.

Management in the NICU involves effective communication at three levels (Fig. 22.2):

1. Amongst the staff in the NICU including nurses, Ayabais and doctors (senior and junior)
2. Between the referring obstetrician and the consultant neonatologist, and
3. Between the treating neonatologist and the baby's relatives.

CAUSES OF CONFUSION IN COMMUNICATION

In the NICU, common reasons leading to gaps in communication are enumerated in Table 22.1.

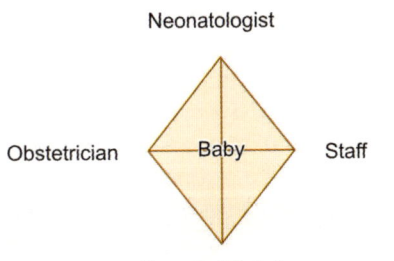

Fig. 22.2: The communication rhomboid in the NICU

In order to prevent problems, NICU should work as a coherent unit. This can happen only if all the cadres communicate well with each other and one another: The housekeeping staff, nurses, junior doctors and consultants. It starts with organizing the structure of the NICU.

We start with the people already working in the NICU, and those who will be hired.

1. Two basic requirements for all the staff members are proper qualification and the ability to communicate.
2. When employing people to work in an NICU, it should be mandatory to have an induction to the workings of the department, as well as certain qualifications like certification in basic neonatal resuscitation, etc.
3. Once working in the NICU, it should be a mandatory requirement to continue to keep up with CMEs/drills regarding important topics, so that the whole staff learns and has a better understanding of what is being done at the workplace.
4. *Delegation of duties and teamwork:* As a result, the work gets done in a more smooth fashion. The best method to inculcate team spirit is by mock drills and training sessions.
5. Use of protocols/mock drills/training
6. Periodic assessments
7. Regular staff meetings

Medical problems in the NICU are 'real' and are tackled by the doctors on their merit, but most problems in the NICU are 'perceived', and are a result of improper understanding/comprehension by the relatives of the baby. The solution lies in using effective communication. These problems will decrease if the parents and relatives are mentally prepared for possible problems/outcomes and aware of the process. The counselling should happen not just at that moment, but much earlier before the delivery.

Table 22.1	Causes of problems in communication	
Doctors	*Staff (Nurses, Ayabais)*	*Relations*
• Availability (senior consultant not always available)	• Absence of protocols	• Situation not anticipated
• Consistency (the same person should be counselling the relatives)	• No established chain of commands	• Shock and disbelief at the condition of the baby
• Delay in the decision-making process (unavailability of doctors and reports in time, etc.)	• Unclear orders	• Problems in understanding/comprehensive
• No daily plan of action and realistic prognostication	• Uncoordinated teamwork	• Unrealistic expectations and demands
• No timely and established liasions with other departments	• Untrained staff	• Unexpected expenditure

ANTENATAL COUNSELING (Fig. 22.3)

Since it is difficult to predict which baby would be admitted to the NICU after birth, effective counseling should start in the antenatal period. We prefer to do this at around 28 weeks in an interactive group session. The benefits would be huge if the husband and the mother in law could be a part of this group counseling session. The obstetrician should introduce this concept of possible neonatal transfer to NICU a little before the time of birth and preferably introduce the neonatologist to the expectant mother/family as the doctor who will look after their baby after birth. Broadly, possible causes of transfer would be classified as: Less than normal birth weight, less than term birth age, problems at birth like did not cry or presence of meconium and congenital defects/anomalies. High risk situations which warrant a neonatal transfer should be enumerated (Table 22.2).

COUNSELING AT THE TIME OF BIRTH, IN THE LABOR ROOM

Every delivery should be supervised. Every birth attendant—be it a doctor, nurse or a trained birth attendant—depending on the infrastructure of the maternity service, should be well trained in neonatal resuscitation. Presence of a neonatologist at the time of delivery is always welcome, but may not be feasible in our country. The relatives should be counseled by the birth attendant about the condition of the baby, possible reasons for the condition, probable course of action and prognosis. The counseling should be done jointly by the obstetrician and the neonatologist (or the seniormost birth attendant). It should be realistic, honest and crisp. Try to include not just the parents but also important relations from the crowd, especially the leaders or potential trouble makers.

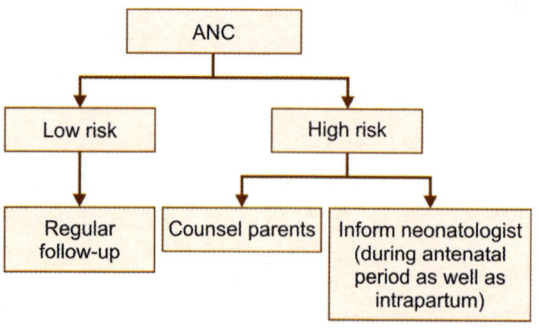

Fig. 22.3: ANC counseling

Table 22.2		Possible causes of neonatal transfer	
Antenatal		*Intranatal*	*Postnatal*
Fetal	*Maternal*		
• Prematurity	• GDM	• Prolonged labor	• Asphyxia
• Anomaly	• Elderly	• Abnormal labor	• Tachypnea
• IUGR	• Post-term	• PROM, PPROM	• Poor suck/failure to feed
• Anemia		• Fetal distress	• Hypoglycemia
• Macrosomic IDM		• Instrumental delivery	• Convulsions
		• CS following a failed instrumental delivery	• Polycythemia
		• Traumatic delivery	

COUNSELING IN THE NICU

The next area where communication is most essential is the NICU itself.

The most important time is the initial contact between the neonatologist and the relatives/parent. Once a strong foundation has been laid, relation based on trust develops which paves the way to minimize problems in the future.

- Seniormost doctor should be communicating with the parents/relatives.
- Preferably, the same person should be communicating, to maintain continuity. If (s)he cannot attend the counseling session, the duty should be delegated to a competent substitute who should have been introduced to the relations earlier by the first doctor.

This will help in providing a uniform statement when talking to any relative, and also helps everyone remain on the same page, reducing errors borne out of any misunderstandings and miscommunications that may arise.

- Counselling should be done at a specific predetermined time and in a specific place, preferably the office.
- The session should be attended by the same set of relations.
- *Each counselling session should be documented:* A format/protocol used for documenting these counselling sessions may be found as Appendices at the end of this chapter.
- Each session should include the following.

Initial counseling: 1st consultation
- Introduction, establishing relationships
- Current condition of the baby
- Vitals
- Any change in condition (for the better or worse) during transfer
- Baby weight
- Risk assessment
- Reason for admission (if not already mentioned earlier, or in greater detail if briefly mentioned earlier)
- *Plan of action:* Investigations, specialized consultations, medications, procedures
- *Prognosis:* Short-term and long-term
- Possible complications and outcomes with statistical data
- Estimated cost to be borne, additional costs that might be borne depending on the case.
- Always ask for any concerns and questions that the relatives might have, thus checking their understanding of the situation
- Confirm their understanding/comprehension.
- NEVER give false assurances and hopes.
- Communicate in a language the relative understands.
- Let at least one relation see the baby
- Always maintain a policy of full disclosure. Remember that the baby belongs to the parents and they are entitled to complete case records by Law.
- *Transparency:* Do not hide facts. Discuss the important investigation results and the

change in plan of treatment in view of the results, more so in case of complications. At no stage the relatives should feel that you are trying to suppress facts.

- In the litigation-philic society where 'distrust' is the driving force, mishaps/accidents are not acceptable by patients and the doctor is blamed for everything that goes wrong. Every counseling session must be documented (Appendix 1 and Appendix 2) and may be recorded on camera.

Follow-up counseling
- This should be done at least once or twice a day, after rounds
- To be done by the same doctor, to the same relatives
- Extra session could be called for in case of a serious problem or drastic change in condition of the baby: Worsening of condition or addition of any expensive medication/investigation/therapy modality.

Fresh consent to be taken at such time
- Session should include
- Condition of the baby
- Progress since the last update
- Feeding
- Review of the investigations/specialist consultations
- Any additional investigations or procedures to be done
- Relative cost that might be incurred due to the above
- Reiterate short-term goals, review prognosis
- Encourage the relatives to ask questions, answer truthfully
- If the baby turns serious, do not hide facts, do not give false hopes, go through the reasons leading to this condition and do not push the blame to colleagues or subordinates. Offer 2nd opinion from a senior colleague and discuss transfer to a tertiary care/higher center. Do not compromise on investigations and expert opinions.

- Check their understanding.
- Update the referring obstetrician about not only the condition of the baby but also the relative's responses.
- Loose talk between the staff and junior doctors should never be allowed.
- Apart from this, other factors which promote effective communication in the NICU (Fig. 22.4) are:
  i. *Uniformity:* In appearance, method and content of communication
  ii. *Body language:* Most of the times, it is not what you say that has the most impact, rather HOW you say it that makes all the difference. Whether it is breaking bad news or just giving an update, it is the way it is put across that determines the opposite person's response, and probably what causes most lawsuits to be filed against people from the profession.
  iii. Ease of communication between the doctors, staff.
  iv. *Work ethics:* While working, there should be a strict code of conduct to be followed at work, the most important being the behavior at the workplace which includes NOT playing a blame game, NOT having loose talk between staff including doctors, doing the assigned work and working as a part of the team.
  v. *Chain of command:* To be defined, thus minimising the gaps in communication.
  vi. Staff members should not be allowed to update the relatives about the condition of the baby—that is the job of the doctor.

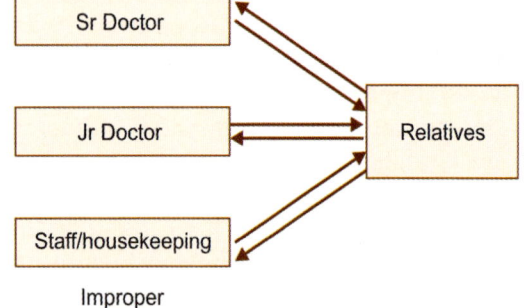

Fig. 22.4: Communication in the NICU

Their counseling relates to proper feeding and caring techniques to be taught to the mother, and should remain confined to those areas only.

vii. When any staff member, including a doctor finds an abnormality in the baby's vitals and wants to escalate the treatment, it is best to have a common protocol for informing the relatives—thus standardising the information given and reducing any disparity and omissions which might occur otherwise and prove detrimental later.

In such cases, it is best to follow the SBAR format.

S—Situation

B—Background

A—Assessment

R—Recommendation

There is a dual advantage in using the SBAR format. First, it helps standardise the way of informing and communicating essential information which everyone is familiar in receiving. Second, it gives the person an opportunity to learn and think, and at the same time not commit mistakes because of faulty communication.

(For example, if a baby has not passed stools for the past 6 hours, and you are worried and inform the senior, the pattern of informing should be such:

S—Situation: Baby of ABC, D3 of birth, has not passed stools over the last 6 hours.

B—Background: He/she was term/preterm, admitted for observation on D0 birth. Vitals are stable, abdominal distension is noted.

A—Assessment: Suspecting a probable case of intestinal obstruction

R—Recommendation: Intend passing an NG tube to decompress the intestine, and confirm with a USG. Is there anything else that should be done?).

BEREAVEMENT COUNSELING

Sometimes, we do have adverse results, and that unfortunately means the loss of a life. It is a sad time for the parents, and they are emotionally not very stable. Understanding that and trying to help them come to terms with it and provide a form of closure is important, both on humanitarian grounds as well as medicolegally.

CONCLUSION

NICU is a super charged high-risk area where problems are bound to occur and outcomes are unpredictable (Figs 22.5 to 22.7). However, some of these problems may escalate and end up in a medico-legal suit in the future. The most effective method of preventing these complications is nothing else but excellent communication.

For communication to be most effective, a channel should be established early in pregnancy, rapport established with the patient and the relations, effective counseling done for all pregnancies—especially the high risk cases, to outline not only the management plans but also to discuss strategies in case of complications and possible neonatal transfer to NICU.

If the baby is shifted to the NICU, the relatives should be handled gently with all the empathy, keeping all the principles of effective

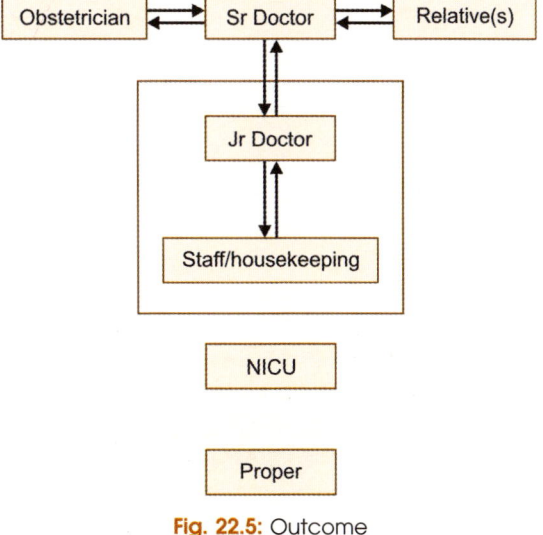

Fig. 22.5: Outcome

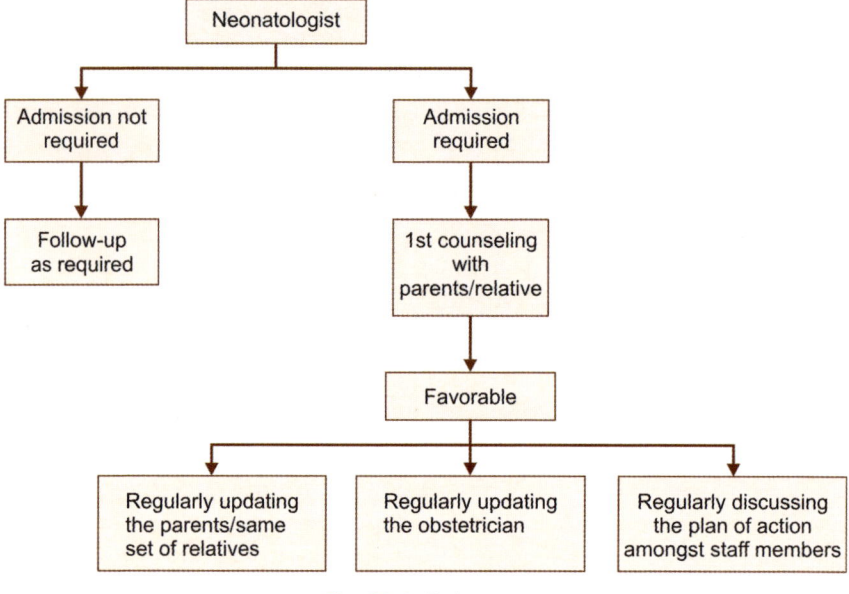

Fig. 22.6: Outcome

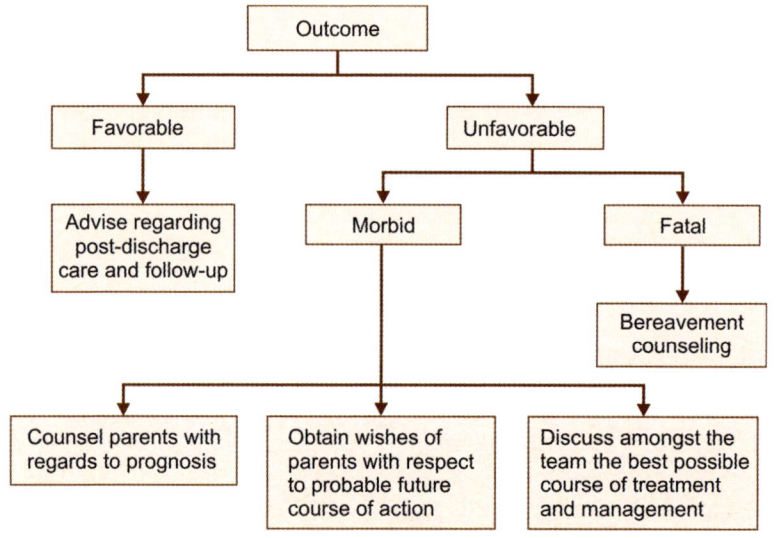

Fig. 22.7: Outcome

communication always in mind. While communicating with the baby's relatives, always be honest, never lie, disclose all the facts without prejudice, be open to suggestions and criticism, accept your limitations and welcome a 2nd opinion whenever required. Remember, you are dealing with parents of a very sick neonate who was to bring a lot of joy to the family. Since the outcomes are not always predictable and may not always be favorable, the least you can do is empathise with the relatives. Though the outcomes are not in your hands, proper communication in the NICU will go a long way and eliminate or definitely minimize some medico-legal problems in future.

REFERENCES

1. Wikipedia

2. Acute care toolkit 6: the medical patient at risk: recognition and care of the seriously ill or deteriorating medical patient (PDF). Royal College of Physicians of London. May 2001.

Appendix 1

Initial Assessment: High Risk Factors

Baby of ....   Gender ....   Age ....

Name of Parents/Father ....

Mother ....

Brought by   ....   Relation ....   History given by...................

Details of Delivery

Date of Birth .... Time ....

Place of Delivery ....   Mode of Delivery : Normal/vacuum/forceps/CS

Apgar Score ....

Risk Factor Assessment   Date........

- Low/very low/extremely low birth weight
- Less than term gestational age
- Small/ large for gestational age
- Male gender
- Congenital malformations
- Low min Apgar score
- Below normal temperature on admission
- Use of respiratory support
- Poor respiratory status
- Maternal risk factors/antenatal conditions
  - GDM/ PIH/anemia/hypothyroidism/other diseases
  - IUGR/Polyhydramnios/Macrosomia
  - Use of drugs during pregnancy:

Risk factors added after admission to the NICU

-
-

Signatures: Parents ....   Relative/ Witness ....   Relation ....

Baby's Footprint

Signature of the Parent/ Relative

Appendix 2
Counseling Sheet

Name of the Patient ....   Age ....   Sex ....

Admission ....   Date ....   Time ....   Day .... of Admission

Admitted by   Self/ Relation
                     Referring Doctor
                     Transferred from

Name of the Parent/ Guardian present

1. ....   Relation
2. ....   Relation
3. ....   Relation

Counseling done by   Dr
                            Date ....   Time .... to ....

Language of counseling ....

 Points covered during counseling session
• Diagnosis: Provisional/final
• Reason for transfer/admission
• Condition of the baby
• Plan of action: Investigations, specialized consultations, medications, procedures
• Prognosis: Short-term and long-term
• Expected hospital stay
• Possible complications and outcomes
• Estimated cost to be borne, additional costs that might be borne depending on the case.
• Notes/additional points

Dr .... has informed us about the condition of our child .... I/We understand the information provided by the Doctor and the condition of our child. Our concerns and questions have been adequately addressed by the Doctor.

We also understand that our child needs further investigations/medical or surgical treatment in the form of ....

 I/we hereby give consent to carry out the investigation/medical or surgical treatment on our child ....

Name ....   Relation ....   Sign

Stillbirth Problem that is not Counted or Accounted

Ganesh Shinde, Nihita Pandey, Hemlata Kuhite

Stillbirth is a largely unstudied and unspoken complication in obstetrics. Despite the progress in antenatal and intrapartum care, stillbirth remains an important complication which is largely missing in the discourse on maternal and fetal well-being.[1, 2]

The debate on stillbirth starts with the definition itself. WHO defines stillbirth as the death of a fetus either antepartum or intrapartum above 28 weeks of gestation or >1000 g of weight, for international comparison. In cases of discrepancy, the 28 weeks mark is given precedence over the 1000 g mark. But on national level, the definition of stillbirth varies. The ACOG considers all fetal deaths above 20 weeks of gestation as stillbirths.[3] The Lancet review and Silver, et al. consider stillbirth as fetal demise above 22 weeks of gestation.[1,2] The RCOG considers fetal demise above 24 weeks as stillbirth.[4] In India, the WHO definition of stillbirth is followed. This leaves a grey area for fetal demise between 20 weeks and 28 weeks, which is often labeled as abortion and not presented in the audit data of perinatal mortality.

A proper assessment of the prevalence of stillbirth and its causes is mandatory for development of preventive protocols and health facilities to deal with the burning issue of stillbirth.

EPIDEMIOLOGY

It is estimated that 2.6 million stillbirths occur annually, of which 75% occur in sub-Saharan Africa and south Asia. Around half of all stillbirths occur intrapartum. The worldwide distribution of countries with highest rates and number of stillbirths in the year 2015 is represented in Fig. 23.1.

India has the highest stillbirth number amongst all nations with a total of 592,000 stillbirths in the year 2015. The current stillbirth rate is 22.1 per 1000 births in India.[6]

CLASSIFICATION

In most cases, it is difficult to ascertain the etiology of stillbirth. Also, at times more than one etiology contributes to stillbirth. In spite of an array of modern diagnostics available, most cases of stillbirth are largely unexplained. Many classification systems are available but none is universal. Even in the available system of classification, the lack of a universal definition of stillbirth acts as a source of confusion. Also, in a few systems both stillbirth and neonatal deaths are included. Hence, no classification system should be considered a benchmark and should only be used to aid in evaluation of a case.[1]

The Wigglesworth classification is the most commonly used for reporting of perinatal mortality rates. The nine point classification is presented in Table 23.1.

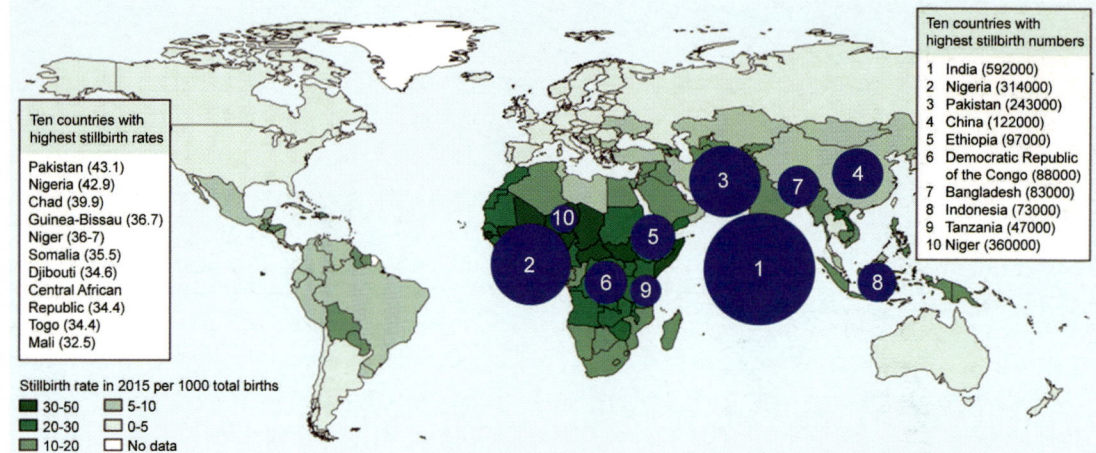

Fig. 23.1: The countries with the highest stillbirth rates and the highest stillbirth number[5]

DIAGNOSIS

The definitive diagnosis of intra-uterine fetal demise can be made only on an ultrasonography.[4] Auscultation and cardio-tocography can give false reassurance and should not be used for investigating a suspected case of intrauterine fetal demise.

On confirmation of diagnosis, it should be ascertained that the patient is accompanied. Else, an immediate offer must be made to call her relatives or partner.

INVESTIGATION

Patients must be counseled about the need of detailed investigation since an abnormal test may have a bearing on future pregnancies. At the same time they must be informed that in almost half of the cases no cause is determined. Also, when an abnormal test is obtained, there is a probability that it is unrelated to the cause

of IUFD. Hence, proper counseling of the couple is mandatory.

Following tests are advised (Table 23.3):[4]

LABOR AND BIRTH

The decision to induce labor should be taken considering the maternal risk factors and also the preferences of the mother.

Indications for immediate induction are:[4]
- Pre-eclampsia/hypertensive disorder of pregnancy
- Rupture of membranes
- Sepsis
- Abruptio placentae

Women who do not have these risk factors may consider waiting for spontaneous onset of labor after proper counseling. DIC profile needs to be repeated twice a week in case pregnancy is prolonged for more than 48 hours.

Table 23.1	Wigglesworth classification
1. Congenital defect/ malformation (lethal or severe)	6. Death due to other specific causes
2. Unexplained antepartum fetal death	7. Death due to accident/non-intrapartum trauma
3. Death from intrapartum 'asphyxia', 'anoxia' or 'trauma'	8. Sudden Infant death, cause unknown
4. Immaturity	9. Unclassifiable
5. Infection	

Gardosi, et al. have proposed a new classification system which excludes neonatal death and focuses on the relevant conditions present at the time of death *in utero*. The ReCoDe classification as proposed by Gardosi, et al. is described in Table 23.2.[7]

Table 23.2	Relevant condition at death (ReCoDe) classification

A. Fetus
  1. Lethal congenital anomaly
  2. Chronic, e.g. TORCH
  3. Acute
  4. Non-immune hydrops
  5. Iso-immunisation
  6. Feto-maternal hemorrhage
  7. Twin-twin transfusion
  8. Intrapartum asphyxia
  9. Fetal growth restriction
 10. Other
B. Umbilical cord
  1. Prolapse
  2. Constricting loop or knot
  3. Velamentous insertion
  4. Other
C. Placenta
  1. Abruptio
  2. Praevia
  3. Vasa praevia
  4. Placental infarction
  5. Other placental insufficiency
  6. Other
D. Amniotic fluid
  1. Chorioamnionitis
  2. Oligohydramnios
  3. Polyhydramnios
  4. Other
E. Uterus
  1. Rupture
  2. Uterine anomalies
  3. Other
F. Mother
  1. Diabetes
  2. Thyroid disease
  3. Essential hypertension
  4. Hypertensive disease in pregnancy
  5. Lupus/anti-phospholipid syndrome
  6. Cholestasis
  7. Drug abuse
  8. Other
G. Trauma
  1. External
  2. Iatrogenic
H. Unclassified
  1. No relevant condition identified
  2. No information available

The woman should be warned that increased prolongation of pregnancy may have ill effects on her while also deteriorating the appearance of the baby and the value of post mortem.

Vaginal route is the preferred route for delivery and LSCS should be undertaken only for obstetric reasons or when vaginal route poses a threat to maternal health. NICE suggests the use of misoprost as per the gestational age of the patient:[4]

- <26 weeks: 100 mcg of misoprost 6 hourly for 24 hours
- >27 weeks: 25–50 mcg of misoprost 4 hourly for 24 hours

RCOG recommends the use of combined mifepristone 200 mg with misoprost. Misoprost has been found to be more effective than prostaglandin E2 for induction of labor.

For patients with a single previous LSCS use of prostaglandins is considered safe. The rate of scar rupture is 0.7%. For patients with two previous LSCS the risk of scar rupture with prostaglandins is only slightly higher at 0.9%. Most obstetricians, however, prefer surgical management in cases of previous two LSCS.

PUERPERIUM

Lactation suppression should be given with dopamine agonists such as cabergoline single dose of 1 mg, alongwith tight breast support. Dopamine agonist should be avoided in patients with hypertensive disorders. Pyridoxine 100 mg is given to hypertensive patients for 21 days. Counseling and bereavement services should be offered in each case.

Contraceptive counseling should be done and patient should be explained the need for pre-pregnancy check up and evaluation. These patients tend to have increased anxiety due to this mishap and enter their subsequent pregnancy fearing that they will lose it. Such women need to be explained and reassured that one bad outcome does not mean that other pregnancies will also end in a mishap.

Table 23.3		Relevant investigations
Tests	Reasons	
1.	Maternal hematology and biochemistry	To rule out complications of pre-eclampsia and multi-organ failure
2.	Bile salts	To rule out obstetric cholestasis
3.	Coagulation profile	To rule out onset of DIC
4.	Blood culture, urine culture, cervical and vaginal cultures	Indicated in cases of maternal fever and prolonged rupture of membranes prior to IUFD.
5.	TORCH test, parvovirus infection and VDRL	Parvovirus infection may lead to hydrops and IUFD.
6.	Random blood sugar and HbA1C	To rule out gestational diabetes
7.	Thyroid function tests	To rule out maternal thyroid dysfunction
8.	Thrombophilia screen	Indicated in cases of intrauterine growth restriction or placental disease
9.	Anti-RBC antibody, anti-Ro, anti-La antibody, anti-platelet antibody	Indicated if there is evidence of hydrops in baby, presence of AV valve calcification and intra-cranial hemorrhage on postmortem examination of fetus
10.	Parental karyotype	Indicated in cases of fetal unbalanced translocation and fetal aneuploidy
11.	Fetal and placental karyotype	To detect fetal aneuploidy and single gene defects. It is absolutely contraindicated if the parents are unwilling.
12.	Postmortem examination: External, autopsy, microscopy, X-ray, placenta and cord	Weight and length measurement should be included. A written informed consent should be obtained prior to procedure. Absolutely contraindicated if parents not willing.
13.	Fetal and placental microscopy	For detecting fetal infections

THE HIDDEN PROBLEM

Stillbirth in many societies remains a stigma and the total magnitude remains hidden. The grief felt by parents is often not acknowledged by family or health professionals. It is a common advice given to such unfortunate couples that they should forget their stillborn child and have another one. The mothers of stillborn feel stigmatized and isolated from society. At times this becomes the cause for their abuse and mistreatment. Many a marriages are broken because the woman is blamed as the sole cause of the stillbirth.

Apart from the parents, stillbirth has wide reaching effects on the society and also on the health care providers. An estimated 4.2 million women worldwide are living with depression as a result of stillbirth.[2] The health care professionals are also negatively affected by this event. Many care providers experience anger, sadness, guilt, anxiety, blame and also fear of litigation. In a study conducted by the RCOG, 75% of care providers reported that caring for a patient with stillbirth took a huge emotional toll on them personally. Nearly one in ten obstetricians reported having considered giving up practice because of the emotional difficulty in caring for a woman with stillbirth.[4] RCOG recommends that there should be a system in place to give psychological and clinical support for staff involved with an IUFD.[4]

Apart from the psychological problems, stillbirth also takes an economic toll on the system. The cost of treating a stillbirth was estimated to be 10–70% higher than the cost for a normal pregnancy.[2] The reduced productivity of the parents and time taken off work as a result of the psychological impact of this event adds further to the total cost of management. These immense costs should be taken into account while evaluating the effectiveness of a new intervention for the management of stillbirth.

SETTING STANDARDS OF CARE

In order to streamline the efforts in improving care for prevention of stillbirth, we need to set standards for audit and also need to improvise training and sensitize the care providers regarding the approach to such a case. RCOG recommends the following auditable standards to be followed for the evaluation of interventions:[4]

a. Proportion of stillbirths reported as a clinical incident
b. Completion of investigations for the evaluation of IUFD
c. Proportion of parents offered postmortem evaluation
d. Proportion of parents declining full post-mortem who were offered alternative tests
e. Proportion of parents who have post-mortem consent undertaken by an appropriately trained obstetrician or midwife
f. Proportion of women offered suppression of lactation
g. Proportion of women given fertility and contraceptive advice.
h. Proportion of parents offered follow-up with a senior obstetrician
i. Proportion of women and families offered counselling follow-up

The evaluation of the aforementioned parameters will help establish the effectiveness of the system in place and will also help identify the lacunae in our management which need to be strengthened. At the same time, steps need to be taken to improve the training of the care providers. The following steps can be considered for the same:[4]

a. Seminars on the causes and care of late IUFD.
b. Skills training for the ultrasound diagnosis of late IUFD.
c. Training for discussions with parents about late IUFD.
d. Training on the postmortem examination, including consent.
e. Additional training in IUFD for bereavement counsellors.
f. Quarterly multidisciplinary clinic—pathology meetings for critical analysis of stillbirths.
g. Role play of follow-up appointments for obstetric trainees.

TOWARDS 2030: AN INTEGRATED APPROACH TO ADDRESSING STILLBIRTH

A series conducted by The Lancet, emphasized on the call to action for ending preventable stillbirths. This series presents three criteria to ensure that stillbirth is being integrated in the global and national programs for maternal and child health:

i. Has stillbirth been included in relevant summaries of the burden of maternal, neonatal and child mortality?
ii. Is high quality antenatal and antepartum care included with specific interventions to prevent stillbirth?
iii. Are stillbirths monitored through use of a target, outcome indicator, or both?

The series recommends the mortality targets to be achieved by 2030:

1. 12 stillbirths or fewer per 1000 total births in every country
2. All countries set and meet targets to close equity gaps and use data to track and prevent stillbirths

The steps mentioned can be summarized in Fig. 23.2.

To conclude, stillbirth is a largely neglected problem which is a major impediment in

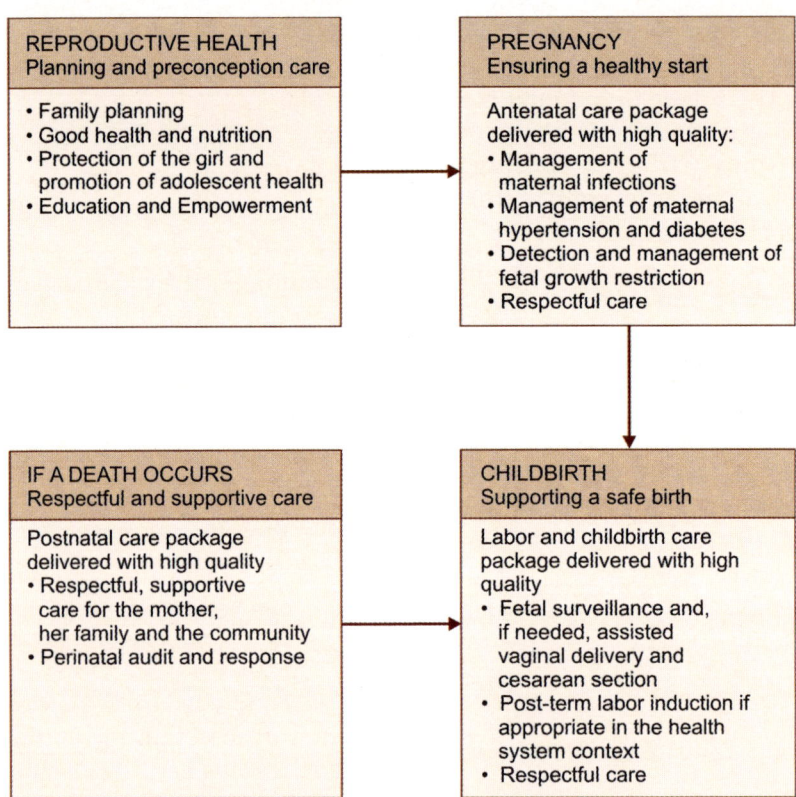

Fig. 23.2: Integrated approach to stillbirth

maternal and neonatal care. Urgent steps need to be taken to formulate and implement policies aimed at ending preventable stillbirths.

REFERENCES

1. Silver et al. Work-up of stillbirth—a review of the evidence. Am J Obstet Gynecol. 2007 May; 196(5): 433–444.
2. Executive summary for The Lancet series. Ending Preventable Stillbirths. The Lancet, Jan 2016.
3. ACOG. Practice Bulletin #102, "Management of Stillbirth". Obstetrics & Gynecology. March 2009.
4. RCOG. Late intrauterine fetal death and stillbirth. Green-top guideline no. 55. October 2010.
5. Lawn JE, Blencowe H, Waiswa P, et al, for The Lancet Ending Preventable Stillbirths Series study group with The Lancet Stillbirth Epidemiology investigator group. Stillbirths: rates, risk factors, and acceleration towards 2030. Lancet 2016; published online Jan 18. http://dx.doi.org/10.1016/S0140-6736 (15)00837-5.
6. ChartsBin statistics collector team 2011, Current Worldwide Stillbirth Rate (per 1000 births), ChartsBin.com, viewed 26th June, 2016, <http://chartsbin.com/view/1445>.
7. Gardosi J, Kady SM, McGeown P, Francis A, Tonks A. Classification of stillbirth by relevant condition at death (ReCoDe): population based cohort study. Bmj. 2005;331(7525):1113–7.

Medico-legal Angle—Cerebral Palsy

Rahul Wani, Archana Wani

INTRODUCTION

The fear of medico-legal litigation involving cerebral palsy (CP) is one which justifiably haunts modern obstetric practice. 73.6% of US obstetricians have faced litigation at some time—most often for alleged causation of fetal neurological impairment. Obstetric malpractice insurance premiums are as high as to $200 000 per year in some states. These high amounts have deterred doctors from entering the specialty and encouraged numerous others to leave OBGYN or limit themselves to gynecology.

In a case of cerebral palsy in India compensation of more than one crore was awarded.

CEREBRAL PALSY

The process of birth in cases of premature birth, difficult labors, mechanical injuries to head and neck during delivery, where the baby survives, may cause hypoxic damage to brain, leading to spastic rigidity of limbs, neurological and muscular impairment leading to cerebral palsy.

In 1999, an international consensus statement was published to provide an agreed reference for use by the courts and expert witnesses in birth injury litigation to establish the cause of cerebral palsy.

Conclusions from the International Cerebral Palsy Task Force Consensus Statement

A criteria to define cerebral palsy caused due to an acute intrapartum hypoxic event.

Essential Criteria

1. Evidence of a metabolic acidosis in intrapartum fetal umbilical arterial cord or very early neonatal blood samples (pH < 7 and base deficit ≥ 12 mmol/L).
2. Early onset of severe or moderate neonatal encephalopathy in infants ≥ 34 weeks of gestation.
3. Cerebral palsy of the spastic quadriplegic or dyskinetic type.

Criteria that together suggest an intrapartum timing but by themselves are non-specific

1. A sentinel (signal) hypoxic event occurring immediately before or during labor.
2. A sudden, rapid and sustained deterioration of the fetal heart-rate pattern, usually after the hypoxic sentinel event where the pattern was previously normal.
3. Apgar score of 6 for longer than five minutes.
4. Early evidence of multisystem involvement.
5. Early imaging evidence of acute cerebral abnormality.

Examples of sentinel hypoxic events
- Ruptured uterus
- Placental abruption
- Cord prolapse
- Amniotic fluid embolism
- Fetal exsanguination (from vasa praevia or fetal maternal hemorrhage).

Factors that suggest a cause of cerebral palsy other than acute intrapartum hypoxia

1. Umbilical arterial base deficit less than 12 nmol/L or pH >7.
2. Infants with major or multiple congenital or metabolic abnormalities.
3. Central nervous system or systemic infection.
4. Early imaging evidence of longstanding neurological abnormalities.
5. Infants with signs of intrauterine growth restriction.
6. Reduced fetal heart rate variability from the onset of labor.
7. Microcephaly at birth.
8. Major antenatal placental abruption.
9. Extensive chorioamnionitis.
10. Congenital coagulation disorders in the child.
11. Presence of other major antenatal risk factors for cerebral palsy, for example, preterm birth less than 34 weeks of gestation, multiple pregnancy or auto-immune disease.
12. Presence of major postnatal risk factors for cerebral palsy, for example, postnatal encephalitis, prolonged hypotension or hypoxia due to severe respiratory disease.
13. A sibling with cerebral palsy, especially of the same type.

Cerebral palsy caused by intrapartum hypoxia is always associated with neonatal encephalopathy and seizures. The inclusion of fetal acidemia as an essential criterion in the consensus statement has been criticized, as pH measurements are often not available. The onus is now on maternity units to obtain that information at delivery.

Acidemia is defined as a pH <7 or base deficit >12. A normal pH excludes hypoxic encephalopathy. By contrast, a pH <7 is associated with encephalopathy in only 10–20%. The majority of severely acidotic infants, born with a base deficit >16, are also normal.

All the above findings suggest that intrapartum hypoxia accounts for only 10% of cases of cerebral palsy. Just focusing on intrapartum factors for the cause cerebral palsy is a misplaced approach in cases of CP litigation.

Analysis in CP Litigation

Intrapartum management of labor could be flawless but antenatal care analysis may reveal missed diagnoses of conditions like SLE, factor V Leiden, polycythemia, thrombophilic disorders, autoimmune thrombocytopenia, chorioamnionitis, mismanagement of IUGR baby, hydramnios leading to preterm birth which could be the cause of cerebral palsy.

Some of these conditions which may be associated with CP later on include:
Low birthweight especially if below 1.5 kg.
- Prematurity especially before the 32nd week of pregnancy.
- Multiple births, particularly with an *in utero* death of a co-twin.
- Assisted reproductive technology conceptions possibly associated with preterm delivery or multiple births or both.
- Antenatal infections such as chickenpox, rubella, CMV virus, bacterial infections associated with chorio-amnionitis.
- Severe and prolonged fetal jaundice such as that resulting from ABO or Rh sensitization may lead to brain damage including CP. Prolonged fetal exposure to elevated maternal serum bilirubin levels may not necessarily result in developmental or neurologic handicap of the fetus.
- Maternal medical conditions of the pregnant mother such as ignored (or undiagnosed) thyroid dysfunction or uncontrolled epilepsy.

Intrapartum Analysis

Labor is the usual focus of CP litigation though it forms only 10% of causes of cerebral palsy. Invariably, this leads to scrutiny of tracings of electronic fetal monitoring in the form of cardiotocographic tracings. It may be both antenatal or intrapartum but more often than not it is the intrapartum CTG.

To prove liability, the court must firstly be convinced that evidence of intrapartum hypoxia was present, that the hypoxia was actively or passively mismanaged and that by not fulfilling expected duty of care, has subsequently caused cerebral palsy.

Mismanagement may include ignoring or misdiagnosing the signs of intrapartum hypoxia, failing to act on them by shortening the labor process, e.g. by C-section, taking the wrong action or even compounding the hypoxia, e.g. by the use of syntocinon stimulation.

CTG monitoring has long been officially accepted as a useful tool of detecting fetal hypoxia/acidosis especially in high-risk labor. Though this is controversial, the fact remains that there is no alternative to CTG monitoring to date. Using an I-P CTG monitor in labor and then ignoring/misinterpreting it, in the face of affected case of CP, does pose serious defense problems. Gross deviation from accepted norms of managing generally accepted abnormal I-PCTG tracings, certainly contributes heavily to the 10% of practitioners found liable at law. And most I-P CTG proven liability is due to

a. Inability to interpret FHR trace
b. Inappropriate action,
c. Technical aspects and
d. Record keeping.

With this firmly in mind, let us look at some worrying cases of actual court litigation centered on I-P CTG discussion.

One worrying aspect of I-P CTG centered argumentation in some CP trials, rendering the tracing as absolute for assessment of medical management. This aspect of scrutiny more often than not tends to work against the defendant.

Let us see a case which has gone in favor of defendant.

In Gossland v East of England Strategic Health Authority, the defendant's name was cleared of malpractice after the court concluded that:

The cardiotocograph trace was not such as to lead an obstetrician of ordinary competence to take the view that it was a case of 'complicated tachycardia' such as to make it imprudent to administer oxytocin.

And on this basis the court accepted the defendant's view that the plaintiff's CP had not been caused negligent act of the defendant. No matter that this 'commonly occurring CTG showing moderate tachycardia' came from a neonate who never mind the CTG,

Was abnormally large at 5.22 kg at birth.

Underwent a rotational forceps with alleged excessive force.

Suffered shoulder dystocia, and subsequently.

Suffered a fractured clavicle.

Here the plantiffs counsel has completely misled the argument focused on CTG, while if he had focused on factors such as birthweight, difficult forceps and birth injuries it would have pointed that this case was fit to be delivered by cesarean section and the defendant was negligent in management of the case.

I-P CTG Controversies

Clinical use of CTG has not diminished the overall prevalence of CP which has remained stable in the past 40 years at 2–3.5 cases per 1000 live births, much expected diminution in CP incidence through early detection and rectification of intrapartum hypoxia has not happened. May be this is hardly surprising now we know that intrapartum hypoxia causes CP in only 10% of cases.

Although extreme caution and balanced judgment needs to be used in the use of I-P CTG in medico-legal litigation, it still cannot be discarded. Wise and reflective clinical response to non-reassuring I-P CTG tracings can and does save fetal lives.

Quoting one example, in one series of 3600 deliveries at the Middlesex Northwick Park Hospital (UK) between 1996 and 2000, 22% of the management care problems were directly attributed to CTG misinterpretation.

Universally Accepted Classification of CTG Abnormalities

There is a need for the court to functionally recognize both the confusion as well as the shortcomings of the absence of one, truly functional and universal classification of CTG abnormalities.

In Brodie McCoy v East Midlands Strategic Health Authority the Court itself , clearly and justifiably brings out the "internal inconsistency" of one such classification:

... reference was made to the 1987 FIGO Guidelines for interpreting CTG traces. FIGO classification of decelerations in antepartum CTGs, as these state that the "absence of decelerations except for sporadic, mild decelerations of very short duration" is consistent with a normal fetal heart pattern; but "sporadic decelerations of any type unless severe" are part of the definition of "suspicious" fetal heart patterns. Thus in cases such as this, where decelerations are difficult to identify, it is not obvious whether a CTG should be classified as normal or "suspicious".

While the quote refers to antenatal and not intrapartum monitoring, the point is still made.

There exists a "considerable variation in the classification of CTG patterns."

This is a major flaw. Management depends on diagnosis based on these classification. If court cases are decided on such flawed classification all affected parties will be at sea.

A few people challenge established practices when the going is good. Who questions I-P CTG science and its medico-legal echoes when babies are born alive and healthy? Though an unnecessary C-Section on a misinterpreted I-P CTG may have been done, but who goes to court when baby is well? But, questions are asked when outcome is otherwise.

If we look at Ludwig (by her mother and litigation friend Della Louise Ludwig) v Oxford Radclie Hospitals NHS Trust and another the court quotes the NICE CTG guidelines:

In cases where the CTG falls into the suspicious category, conservative measures should be used. In cases where the CTG falls into the pathological category, conservative measures should be used and fetal blood sampling is undertaken where appropriate/feasible.

This looks simple on paper but considering the wide intra- and inter-observer variations in interpretation it is not so. In this context, let us re-visit an important and key principle of medical jurisprudence—the Bolam test, Bolam v. Friern Hospital Management Committee:

The court held that there is no breach of standard of care if a responsible body of similar professionals supports the practice judged even if this did not comply with the established standard of care.

Now the question arises as to what is to be concluded if in a CP Court trial for alleged malpractice, one responsible body of peers supports one classification and another supports an equally valid one.

To this add the existent lack of a universally accepted classification of CTG abnormalities.

Court opinions may be influenced (either way) by I-P CTG analysis in the absence of internationally agreed practice recommendations. Action is needed now to rectify the poor standardisation in the interpretation of CTGs and disagreement about appropriate interventions.

Inherent Problems of I-P CTG Monitoring

As already referred to, I-P CTG interpretation does exhibit worrying features including:

a. High specificity but low sensitivity, as well as

b. High intra- and inter-observer errors.

a. *Sensitivity may be as low as 99.8%, with only 0.19% of abnormal CTG tracings being (truly) associated with moderate or severe cerebral palsy.*

Taking a scenario where the I-P CTG tracing is as bad as it can get, with absent baseline variability, late decelerations. True fetal hypoxaemia and acidosis will only be present in 50–60% of cases. Rest of the cases would undergo an unnecessary C-section if the practitioner, resting solely on the non-reassuring I-P CTG tracing, proceeds to prevent "fetal distress". Traditionally, FBS was employed to distinguish which non-reassuring CTG was showing true fetal hypoxia and acidosis.

b. *There is a vast degree of inter-observer and intra-observer variation in pattern recognition.*

In other words, experts may disagree between themselves on a tracing but even far worse, the same expert may disagree with himself at different times. Medico-legally one must accept that this is frightening.

Knowing which interpretation of I-P CTG is correct in, may also be resolved by FBS.

...it must be recognized that FBS is a 'snapshot' test and is not useful if hypoxia evolves rapidly during second stage of labor because the values may not represent the actual fetal condition.

This problem may be partly set of by repeating FBS as necessary. However, the real current problem with FBS is that it has not withstood the scrutiny of modern evidence-based practice. Evidence-based publications do not recommend the use of FBS during second stage of labor.

Two other factors have to be taken in full consideration at this juncture:

1. FBS is still officially recommended by NICE guidelines as well as the Obstetric Colleges such as RCOG and RCPI. Evidence based practice says otherwise.

2. There are other investigations on the horizon which may offer better answers than FBS. One of the most promising is STAN (ST analysis of the fetal electro-cardiogram).

Evidence-based medicine takes time to seep into medical minds and change practice. When it comes to the law courts, it takes even longer.

In Milkhu v North West Hospitals NHS Trust, lack of performance of FBS led to a ruling against the defendant:

If fetal blood samples had been taken at 22.00 and 23.00 when they should have been, then the patient would have been delivered by immediate cesarean section and would have escaped the brain damage he suffered in the 15 minutes that preceded his delivery. It follows that the defendant is liable for that damage.

Confirmation of fetal hypoxia and acidosis, is right at the cross-roads of obstetric science evaluation. Medico-legal cross-roads are likely to come later. In the meantime, the burning question remains: In a CP trial, will the court favors the latest evidence-based practice or bow to the time honoured and still officially recognized FBS?

In India, the situation is still grave. There are no local guidelines. FBS sampling expertise and availability is negligible. The courts will depend more on foreign literature, which would more often, then not go against the obstetrician.

CONCLUSION

Cerebral palsy will always be with us. Every pregnancy is precious and must be handled with utmost care antenatally as well as intrapartum. Good and timely communication in a honest manner may be one step in minimising litigation.

There is an urgent need for universal and uniform guidelines on intrapartum CTG monitoring and management based on its findings and further research in developing a test which would reflect true hypoxic state of the fetus.

Peer review of our practices will help us identify medico legally deficient practices which can be corrected.

CASE SCENARIO

Dr Indu Sharma *VS* Dr Sohini Verma

Facts of the Case

The complainant patient, Dr Indu Sharma, previously treated for infertility for 4½ yrs was under care of Dr Sohni Verma, for her pregnancy.

On 10-6-1999, after midnight, due to rupture of membranes, she got admitted in Apollo Hospital for her delivery. Resident doctor examined and admitted her. In the morning, Dr Sohini Verma examined her and advised her medicines, started IV fluid with 1 ampule of syntocinon.

CTG machine showed that the heart rate of the child began to drop (80/min), during the midnight of 11/12-6-1999. It was alleged that none attended the patient immediately.

Patient was shifted to operation theatre at 2.00 am for emergency cesarean (LSCS), and at 3.36 a.m. a female baby was delivered by LSCS, weighing 3.7 kg.

The baby did not cry immediately after birth and it took almost five minutes. The baby was kept on ventilator in NICU. The baby had a stormy course latter developed tonic clonic seizures, milestones were delayed, investigations revealed severe atrophy of the brain due to hypoxia leading to mental retardation and cerebral palsy.

The Disability Board of AIIMS, New Delhi certified the baby as '95% disability'. Baby survived for 12 years with disabilities and with mental retardation.

Allegations of Medical Negligence

Allegations are mainly related to protocol failure, manipulation in medical record and not supplying the medical record to patient.

Specific allegations are as follows:
The complainant alleged that doctor failed to perform LSCS within 12 to 18 hours after rupture of membrane. It was abnormally delayed for about 27 hours.

The doctor advised excessive dose of syntocinon, which caused fetal distress and cerebral anoxia palsy.

The doctors/hospital made number of corrections/interpolations on the case sheets. The neonatal record was also tampered. The hospital purposely concealed cardiotocograms (CTG) tracings, which was the vital document in this case.

The doctor failed to take proper care during delivery, which resulted in birth of an asphyxiated baby.

The hospital did not issue entire medical record, CTG graphs, etc.

Compensation Claimed

The complainant filed this complaint of medical negligence and has prayed total compensation of Rs. 2.5 crores plus Rs. 5 lacs for the mental agony and Rs. 25000/- as costs of litigation. The complainant paid approximately 2.5 lakhs towards hospitalisation.

Findings of NCDRC

The patient had pregnancy after 4½ years of infertility, thus it was a precious pregnancy, hence should have taken prudent approach to deliver baby with utmost care and caution. After spontaneous rupture of membranes and administration of Syntocinon she should not have waited for more than 8 hours to take decision of C-section.

The nursing notes clearly establish hypertonic contractions fetal distress; which was not acted upon. It was an act of omission, thus negligence.

Excessive use of syntocinon and delay in decision to perform C-section, which caused birth asphyxia to baby.

In addition there is unflappable evidence that, the medical record of baby and mother are tampered in several places.

CTG record was not provided to the patient.

Therefore, the doctor and its nursing staff failed in a duty of care to accord the obstetric and paediatric care with the reasonable skill and diligence prevailing in the medical profession in order to the safe delivery of the baby.

Compensation Awarded

Parties were held responsible for medical negligence in this case, NCDRC, therefore fixed total compensation of Rs. one crore.

Further, NCDRC imposed Rs. 10 lacs as punitive cost which Apollo Hospital shall deposit in the Consumer Legal Aid Account.

SUGGESTED READING

1. Cerebral Palsy: Medico-Legal Issues, Buttigieg GG Phys Med Rehabil Int . 2016; 3(2): 1084.
2. The use and interpretation of cardiotocography in intrapartum fetal surveillance, Evidence-based Clinical Guideline, Number 8.
3. Cerebral palsy—medicolegal aspects, Ivan Blumenthal, MRCP DCH, J R Soc Med. 2001 Dec; 94(12): 624–627.
4. 2015 ACOG Ob-Gyn Professional Liability Survey Results. The American Congress of Obstetricians and Gynecologists.
5. WJ. On the influence of abnormal parturition, difficult labors, premature births, and asphyxia neonatorum, on the mental and physical condition of the child, especially in relation to deformities. Trans Obstet Soc Lond 1862;3: 293–344 [PubMed].
6. Schifrin B, Longo L. William John Little and cerebral palsy. A reappraisal. European J Obstet Gynecol2000;90: 139–44.
7. Myers R. Two patterns of perinatal brain damage and their conditions of occurrence. Am J Obstet Gynecol1972;112: 246–76.
8. Colver A, Gibson M, Hey E, et al. Increasing rates of cerebral palsy across the severity spectrum in north-east England 1964–1993. Arch Dis Child Fetal Neonatal Edn 2000;83: F7-F12 [PMC free article].
9. Medico-Legal Aspect of Pregnancy and Delivery: A Critical Case Review. J Indian Acad Forensic Med. October, December 2015, Vol. 37, No. 4.
10. JM Malik, J Presiding Member, NCDRC, Mr Dr (Mr.) SM Kantikar, Member, NCDRC. Dr (Mrs) Indu Sharma vs. Indraprastha Apollo Hospital and Ors.O.P./104/2002, Date of Judgment: 24-04-2015.

Timely Referral Makes a Difference

Pradnya Supe, Mangala Gomare

A referral can be defined as a process in which a health worker at one level of the health system, having insufficient resources (drugs, equipment, skills) to manage a clinical condition, seeks the assistance of a better or differently resourced facility at the same or higher level to either assist or take over the management of the patient.

The reasons for referral are normally

- To seek expert opinion
- To seek additional or different services
- To seek admission and management
- To seek use of diagnostic and therapeutic tools[1]

An effective referral system ensures a close relationship between all levels of the health system and helps to ensure that people receive the best care available. It also involves making effective use of hospitals and primary health care services. A good referral system provides support to health care centers and outreach services from experienced staff at secondary and tertiary level centers. It thus helps in building capacity and enhances the access to better care at every level.[2]

A good referral system should ensure:

1. Optimal care at appropriate level and not unnecessarily costly.
2. Hospital facilities are used optimally and cost effectively.
3. Patients who need specialist services will get them in a timely manner.

4. Primary Health services are well utilized and their reputation is enhanced.

A referral system will only function effectively if all service providers adhere to the referral discipline; to refer appropriately, and to follow the agreed protocols of care.[1] It is the role of the supervising organization and facility supervisors to monitor referral statistics and to provide feedback as appropriate.

Pregnancy and childbirth is associated with health risks for both the mother and child. Timely and prompt referral service has been identified as one of the effective strategies to combat related risks and adverse outcomes. In rural areas, this problem is compounded by multiple factors and referral often plays a key role to ensure favorable outcome.[3]

The Reproductive and Child Health Program Phase-II "A flagship programme" within the National Rural Health Mission, aimed to reduce maternal mortality ratio to less than 150 by 2010.[4] India is definitely lagging behind in this aspect.

The current Maternal Mortality (MMR) of India is 1675 per one lakh live births, where as per the Millennium Developmental Goal (MDG) for India was to reduce the MMR by 3/4th the value of 1990 (i.e. 560)[6] by 2015. The National Health Mission aims to reduce the MMR to 100/1 lakh live births by 2017.[7]

Emergency obstetric care (EmOC) includes urgent services to prevent maternal death. It was started in 1997 as a joint venture by the WHO, UNICEF and UNFPA to treat major direct obstetric complications.

Basic emergency obstetric care (BEmOC) is defined as seven essential medical interventions or "signal functions" which include the following:

1. Antibiotics to prevent puerperal infection
2. Anticonvulsants for treatment of pre-eclampsia/eclampsia
3. Uterotonic drugs for postpartum hemorrhage
4. Manual removal of placenta
5. Assisted or instrumental delivery
6. Removal of retained products of conception
7. Neonatal resuscitation

Comprehensive emergency obstetric and newborn care (CEmOC) in addition to the above includes blood transfusion, surgery (e.g. cesarean section), neonatal intubation and advanced resuscitation.

Referral systems should basically include primary centres which categorize under BEmOC and referral centers which categorize under CEmOC.[8]

The importance of effective and timely referrals in obstetric emergency is related to the unpredictability of pregnancy complications and their potential to progress rapidly to become severe and life threatening. For example, a serious hemorrhage can lead to the death of a woman and the unborn child within minutes or hours without timely intervention.[9]

Health Care Delivery Systems in India

A. Public Sector

1. Primary health care: Primary health center; subcenters
2. Hospitals—Community health centers; rural/district hospital; specialist/teaching hospital
3. Health Insurance Schemes: Employees State Insurance Scheme; Central Government Health Scheme.

B. Private Sector

Nursing homes, hospitals.

Commonest Reasons for Referral

a. *Antenatal*
   - Severe anemia
   - Pre-eclampsia and eclampsia
   - Gestational diabetes
   - Severe vomiting
   - Missed/threatened abortion
   - Ultrasound
   - Medical termination of pregnancy

b. *Intranatal*
   - Non-progress of labor (most common)
   - Pre-eclampsia
   - Malpresentation
   - Post-term
   - Short stature
   - Previous LSCS
   - Premature labor/PROM
   - Fetal distress
   - Severe anemia
   - Intrapartum hemorrhage
   - Multiple gestation
   - Placenta previa
   - Previous stillbirth
   - Non co-operative mother
   - Medical disorders
   - Need for intensive care unit and blood transfusion

c. *Postnatal*
   - Postpartum hemorrhage (2nd most common)
   - Postpartum psychosis
   - Postpartum sepsis

Other common reasons seen in developing countries is unavailability of doctors and other health professionals and paucity of resources at peripheral centers.

Delays during Referrals

According to Thaddeus' and Maine's[10] "3 delays model", there are three main delays which affect the timely delivery of obstetric care to a pregnant women in a facility.

They are classified as follows:

Type (I): Delay in seeking care

Type (II): Delay in identifying and reaching the appropriate facility

Type (III): Delay in receiving adequate care at the referral facility

Distance, cost and quality; all three matter as far as the delays are concerned. These delays may not be just because of transportation problems, but also due to shortages of qualified staff, essential drugs and supplies coupled with administrative delays and clinical mismanagement. Also in rural areas have access to only basic care at the nearest available center.

Transport during Referral

The transport during referral has to be provided by either the government (in case of public sector) or by the referring hospital (in case of the private sector). Depending on the condition of the patient in question, this transportation system should include trained and experienced staff and attendants and adequate resuscitation and delivery facilities.

The Government of India has introduced "108" which is the free telephone number for emergency services. When an emergency is reported through 108, the call taker gathers the needed basic information and dispatches appropriate services.

Basic information obtained includes:

- Where the call is placed from (district/taluka/city/town/exact word location/landmark)
- The type of emergency.
- Number of people injured and the condition of the injured.
- The caller's name and contact number: For location guidance if required.

Emergency help dispatched through this process is expected to reach the site of the emergency in an average of 18 minutes. Pre-hospital care will be given to patients being transported to the nearest hospital.

Documentation during Referral (Fig. 25.1)

The patient who is being referred should always be accompanied by a referral chit/note.

An ideal referral note should include the following:

1. The complete name and details of the patient
2. Name of the facility from which the patient is being referred
3. Name of the facility to which the patient is referred
4. Complete history and clinical findings of the patient with timely details mentioned
5. Investigations and treatment given to the patient at the referring facility
6. Time lag between receiving and referring the patient
7. Reason for referral
8. Details of the doctor who is referring the patient

It is important for the referring center to keep a copy of the referral note to preserve as evidence. Also any other related documents should also be kept.

Communication to the referral center before transferring the patient is also extremely important. This helps for the doctors and other health care professionals at the referral center to be prepared for the emergency being presented to them. For example, if a patient of bleeding placenta previa is being referred, the referral center can arrange for blood and blood products and also for the operation table to be vacant at the said time.

Also it is equally important that only those patients who require referral essentially should be referred. Unnecessary referrals will not just increase the cost and use of resources but it will also cause increased workload and unnecessary use of resources at the referral centers.

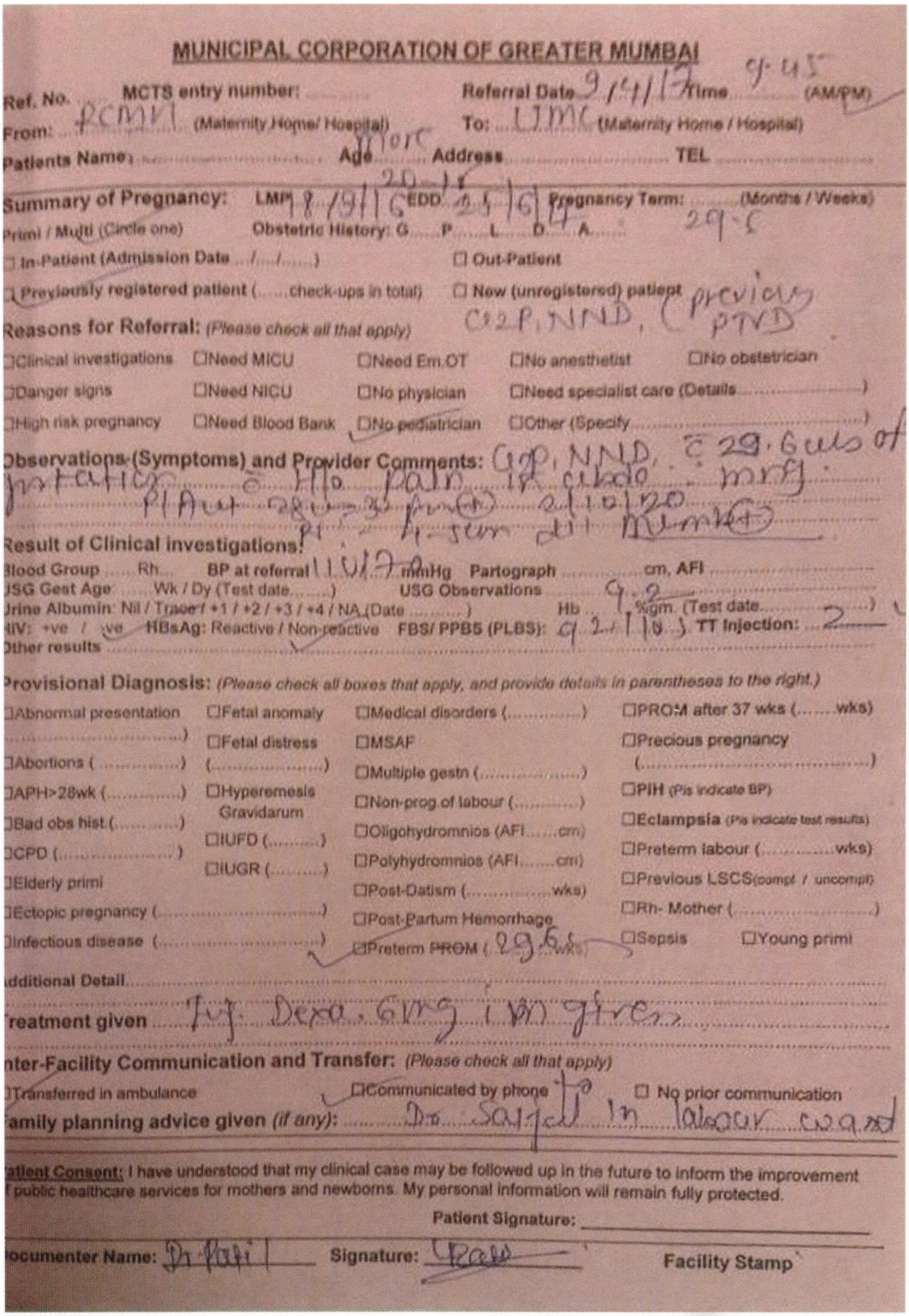
MUNICIPAL CORPORATION OF GREATER MUMBAI

Ref. No. MCTS entry number: .... Referral Date 9/1/17 Time 9.45 (AM/PM)
From: PCMH (Maternity Home/ Hospital) To: LTMG (Maternity Home / Hospital)
Patients Name: .... Age .... Address .... TEL ....

Summary of Pregnancy: LMP 18/9/16 EDD 25/6/17 Pregnancy Term: .... (Months / Weeks) 29.5
Primi / Multi (Circle one) Obstetric History: G.... P.... L.... D.... A....
☐ In-Patient (Admission Date ..../..../....) ☐ Out-Patient
☐ Previously registered patient (......check-ups in total) ☐ New (unregistered) patient

Reasons for Referral: *(Please check all that apply)*
☐ Clinical investigations ☐ Need MICU ☐ Need Em OT ☐ No anesthetist ☐ No obstetrician
☐ Danger signs ☐ Need NICU ☐ No physician ☐ Need specialist care (Details ....)
☐ High risk pregnancy ☐ Need Blood Bank ☐ No pediatrician ☐ Other (Specify ....)

Observations (Symptoms) and Provider Comments: G2P1 NND, c 29.6 wks of gestation ...

Result of Clinical investigations:
Blood Group ...... Rh...... BP at referral 110/70 mmHg Partograph .... cm. AFI ....
USG Gest Age: ...... Wk / Dy (Test date........) USG Observations ....
Urine Albumin: Nil / Trace / +1 / +2 / +3 / +4 / NA (Date ....) Hb 9.2 %gm. (Test date........)
HIV: +ve / -ve HBsAg: Reactive / Non-reactive FBS/ PPBS (PLBS): 9/2/16 TT Injection: 2
Other results ....

Provisional Diagnosis: *(Please check all boxes that apply, and provide details in parentheses to the right.)*
☐ Abnormal presentation (........) ☐ Fetal anomaly ☐ Medical disorders (........) ☐ PROM after 37 wks (.......wks)
☐ Fetal distress ☐ MSAF ☐ Precious pregnancy (........)
☐ Abortions (........) (........) ☐ Multiple gestn (........)
☐ APH>28wk (........) ☐ Hyperemesis Gravidarum ☐ Non-prog.of labour (........) ☐ PIH (Pls indicate BP)
☐ Bad obs hist (........) ☐ Oligohydromnios (AFI......cm) ☐ Eclampsia (Pls indicate test results)
☐ CPD (........) ☐ IUFD (........) ☐ Polyhydromnios (AFI......cm) ☐ Preterm labour (........wks)
☐ Elderly primi ☐ IUGR (........) ☐ Post-Datism (........wks) ☐ Previous LSCS (compl / uncompl)
☐ Ectopic pregnancy (........) ☐ Post-Partum Hemorrhage ☐ Rh- Mother (........)
☐ Infectious disease (........) ☐ Preterm PROM (29.6 wks) ☐ Sepsis ☐ Young primi

Additional Detail ....
Treatment given Inj. Dexa. 6mg i.m. given

Inter-Facility Communication and Transfer: *(Please check all that apply)*
☐ Transferred in ambulance ☐ Communicated by phone ☐ No prior communication
Family planning advice given *(if any)*: To Dr. Sajid in labour ward

Patient Consent: I have understood that my clinical case may be followed up in the future to inform the improvement of public healthcare services for mothers and newborns. My personal information will remain fully protected.
Patient Signature: ________

Documenter Name: Dr. Patil Signature: ________ Facility Stamp

Fig. 25.1: Documentation during referral

Current Scenario in Mumbai

The city of Mumbai is broadly divided into two parts; city and suburban Mumbai. However, being the metropolitan city that it is, all public hospitals in Mumbai get referrals not only from smaller peripheral hospitals and maternity homes located in Mumbai, but also from public and private hospitals from outside

Mumbai, i.e. from Thane and Virar areas (which include neighboring 7 corporations, Thane district, Raigad and Palghar).

Although the referral system in Mumbai is existing since beginning; it became organized in 2014.

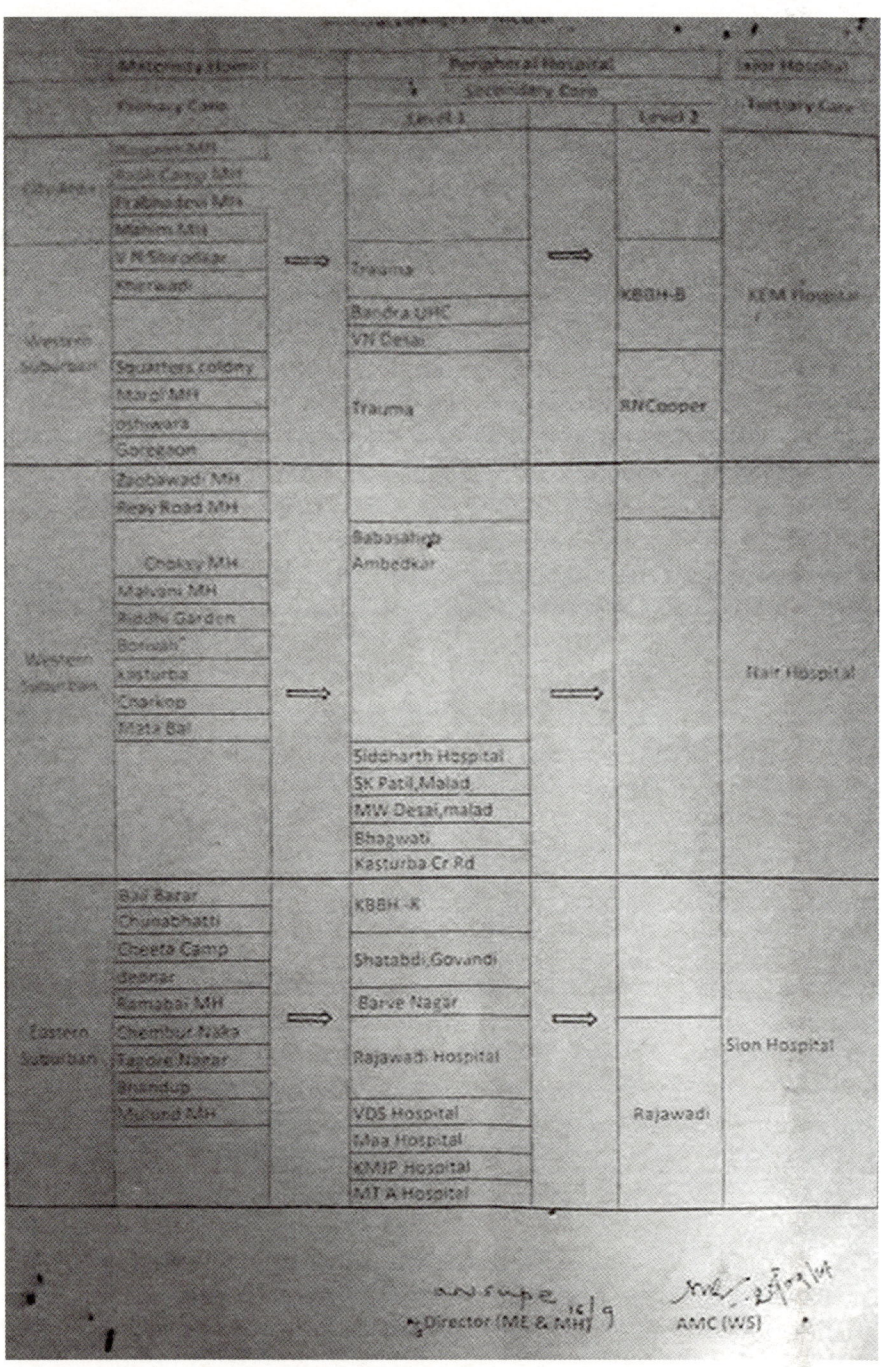

Fig. 25.2: Documentation during referral

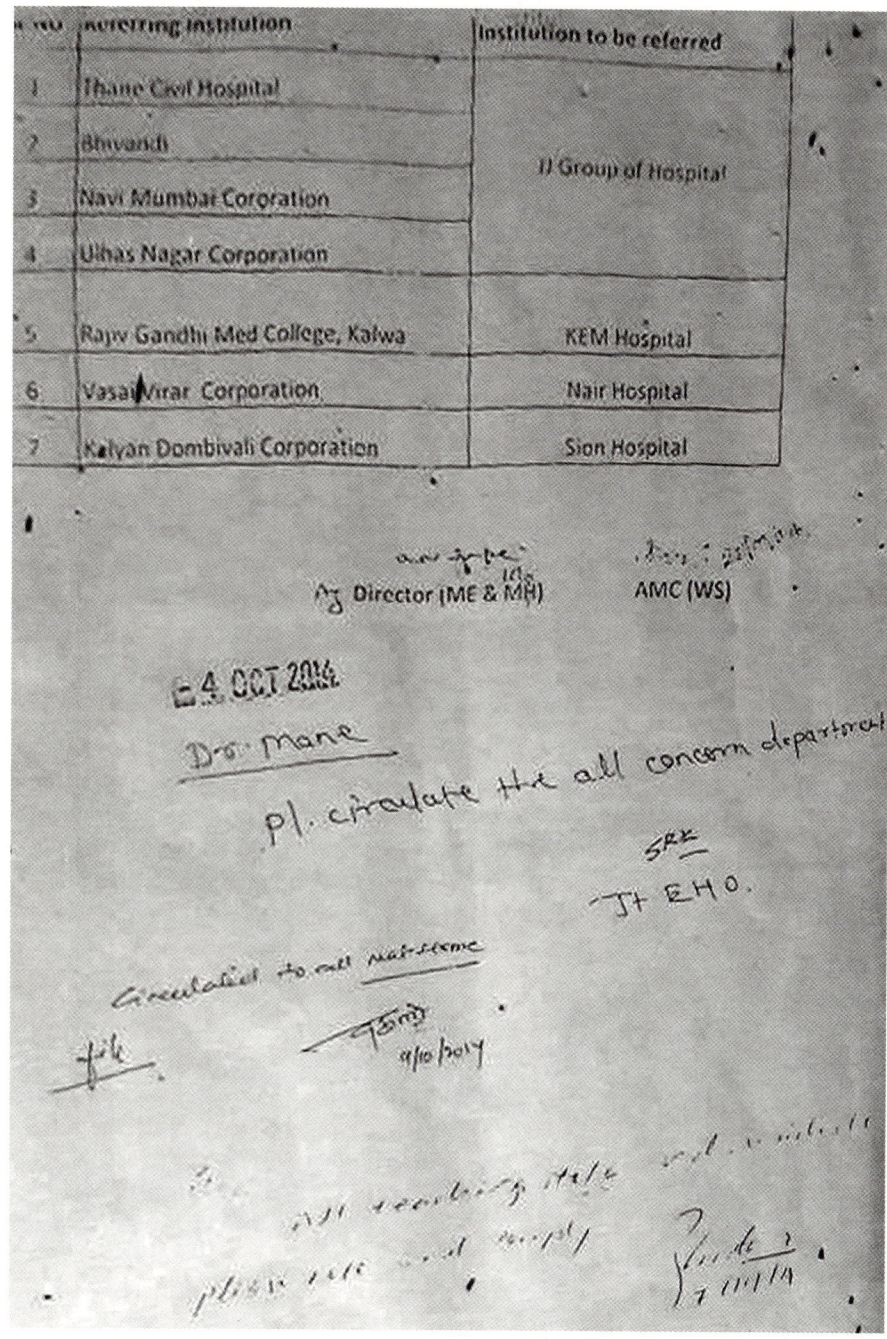

No	Referring institution	Institution to be referred
1	Thane Civil Hospital	I) Group of Hospital
2	Bhiwandi	
3	Navi Mumbai Corporation	
4	Ulhas Nagar Corporation	
5	Rajiv Gandhi Med College, Kalwa	KEM Hospital
6	Vasai Virar Corporation	Nair Hospital
7	Kalyan Dombivali Corporation	Sion Hospital

Fig. 25.3: Documentation during referral

A referral linkage system was created, which linked the primary health care centers, secondary level hospitals to the tertiary hospitals depending on the geographical location. It also included hospitals from Thane, Bhiwandi, Virar, Kalyan, Dombivali and Vashi which were linked to one of the tertiary level teaching hospitals. This not only reduced the confusion in referrals, but also made the workload well distributed. This referral system was further strengthened by having quarterly or six monthly referral meetings of a particular area, which were attended by representatives from all concerned health facilities. In these meetings all the problems regarding referrals were ironed out and also communication details were provided which have helped in bridging the communication gap to a great extent (Figs 25.2 to 25.4).

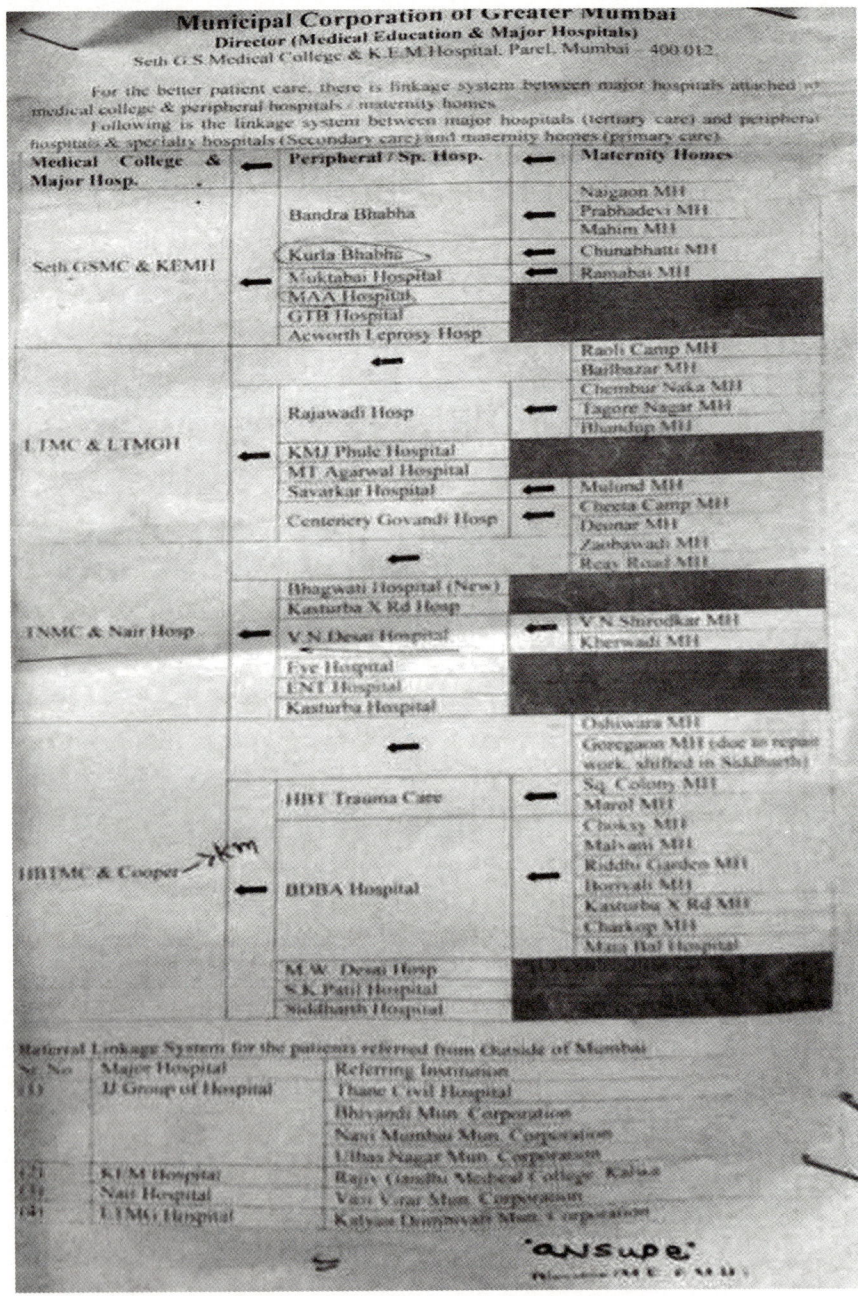

Fig. 25.4: Documentation during referral

CONCLUSION

The most common reason for referrals is obstetric indication. However, approximately 25–30% of the patients are referred for medical conditions. Therefore, strengthening the referral system will play an important role in timely manangement of high-risk obstetric cases and overall improve the maternal and fetal outcome.

SUGGESTED READING

1. 12th Five-Year Plan document, NHM, MOHFW
2. Afari, Henrietta A.O. 2015. Improving Emergency Obstetric Referrals: A Mixed Methods Study of Barriers and Solutions in Assin North, Ghana. Doctoral dissertation, Harvard Medical School
3. Maternal Health Division, Ministry of Health and Family Welfare Government of India. Trainees' Handbook for Training of Medical Officers in Pregnancy Care and Management of Common Obstetric Complications. August 2009. New Delhi, India.
4. Ministry of Health 2012. Guidelines for consultation with Obstetric and Related Medical Services (Referral guidelines). Wellington: Ministry of Health
5. Patel HC, Singh BB, Moitra M, Kantharia SL. Obstetric Referrals: Scenario at a Primary Health Centre in Gujarat. Natl J Community Med. 2012; 3(4):711–4.
6. Press information Bureau/GOI/MOHFW
7. SRS 2011–2013 Bulletin, Maternal Mortality of India.
8. Thaddeus S, Maine D. Too far to walk: maternal mortality in context. Soc Sci Med. 1994;38(36): 1091–1110.
9. UNICEF, WHO, UNFPA. Guidelines for Monitoring the Availability and Use of Obstetric Services.; 1997:1-103. http://www.childinfo.org/files/maternal_mortality_finalgui.pdf.
10. WHO; Management of Health facilities:Referral systems.

What a Neonatologist can do?

A Healthy Newborn: What Obstetricians and Neonatologists can do?

Madhuri Mehendale, Neha Singh

BACKGROUND

A newborn or neonate is defined as an infant till 28 days since birth. Every year nearly 45% of all under-five child deaths takes place in neonatal period. Three-fourths of all neonatal deaths occur in the first week of life. Major causes of neonatal deaths are infections (33%) such as pneumonia, septicemia and umbilical cord infection; prematurity (35%) and asphyxia (20%). In developing countries approximately half of all mothers and newborns do not receive skilled care during antepartum, intrapartum and immediate postpartum period.[1] Up to two-thirds of newborn deaths can be prevented by effective health measures provided during pregnancy, at birth and in first week of life. Thus, it is crucial that appropriate antenatal, intranatal and postnatal care is provided to improve child's chances of survival and to lay the foundations for a healthy life. To achieve this goal, roles of both obstetrician and neonatologist are very important by identification of high-risk pregnancies, intense antepartum, intrapartum surveillance, and providing essential newborn care respectively.

INDIAN SCENARIO

In India, 57% of under-five deaths take place within first one month of life which accounts for 7.3 lakh neonatal deaths every year in the country. As per WHO 2012 estimates, causes of under-five mortality in India are neonatal (53%), pneumonia (15%), diarrheal diseases (12%), measles (3%) and others (14%). As per sample registration system 2013, current neonatal mortality rate in India is 28 per 1000 live births.[2]

ROLE OF AN OBSTETRICIAN

A continuum of care is essential for the mother in order to achieve ultimate goal of healthy mother and newborn, starting from long before pregnancy (during childhood and adolescence) through pregnancy and childbirth. The traditional approach to provide antenatal care

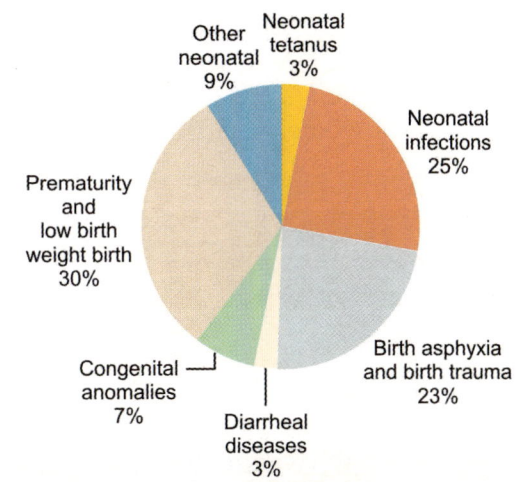

Fig. 26.1: Causes of neonatal deaths. *Source:* World Health Organization. The global burden of disease: 2004 update. World Health organization, Geneva, 2008

includes prevention and identification of potential fetomaternal complications, intense surveillance of high-risk pregnancies and appropriate management of such pregnancies to minimize maternal and perinatal morbidity and mortality.

PRECONCEPTIONAL CARE

A woman's body undergoes significant changes in pregnancy, with the developing fetus making increasing demands. Preparation for pregnancy should begin before conception, as fetal development begins from the third week after the last menstrual period. Damaging effects (e.g. exposure to drugs) may occur before the woman is even aware she is pregnant. Being as fit and healthy as possible before conception maximizes chances of a healthy pregnancy, but not all poor obstetric outcomes can be avoided. Pre-pregnancy counseling by a specialist team is recommended where specific risks and diseases are identified. Women above 35 years of age have reduced fertility rate and increased risk of chromosomal abnormalities in baby. Older mothers are more likely to develop complications in pregnancy, e.g. preeclampsia and diabetes mellitus.

Exercise and Stress

Moderate exercise should be encouraged, as it improves a women's cardiovascular and

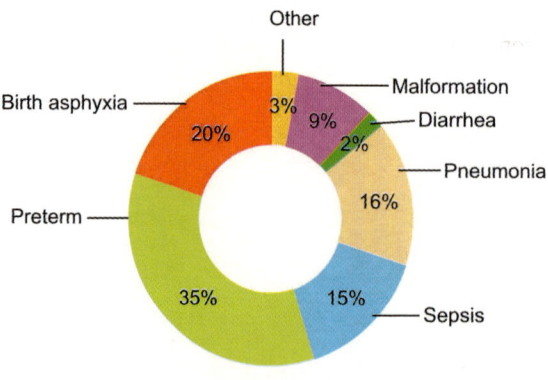

Fig. 26.2: Causes of neonatal deaths in India

muscular fitness and they should be reassured that beginning or continuing such exercises during pregnancy is not associated with adverse outcome. Relaxation and avoiding stress should be encouraged when planning for pregnancy.

Stopping Contraception

It is often recommended that women should wait for 3 months after stopping the pill before trying to conceive.

Nutritional Supplements and Lifestyle Advice

Folic acid 400 microgram daily has been shown to reduce the recurrence of neural tube defects. For women at higher risk (e.g. previous affected child, women with epilepsy, diabetes, and obesity), a dose of 5 mg/day is recommended.

Iron should be prescribed only when it is medically indicated. Calcium supplementation may be necessary if there is low intake. Iodine deficiency can result in cretinism and neonatal hypothyroidism. Supplementation with iodinized salt or oil should be considered preconceptionally. Low serum zinc levels have been associated with an increased risk of preterm labor and growth restriction, but increased intake from dietary sources, such as milk and dairy products, should be sufficient. Vitamin A supplementation (>700 micrograms/day) should be avoided as it might be teratogenic.

Alcohol, Smoking and Recreational Drugs

Excessive alcohol intake has been shown to cause fetal malformations. Smoking during pregnancy has an adverse effect on the developing fetus (e.g. preterm labor, low birth weight). Women should be encouraged to stop and supported through smoking cessation. Recreational drugs cause significant problems including miscarriage, preterm birth, poor fetal development, and intrauterine death.

Weight and Diet

Fertility may be reduced in women who are underweight (BMI <18.5) or overweight (BMI >30). Obesity is the most common nutritional disorder in the industrialized world, with increased risks including gestational diabetes and hypertension. Malnutrition, on the other hand, is a major life hazard in the developing world and is a cause of other problems such as anemia that has its own inherent risk for both mother and fetus. Poor nutrition in pregnant women is associated with the delivery of low birth weight (<2500 g) babies, and improving the nutritional status and maternal weight can have a positive effect on the birth outcome.

General Health Check

Planning a pregnancy provides a good opportunity for a general health check by the GP and may identify any potential obstetric risk factors well in advance. Thus, it provides an opportunity to anticipation of complications and their appropriate timely management.

Pre-pregnancy general health check may include general and systemic (e.g. cardiovascular and respiratory) examination, family history of inherited disorders or congenital anomalies, rubella vaccination status (if not immune, should be vaccinated and avoid conception for 3 months), HIV screening if at risk and dental examination.

Pre-pregnancy Medical Disorders

Pregnancy can have an adverse effect on preexisting medical disorders. For a woman with pre-existing disorder contemplating pregnancy, the advice of a specialist should be sought early. If the risk is very high, pregnancy may be discouraged altogether (e.g. Eisenmenger's complex—ventricular septal defects with pulmonary hypertension). Optimal control of certain diseases before conception may be very important to avoid the risk of fetal malformation or adverse outcome (e.g. diabetes mellitus). Some medications may be changed before conception to reduce the risk of teratogenesis (e.g. antiepileptics).

Pregnancy undertaken when the illness is in remission, stable, or cured will ensure a better outcome.

Working during Pregnancy

Women should be reassured that it is safe to continue working before and during pregnancy. Some workplaces are more likely to present hazards like chemical factories, operation theatres, X-ray departments; hence, precautions may be necessary.

ANTENATAL CARE

The needs of each pregnant woman should be assessed at the first appointment and a plan of care made for her pregnancy. This should be assessed at each appointment as new problems can arise at any time. There should be continuity of care throughout the antenatal period.

Booking Visit

Booking should ideally be early in pregnancy (before 12 weeks) in order to take full advantage of antenatal care. Children born to very late bookers or unbooked women have a 4–5 fold higher risk of perinatal mortality and morbidity, with an attendant increase in maternal morbidity and mortality.

Clinical History

A comprehensive history should be elicited and a full physical examination undertaken. Risk factors from past history should be highlighted. Past obstetric history, history of inheritable diseases, addiction, identifying women at risk of postnatal depression are very important to decide management plan for current pregnancy.

Screening for Chromosomal and Structural Abnormalities

Ideally, screening should be offered to all women at the time of booking. Detailed, unbiased, written information should be provided about the conditions being screened for, types of tests available, and the implications of the results. It is important for a woman to understand that a negative result in any screening test does not guarantee that her baby does not have that or other abnormality.

Routine Antenatal Care

Recent NICE antenatal care guidelines have been described stating schedule of antenatal visits and what needs to be done at each visit. These guidelines are summarized as follows:

Second Trimester

- *16 weeks:* Discuss prenatal screening results, investigate if Hb level< 11 g, offer information and anomaly scan (18–20 weeks)
- *25 weeks:* Nulliparous woman only: BP, urine dipstick, symphysiofundal height
- *28 weeks:* Screening for anemia, anti-D prophylaxis to rhesus negative women, BP, urine dip, plot SFH.

Third Trimester

- *31 weeks:* Nulliparous women only: BP, proteinuria, and plot SFH.
- *34 weeks:* Discuss labor and birth (including labor analgesia and birth plan)
- *36 weeks:* Discuss breastfeeding, vitamin K prophylaxis, postnatal self care, awareness of baby blues and postnatal depression, BP, proteinuria, plot SFH
- *38 weeks:* BP, proteinuria, plot SFH
- *40 weeks:* BP, proteinuria, plot SFH, information about prolonged pregnancy.

Sweeping of membranes at 41 weeks and induction of labor at 42 weeks.

Antenatal Fetal Surveillance

Fetus is vulnerable to any strain in the uteroplacental unit. There are no treatments to reverse uteroplacental insufficiency but potentially delivering the baby at the right time can prevent death and disability.

Stages in Fetal Surveillance

Stage 1: Assigning risk: Finding normal babies developing in an abnormal situation.

Stage 2: Timing delivery:
- Preterm babies should be delivered only if they show signs of distress, ensuring maximum maturity while avoiding any harm.
- After 36 weeks, babies at high risk should be delivered.

Identifying the High Risk Fetus

Symphysis—Fundal Height

Sequential measurements can reveal changes in fetal growth. Therefore, detection rate of small for gestational age babies is improved. If growth is suspected to be abnormal, a ultrasound should be done to confirm.

Ultrasound Assessment of Fetal Growth

It should be advised in cases of clinically suspected growth abnormality or in patients with high risk of uteroplacental insufficiency. Serial measurements are useful to assess growth. Color Doppler helps in monitoring of high-risk fetuses like IUGR, Rh isoimmunized pregnancy.

NEWBORN SCREENING AND ROLE OF OBSTETRICIAN

Newborn screening is the largest genetic program in the United States which has essential goal of decreasing neonatal morbidity and mortality by screening for disorders in which early intervention will improve neonatal and long-term health outcomes. There is a panel of genetic disorders which are

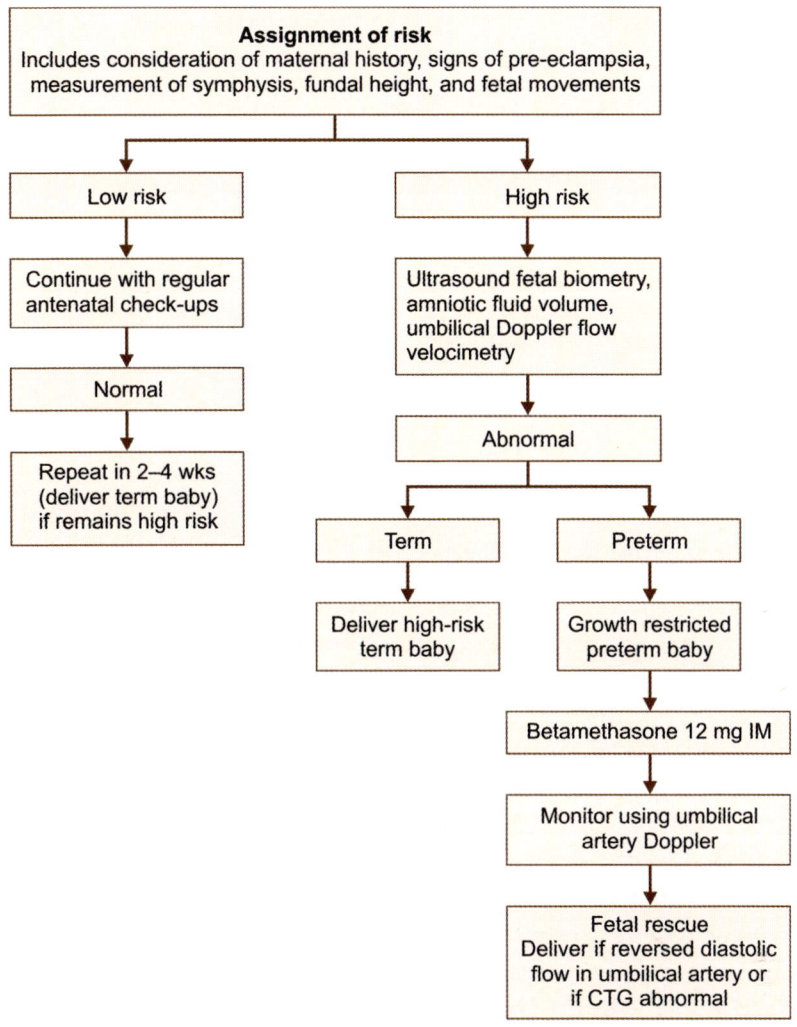

Fig. 26.3: Assigning risk in antenatal monitoring

screened by tests performed within 24–48 hours after birth. This panel has five main categories of disorders:

1. Hemoglobinopathies
2. Organic acid disorders
3. Aminoacid disorders
4. Fatty acid oxidation disorders, and
5. Miscellaneous group which includes cystic fibrosis, hypothyroidism, and hearing loss. The American college of obstetricians and gynecologists recommends that obstetric care providers review and make resources about newborn screening available to patients during pregnancy during their prenatal care visits.

ROLE OF A NEONATOLOGIST

A pediatrician can support development of a healthy newborn in several ways. A prenatal pediatric visit allows pediatricians to identify high-risk fetuses and to decide their management in postnatal period, to counsel parents antenatally about neonatal outcome of high-risk fetuses, to assess potential threats to bonding (a tense spousal relationship) and sources of social support.

WHO RECOMMENDATIONS FOR NEWBORN HEALTH

They can be summarized as below: Care of the newborn immediately after birth

1. Immediate drying and additional stimulation, if required
2. Suction in newborns who start breathing on their own; routine oral/nasal suction should not be done in babies who start breathing on their own after birth with clear amniotic fluid and those who are born with meconium and start breathing on their own, intrapartum, post-delivery or tracheal suctioning is not recommended.
3. Cord clamping; late cord clamping (performed after 1–3 minutes after birth) is recommended for all births while initiating simultaneous essential newborn care and early clamping (<1 min) is done only when neonate is asphyxiated or needs resuscitation.
4. Skin to skin contact in first hour of life; for all low-risk newborns to prevent hypothermia and promote breastfeeding.
5. Initiation of breastfeeding; as soon as possible after birth including low birth weight babies who are able to feed.
6. Vitamin K prophylaxis; all newborns should receive 1 mg vitamin K intramuscularly after 1 hour of birth.

Postnatal Care

1. Timing of discharge from health facility; hospital care should be provided for at least 24 hours after an uncomplicated vaginal birth.
2. Timing and number of postnatal visits; if birth in a hospital, postnatal care should be provided for at least 24 hours after birth or if it is at home, first visit should be in first 24 hours of birth, if possible. At least three postnatal visits are recommended, on day 3 (48–72 hours), between day 7–14 and after 6 weeks postpartum.

3. Assessment of the newborn; signs to be assessed during each postnatal contact and newborn to be referred for further evaluation if any of these danger signs is present: Stopped feeding well, history of convulsions, fast breathing (breathing rate >60/min), severe chest in-drawing, no spontaneous movement, fever (>37.5°C), low body temperature (35.5°C), any jaundice in first 24 hours of life or yellow palms/soles at any age.
4. Exclusive breastfeeding; all mothers should be counseled for exclusive breastfeeding until six months.
5. Cord care; daily chlorhexidine application to the umbilical cord stump during the first week of life is recommended for all newborns born at home in settings with high neonatal mortality (> neonatal deaths per 1000 live births). Clean, dry cord care is recommended for newborns born in low mortality settings.
6. Keeping the newborn warm; bathing should be delayed until after 24 hours of birth. Appropriate clothing (1–2 layers more than adults and use of hats/caps) for ambient temperature is recommended. The mother and baby should not be separated and should stay in the same room 24 hours a day.

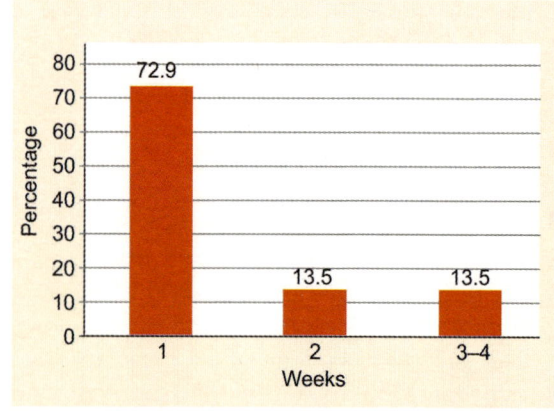

Fig. 26.4: Distribution of neonatal deaths by time since birth

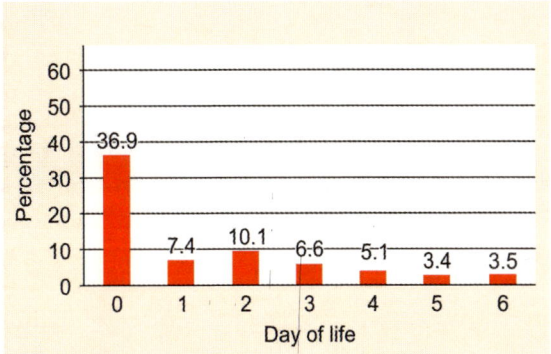

Fig. 26.5: Proportion of infants dying in the first week of life

Newborn Immunization

1. All newborns should receive first dose of hepatitis B vaccine as soon as after birth preferably within 24 hours.
2. Birth or zero dose of oral polio vaccine should be given at birth or as soon as possible after birth in all polio endemic countries.
3. Single dose of BCG vaccine should be administered to all infants in settings where tuberculosis is highly endemic or where there is high risk of exposure to TB.

Newborn Resuscitation

1. Immediate care after birth; in term/preterm newborn babies who do not require positive pressure ventilation, the cord should not be clamped earlier than one minute after birth, whereas those who require PPV, cord should be clamped and cut immediately to allow effective ventilation. Suctioning should be done only if nose or mouth is full of secretions.
2. Positive pressure ventilation; newborns who do not start breathing despite thorough drying and additional stimulation, PPV should be initiated within one minute after birth. Term or preterm (>32 weeks) neonates requiring PPV, ventilation should be initiated with air using self-inflating bag and mask.

3. Stopping resuscitation; in newborn babies with no detectable heart rate after 10 minutes of effective ventilation, resuscitation should be stopped.
4. Post-resuscitation care; head or whole body cooling should not be done outside well-resourced/tertiary neonatal intensive care units, because there is potential for harm from this therapy in low resource settings.

Management of Neonatal Sepsis

1. Prophylactic antibiotics for prevention of sepsis; newborns with risk factors for infection (i.e. ruptured membranes >18 hours before delivery, mother had fever before delivery or during labor, or foul smelling/purulent amniotic fluid) should be given prophylactic antibiotics ampicillin (IM/IV) and gentamicin for at least 2 days. Treatment should be continued only if there are signs of sepsis or positive blood culture.
2. Empirical antibiotics for suspected neonatal sepsis; newborns with signs of sepsis should be started with ampicillin (or penicillin) and gentamicin as the first line antibiotic treatment for at least 10 days. Wherever possible, blood culture should be obtained before starting antibiotics. In cases showing no improvement in 2–3 days of treatment, change of antibiotics and referral should be considered.

Management of Neonatal Jaundice

1. Monitoring jaundice and serum bilirubin; it should be ensured that all neonates are routinely monitored for development of jaundice and that serum bilirubin should be measured in those at risk (i.e. in all babies if jaundice appears on day 1, palms/soles are yellow at any age, preterm babies <35 weeks if jaundice appears on day 2).
2. Phototherapy and exchange transfusion; term and preterm neonates should be treated with phototherapy or exchange transfusion guided by cut-off levels of serum hyperbilirubinemia.

3. Stopping phototherapy; once serum bilirubin is 3 mg/dl or below phototherapy threshold.

NATIONAL HEALTH PROGRAMMES AND GUIDELINES

The Reproductive and Child Health (RCH) programme II under the National Rural Health Mission (NRHM) comprehensively integrates interventions that improve child health and addresses factors contributing to infant and under-five mortality. Further, Twelfth Five-Year Plan (2012–2017) and National Health Mission (NHM) laid down the Goal to Reduce Infant Mortality Rate (IMR) to 25 per 1000 live births by 2017.

One of the thrust areas under child health programme is to improve neonatal health by provision of essential newborn care at every delivery point at the time of birth, facility based sick newborn care at first referral units and district hospitals and home-based newborn care. To address the issues of higher neonatal and early neonatal mortality, facility based newborn care services at health facilities have been emphasized. Setting up of facilities for care of sick newborn such as Special Newborn Care Units (SNCUs), Newborn Stabilization Units (NBSUs) and Newborn Baby Corners (NBCCs) at different levels is a thrust area under NHM.

HOME-BASED NEWBORN CARE

Home-based newborn care is provided by ASHA through home visits to all newborns up to 42 days of life. Newborn's weight recording, ensuring BCG, OPV and DPT vaccination, birth registration and safety of mother and newborn are the various parameters to be looked for during such home visits.

NAVJAT SHISHU SURAKSHA KARYAKRAM (NSSK)

NSSK is a programme aimed to train health personnel in basic newborn care and resuscitation, has been launched to address care at birth issues, i.e. prevention of hypothermia, prevention of infection, early initiation of breastfeeding and basic newborn Resuscitation. Newborn care and resuscitation is an important starting-point for any neonatal program and is required to ensure the best possible start in life. The objective of this new initiative is to have a trained health personal in basic newborn care and resuscitation at every delivery point. The training is for 2 days and is expected to reduce neonatal mortality significantly in the country.

INDIA NEWBORN ACTION PLAN (INAP)

The India Newborn Action Plan (INAP) is India's committed response to the Global Every Newborn Action Plan (ENAP), launched in June 2014 at the 67th World Health Assembly, to advance the Global Strategy for Women's and Children's Health. For the first time, INAP also articulates the Government of India's specific attention on preventing stillbirths. It includes six pillars of intervention packages across various stages with specific actions to impact stillbirths and newborn health. The six pillars are: Preconception and antenatal care; care during labor and child birth; immediate newborn care; care of healthy newborn; care of small and sick newborn; and care beyond newborn survival. Goal 1: Ending preventable newborn deaths to achieve "Single Digit NMR" by 2030, with all the states to individually achieve this target by 2035. India will achieve the target of single digit NMR (NMR less than 10) by 2030. Goal 2: Ending preventable stillbirths to achieve "single digit SBR" by 2030, with all the states to individually achieve this target by 2035.

SUMMARY

Majority of newborn deaths take place in developing countries due to low access to health care. Most of these newborns die at home in absence of skilled health personnel's care that could greatly increase their chances for survival. Skilled health care during

Table 26.1	Interventions under National Health Mission focusing on newborns	
Programme (year)	*Objectives*	*Status*
Janani Suraksha Yojana (JSY) (2005)	Safe motherhood intervention to increase institutional delivery through demand-side financing and conditional cash transfer	• Implemented in all states and union territories (UTs) • Special focus on low-performing states
Integrated management of neonatal and childhood illnesses (IMNCI) at the community level and F-IMNCI at health facilities (2007)	Standard case management of major causes of neonatal and childhood morbidity and mortality	• Operationalised in more than 500 districts • 5.9 lakhs health and other functionaries, including physicians, nurses, AWWs, and ASHAs trained under IMNCI • 26,800 medical officers and specialists placed at the CHCs/FRUs trained under F-IMNCI
Navjat Shishu Suraksha Karyakram (NSSK) (2009)	Basic newborn care and resuscitation training programme	• 1.3 lakh health providers trained to date
Janani Shishu Suraksha Karyakram (JSSK) (2011)	Zero-out-of pocket expenditure for maternal and infant health services through free healthcare and referral transport entitlements	• Implemented in all states and UTs • Assured service package benefits extended to sick children up to age one
Facility Based Newborn Care (FBNC) (2011)	Newborn care facilities at various levels of public health services that includes newborn care corners (NBCCs) at all points of childbirth to provide immediate care; newborn stabilization units (NBSUs) at CHC/FRUs for management of selected conditions and to stabilize sick newborns before referral to higher centers; and special newborn care units (SNCUs) at district/sub-district hospitals to care for sick newborns (all types of care except assisted ventilation and major surgeries)	• 14,135 NBCCs established at delivery points to provide essential newborn care • 1810 NBSUs established at CHCs/FRUs • 548 SNCUs established at district/sub-district hospitals or medical colleges • More than 6300 personnel provided FBNC training • Online reporting system adapted and scaled up in seven states with 245 SNCUs made online and more than 2.5 lakhs newborns registered in the database
Home Based Newborn Care (HBNC) (2011)	Provision of essential newborn care to all newborns, special care of preterm and low-birth-weight newborns; early detection of illness followed by referral; and support to family for adoption of healthy practices by ASHA worker	• Implemented in all states and UTs • Most of the ASHAs trained in newborn care • ASHAs visited more than 12 lakhs newborn in 2013
Rashtriya Bal Swasthya Karyakram (RBSK) (2013)	Screening of children with birth defects, diseases, deficiencies, and developmental delays (including disabilities)	• All children, ages 0 to 18 years, targeted • More than 8 crore children screened and more than 10 lakhs children identified for tertiary care in 2013

Table 26.2	Interventions and tracer/proxy indicators for bottleneck analysis
Interventions	*Tracer indicators*
1. Management of pre-term birth	Antenatal corticosteroids
2. Skilled care at birth	Use of the partograph
3. Basic emergency obstetric care	Assisted vaginal delivery
4. Comprehensive emergency obstetric care	Cesarean section
5. Basic newborn care	Cleanliness including cord care, warmth, and feeding
6. Neonatal resuscitation	Use of bag and mask
7. Kangaroo mother care	Skin to skin, breastfeeding, and feeding support for premature and small babies
8. Treatment of severe infections	Using injectable antibiotics
9. In-patient supportive care for sick and small newborns	IV fluids/feeding support and safe oxygen

pregnancy, childbirth and in the postnatal period prevents complications for mother and newborn, and allows for early detection and management of complications.

SUGGESTED READING

1. http://www.who.int/mediacentre/factsheets/fs333/en/
2. http://nrhm.gov.in/nrhm-components/rmncha/child-health-immunization/child-health/schemes.html
3. INAP
4. Committee opinion role of obstetrician in newborn screening acog 2015
5. NICE 2008 ANC CG 62
6. Rose NC, Dolan SM. Newborn screening and the obstetrician. Obstet Gynecol 2012;120:908–17.
7. CDC Grand Rounds: Newborn screening and improved outcomes. Centers for Disease Control and Prevention (CDC). MMWR Morb Mortal Wkly Rep 2012;61:390–3.

Neonatal Resuscitation—Basics

Alok Sharma, Pancham Kumar

Birth asphyxia is the third commonest cause of the neonatal deaths accounting for almost one-fifth of total neonatal deaths. About 4 million neonatal deaths occur world over due to birth asphyxia every year. Effective neonatal resuscitation in first few minutes of life decreases the mortality significantly and also influence long-term outcome. Hence, science and art of the neonatal resuscitation is very important for all health functionaries who are involved in the delivery of the newborn. All the deliveries should be considered as potential for the need of resuscitation as only 60% of asphyxiated newborns can be predicted antepartum. The aim of resuscitation is to establish rapidly and smoothly respiratory functions which reduces the mortality and morbidity.

CAUSES OF NEONATAL DEATHS

The major causes of newborn deaths in India are pre-maturity (35%); neonatal infections (33%); birth asphyxia (20%); and congenital malformations (9%) (Liu et al, 2012).

Birth asphyxia accounts for about 20% of total neonatal deaths each year. Henceforth, it is essential to reduce the number of neonates suffering from asphyxia so as to achieve the goal of reducing neonatal mortality rate of India to single digit by 2020.

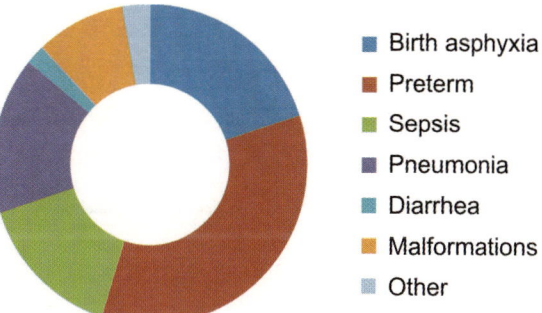

■ Birth asphyxia
■ Preterm
■ Sepsis
■ Pneumonia
■ Diarrhea
■ Malformations
▨ Other

Fig. 27.1: Cause of neonatal deaths

CARDIORESPIRATORY ADAPTATIONS AT BIRTH

During antenatal period placenta is the site for gas exchange and the lung alveoli are fluid filled. In fetus, lungs only receive 15% of the cardiac output due to high pulmonary resistance and rest of the blood flows through low resistance ductus arteriosus into the aorta bypassing the lungs. After the birth site of gas exchange shifts from the placenta to the lungs. In 90% of the cases this transition occurs without much help and only 10% require minimal assistance. Extensive resuscitation is needed in less than 1% of newborn.

As at birth newborn makes effort to breathe and inhale air, the fluid in the alveoli is absorbed by the pulmonary lymphatics and alveoli expand. The pulmonary resistance decreases due to delivery of oxygen through expanded alveoli and systemic resistance increases due to clamping of the cord. Increase in systemic resistance/pressure and decrease in pulmonary resistance/pressure lead to decrease in shunting of blood through ductus arteriosus and increase in flow to the lungs. Thus lungs are established as the site of gas exchange. The increase in the oxygen saturation due to shifting in the site of oxygenation from the placenta to lungs leads to the constriction of the ductus arteriosus functionally it gets closed within 12–24 hours. The blood which was bypassing the lungs in the fetal period now starts flowing through the lungs after the birth. The normal transition occurs in a few minutes of the birth. Oxygen saturation of 90% or more is achieved in 10 minutes in normal newborn.

FETAL CIRCULATION

Birth Asphyxia

Birth asphyxia is constellation of hypoxia, hypercarbia and metabolic acidosis. It has been defined in various ways:

1. National Neonatal Forum of India defines birth asphyxia as when baby has gasping and inadequate breathing at 1-minute.
2. It has been also defined as APGAR of less than 4 at 1 minute.

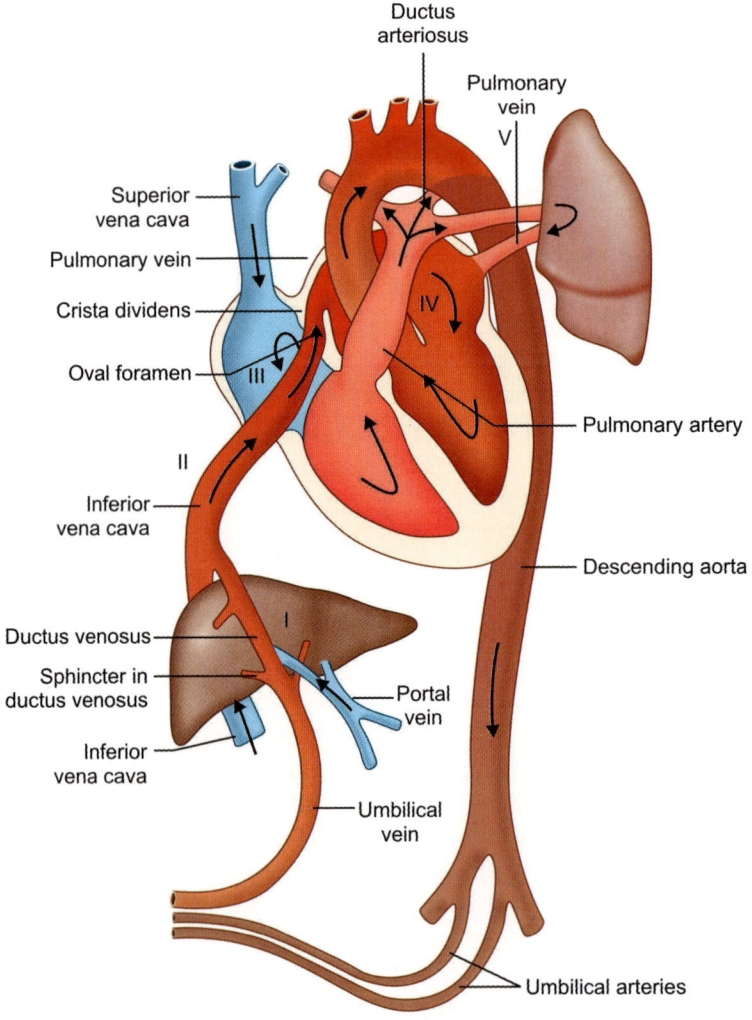

Fig. 27.2: Fetal circulation

3. American Academy of Paediatrics (AAP) defines perinatal asphyxia (term used by AAP) when all of these are present:
   a. Umbilical artery pH of < 7.0 with base deficit of >10 mEq/L
   b. Neurological manifestation of hypoxic ischemic cephalopathy.
   c. Multiorgan dysfunction

NEONATAL RESUSCITATION

Preparedness for Resuscitaion

As it is not possible to predict all the newborns which will require help to initiate breathing. Hence, it is important to have total preparedness for resuscitating a newborn at each delivery.

1. *Equipment:* Ensure that all equipment is in working order
2. *For maintaining desired ambient temperature:* Radiant warmer, heater, plastic bags, prewarmed wraps.
3. *For suctioning:* Suction machine with negative pressure source not to be >100 mmHg, suction catheter (6F, 8F, 10F or 12F) feeding tubes for gastric decompression, meconium aspirator.
4. *For ventilation:* Properly functioning ambu bags, appropriate size mask, oxygen tubings, pulse oximeter, oxygen/gas supply, laryngoscopes and blades, endotracheal tubes of different sizes.
5. *For circulatory support:* Cannula, umbilical catheters, intraosseous needles.
6. Drugs adrenaline (1:1000 and 1:10000 dilutions), atropine, and volume expander.

RISK FACTORS FOR BIRTH ASPHYXIA

Risk factors which are more likely to require some form of resuscitation at birth:

Antepartum

1. Pregnancy in early and late maternal age
2. Lack of antenatal care
3. Maternal infections

4. *Placental insufficiency:* Toxemia, hypertension, diabetes mellitus, anemia, antepartum hemorrhage
5. Polyhydramnios and oligo-hydramnios
6. Multiple gestation
7. Post-term gestation
8. Malformations and malpresentation in fetus
9. Bad obstetrical history

Intrapartum

1. Augmentation of labor with oxytocin
2. Premature labor
3. Antepartum hemorrhage
4. Meconium stained liquor
5. Prolonged labor (>24 hr)
6. Premature rupture of membranes (>12 hr)
7. Cord prolapse

Evaluation of Newborn at Birth

At birth baby is assessed for:
1. Term gestation
2. Respiratory effort: Crying or breathing
3. Good muscle tone.

   If answer is yes to all three, then baby stays with mother with routine care
1. Providing warmth,
2. Maintaining clear airways,
3. Drying
4. Ongoing evaluation
5. Initiation of the breastfeeding.

Initial Steps of Resuscitation

If the answer to any of three is no, one proceeds to the initial steps of resuscitation after cutting the umbilical cord which are as follows:
1. *Warmth:* Baby ideally should be kept under a heat source for maintaining the temperature rather than covering the baby as it helps us to make full and better visual assessment.
2. *Positioning:* Baby should be placed on the back with slight extension (sniffing position) which can be done by keeping a shoulder roll under the shoulders.

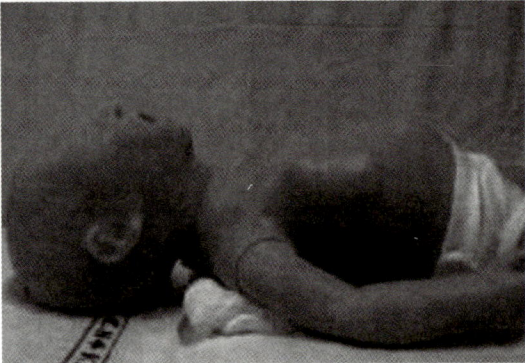

Fig. 27.3: Sniffing position

3. *Clear airways:* Airways are cleared only if required using the method available.
4. *Dry, stimulate and repositioning:* Baby is dried and wet linen are removed to prevent heat loss and wrapped in dry prewarmed linen after drying. Drying also provides stimulation for initiation of breathing. Rubbing the back or flicking the soles are the other methods used for stimulation.

Baby is evaluated again after initial steps with regards to:

   i. *Respiration:* It is assessed by cry or chest movements of the baby.

  ii. *Heart rate (HR):* Of late it has been advisable to use ECG for assessing the HR. If ECG monitoring is not available, as done previously it can be calculated by counting the heart beats or the cord pulsation for 6 seconds and multiplying it by 10.

 iii. *Oxygen saturation:* It is assessed by using the pulse oximetry. Color can also be used for evaluation if pulse oximetry is not available. Color is evaluated by looking for central cyanosis.

After assessment following scenario encountered are:

1. If baby is breathing or crying, HR=/>100/minute and no central cyanosis, then no active intervention is required further except monitoring.
2. In case breathing is labored or central cyanosis is persisting, then oxygen supple-mentation/continuous positive airway pressure (CPAP) is administered. O_2 supplementation and CPAP is titrated to achieve the oxygenation targets as given below:

1. 1 minute   60–65%
2. 2 minutes   65–70%
3. 3 minutes   70–75%
4. 4 minutes   75–80%
5. 5 minutes   80–85%
6. 10 minutes   85–90%

3. If after initial steps baby is not breathing or gasping or HR <100/minute, positive pressure ventilation is initiated.

Positive Pressure Ventilation

Positive pressure ventilation (PPV) is given by self-inflating bag of 250–750 ml capacity with appropriate sized mask (0 size for preterm and 1 for term baby). PPV is given by appropriable sized mask after obtaining a good seal that covers mouth and nose, but not the eyes of the baby. Oxygen supplementation is to be given or not during PPV depends upon whether the oxygen saturation targets are achieved or not.

During PPV 40–60 breaths are delivered per minute calling loudly: "Squeeze, two, three". Breath is delivered by squeezing the bag when you call squeeze and allow the bag to recoil during calling "two-three".

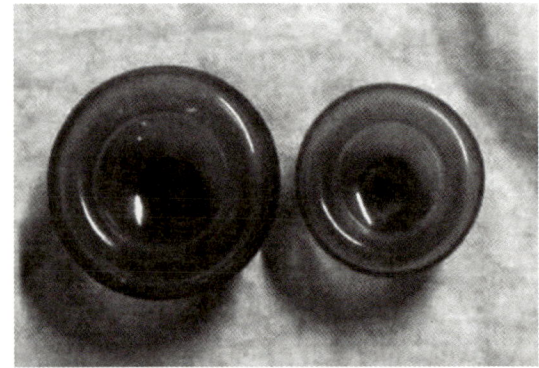

Fig. 27.4: 0 and 1 size mask

PPV is administered for 30 seconds and neonate is assessed for heart rate. If despite effective PPV for 30 seconds heart rate<60, external cardiac message is started while the ventilation is continued.

Cardiac Compression

It is done at the rate of 90 compression per minute and simultaneous ventilation at 30/minute. During cardiac compression chest is pressed to one-third of AP diameter of chest. Endotracheal intubation should be done for delivering the PPV during cardiac compression but if expertise is not available BMV can be continued, cardiac compression can be done by

1. *2-Thumb method:* Sternum is pressed by 2 thumbs while encircling the chest with hands and spine is supported by fingers.
2. *2-Finger method:* In this, tip of middle and index finger are used for pressing the sternum and spine is supported by hand surface or hand.

For performing cardiac compression, xiphisternum is located by sliding the fingers over edge of thoracic cage and cardiac compression is carried out above it. Finger and thumbs should remain in contact with chest during compression and release.

After 30 seconds of cardiac message and ventilation, newborn is again assessed for heart rate.

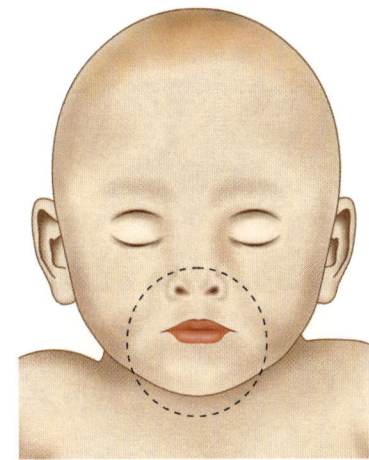

Fig. 27.5: Seal covering mouth and nose

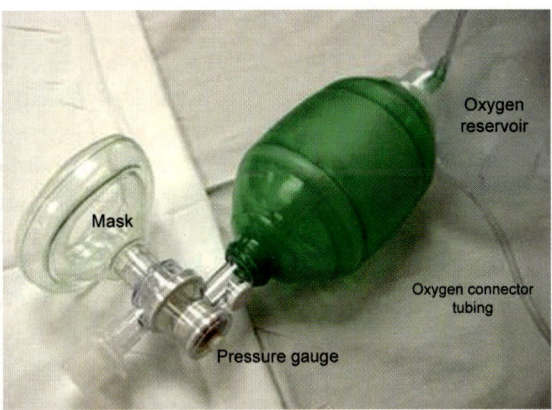

Fig. 27.6: Neonatal ambu bag

While one starts doing PPV, one looks for appropriate chest rise and adequate breath sounds.

If chest rise is not adequate or heart rate and oxygen saturation does not improve, corrective steps are taken:

	Actions
M	Adjust mask to assure good seal on the face
R	Reposition airway by adjusting head to sniffing position
S	Suction mouth and nose of secretion, if present
O	Open mouth slightly and move jaw forward
P	Increase pressure to achieve chest rise
A	Consider airway alternative (endotracheal intubation or laryngeal mask airway)

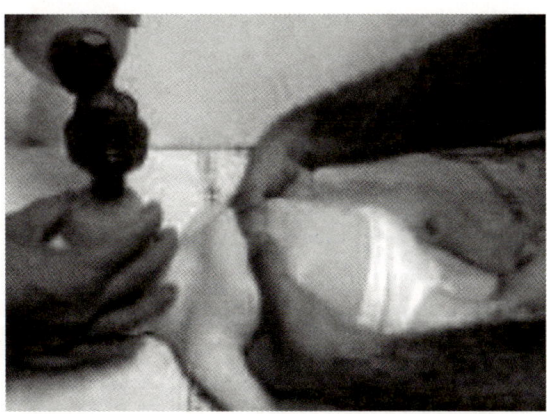

Fig. 27.7: 2-Thumb method

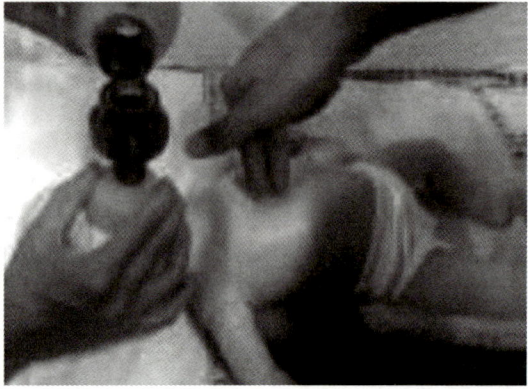

Fig. 27.8: 2-Finger method

1. If HR>60/m, cardiac compression and ventilation is continued, medications have to be used if the HR<60/m
2. If HR>60/m but <100/m chest compression is discontinued but PPV is continued.
3. If HR>100/m then BMV is discontinued if spontaneous breathing is established, and it has to be continued if spontaneous breathing is not established.

Endotracheal intubation can be considered at any step of resuscitation for providing PPV depending upon the expertise available.

Endotracheal Intubation

For intubation neonatal laryngoscope with straight blade 0'size for preterm and 1'size for term is used. Baby's head is kept in midline and neck is slightly extended. Standing at head end and holding laryngoscope in the left hand, blade is introduced in mouth advancing it, so that it is up rest on vallecula. Then blade is lifted to visualise the glottis opening which is surrounded by vocal cords on side. Then endotracheal tube (ET) is introduced from right side of mouth and inserting it though glottis opening until vocal cord guide is at the level of glottis. The size of ET tube depends on the weight or gestation of the baby.

	Size of ET tube	
Diameter of ET tube	Newborn weight in grams	Gestation age of newborn in weeks
2.5	<1000	<28
3.0	1000–2000	28–34
3.5	2000–3000	34–38
4.0	>3000	>38

Medication

1. *Adrenaline:* If baby continues to have heart rate <60/minute after 30 seconds of assisted ventilation and 30 seconds of coordinated ventilation and chest compression, epinephrine in a strength of 1:10000 dilution is administered at a dose of 0.1 ml/kg to 0.3 ml/kg through umbilical vein. The dose of epinephrine can be repeated after every 3–5 minutes as indicated.
2. *Volume expanders:* If shock persists, use of volume expanders like normal saline, Ringer lactate in a dose of 10/kg is administered over 5–10 minutes. O Rh negative blood can rarely be used if there is a history suggestive of fetal blood losses.

One   and   Two   and   Three   and   Breathe and

Person 1:
Chest compression

Person 2:
Positive pressure ventilation

— — — — — — — — — — — 2 second (one cycle):— — — — — — — — —

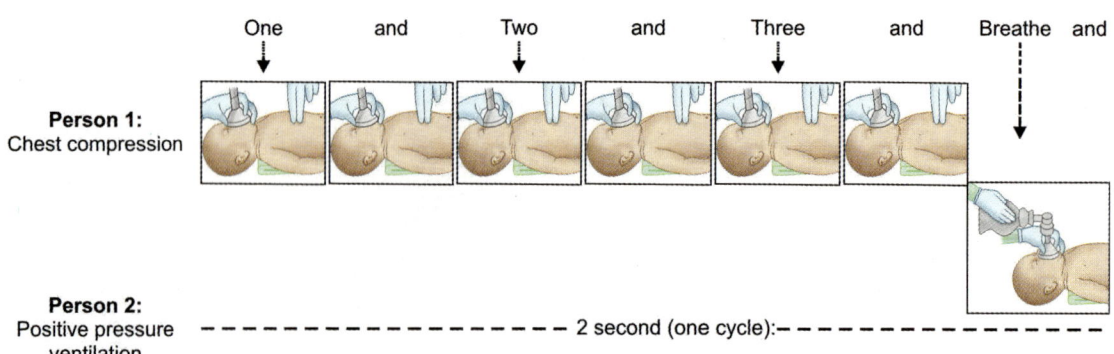

Fig. 27.9: Cardiac compression

3. *Alkali therapy:* Administration of sodium bicarbonate guided by blood acid–base parameters. It can only be given after establishing effective ventilation.

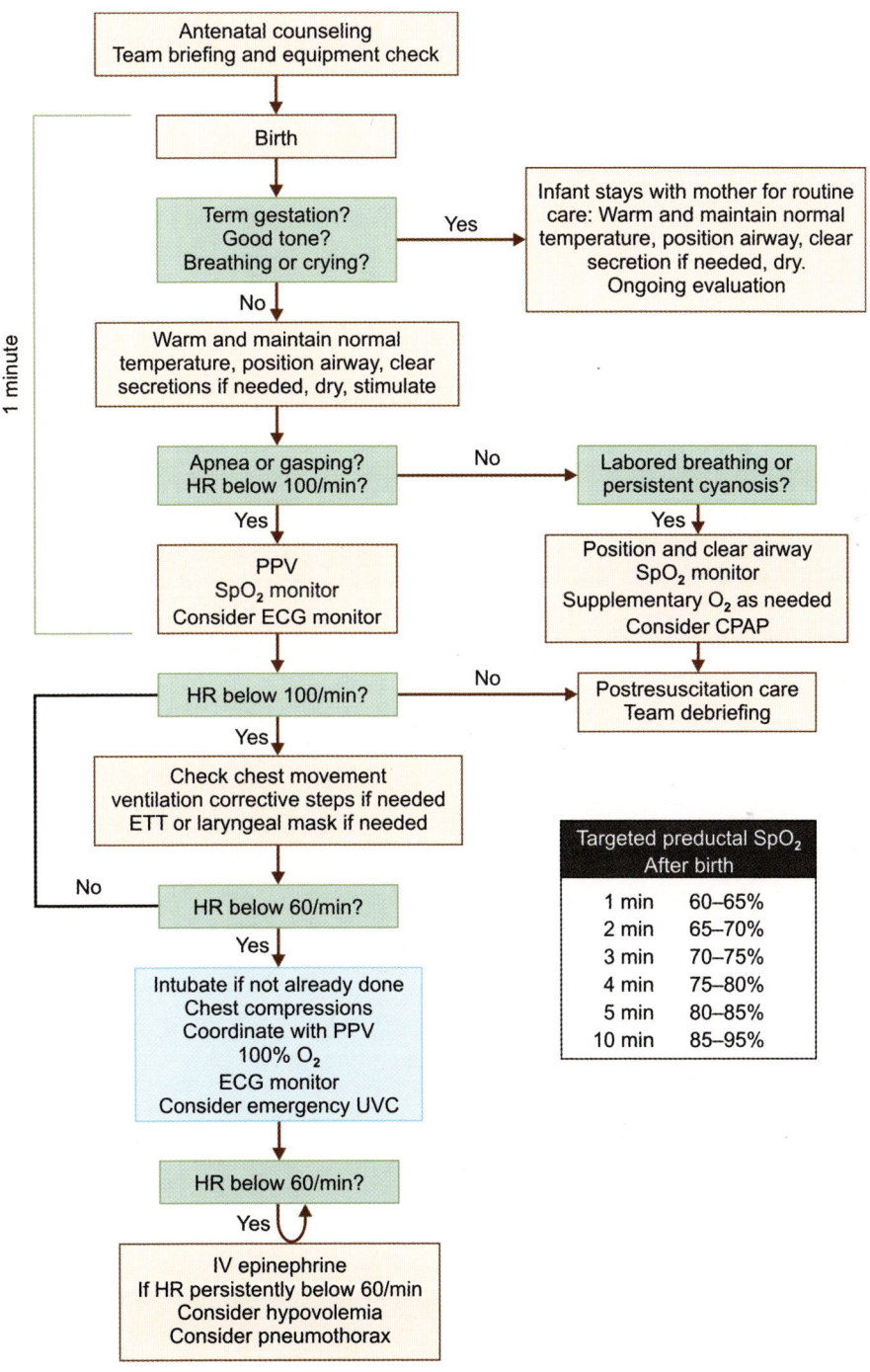

Fig. 27.10: Neonatal resuscitation algorithm, 2015 update

American Heart Association has Suggested following Major Changes in Neonatal Resuscitation in October 2015

1. Tracheal suctioning for non-vigorous newborns born through meconium-stained liquor done in past is not recommended.

2. In vigorous term and preterm newborns cord clamping should be delayed for at least 30 seconds. However, evidence to recommend for delayed cord clamping for newborns that require resuscitation at birth are insufficient. Cord milking is not recommended in routine. Initial steps followed by positive pressure ventilation (PPV) should be done as per routine indications.

3. Newborns < 35 weeks' gestation should be resuscitated with 21% to 30% oxygen. Oxygen concentration is titrated to meet the target oxygen saturation (SpO_2).

4. In infants <32 weeks various strategies like use of radiant warmers, thermal mattress, plastic wrap with cap, increasing of ambient temperature and use of warmed humidified resuscitation gas can be practiced to maintain normothermia. For warming hypothermic newborns rapid ($0.5°C/h$ or greater) or slow rewarming (less than $0.5°C/h$) can be done. In resource-limited settings covering the newborn in a food-grade plastic bag up to level of the neck after drying and skin-to-skin contact or kangaroo mother care can be practiced for maintaining normothermia. Temperature of non-asphyxiated infants should be maintained between $36.5°$ and $37.5°C$.

5. Along with PPV use of approximately 5 cm H_2O of PEEP is suggested in preterm newborns. PPV is done at the rate of 40–60 breaths per minute. Use of CPAP is preferred to routine intubation.

6. Electronic cardiac monitor is recommended for accurate assessment of the heart rate.

7. 100% oxygen is to be used during chest compression.

8. Induced therapeutic hypothermia for birth asphyxia is recommended even in resource-limited settings.

9. Training of the staff is recommended more frequently than the previous interval of 2 years.

SUGGESTED READING

1. Aslam, Hafiz Muhammad; Saleem, Shafaq; Afzal, Rafia; Iqbal, Umair; Saleem, Sehrish Muhammad; Shaikh, Muhammad Waqas Abid; Shahid, Nazish (2014-12-20). "Risk factors of birth asphyxia". Italian Journal of Pediatrics. 40.

2. Kaye, D. "Antenatal and intrapartum risk factors for birth asphyxia among emergency obstetric referrals in Mulago Hospital, Kampala, Uganda". East African Medical Journal. 80 (3): 140–143.

3. Wyckoff MH, Aziz K, Escobedo MB, Kapadia VS, Kattwinkel J, Perlman JM, et al. Part 13: Neonatal Resuscitation: 2015 American Heart Association Guidelines Update for Cardiopulmonary Resuscitation and Emergency Cardiovascular Care. Circulation 2015;132:S543–60.

4. Chettri S, Adhisivam B, Bhat BV. Endotracheal suction for nonvigorous neonates born through meconium stained amniotic fluid: a randomized controlled trial. J Pediatr 2015;166:1208–13.

5. Roehr CC, Hansmann G, Hoehn T, Bührer C. The 2010 Guidelines on Neonatal Resuscitation (AHA, ERC, ILCOR): similarities and differences—what progress has been made since 2005? Klin Padiatr. 2011;223:299–307.

Neonatal Resuscitation—Newer Protocols and Guidelines

Kartikeya Bhagat, Prashant Dixit

A healthy baby—who cries, breathes well and breastfeeds—is the expected outcome of every pregnancy. This depends on good antenatal care, proper intrapartum management with the use of partograms, appropriate implementation of the Neonatal Resuscitation Program, good infection control policies and adoption of the Baby Friendly Hospital Initiative (BFHI) Practices by the Maternity Service.

Out of 100 babies, 90 will breathe well and cry immediately after birth, 10 will need some help to initiate breathing but only 1 out of these 10 will require extensive resuscitation to initiate cry and breathing / begin life. The problem is, we do not know which baby will not cry and it is difficult to predict, especially in the no-risk group.

Since it is difficult to predict which baby will require resuscitation at the time of birth, the maternity service—and every birth attendant—has to be prepared to resuscitate every baby at every birth.

India is the world leader in neonatal births and neonatal deaths.

Since perinatal asphyxia is a major cause of neonatal deaths in our country, accounting for approximately 20–30% of NMR,[2] Neonatal Resuscitation Program becomes the most effective tool which can help address this issue.[3]

Who will Resuscitate?

Since presence of a neonatologist/pediatrician at the time of delivery, though desirable, may not be possible at every birth, whoever attends a birth should be proficient in neonatal resuscitation.

In resource limited settings like those existing in parts of our country, every birth may not be attended by a qualified doctor proficient in neonatal resuscitation. Hence, every birth attendant in our country—be it a doctor, nurse, trained birth attendant or a community worker—should be trained in the basic steps of neonatal resuscitation.

Where will the Neonatal Resuscitation be Carried out?

Every maternity service should have a predefined/dedicated place/corner where neonatal resuscitation would be carried out. This area should be close to the place of delivery, but not in direct sight of the mother. It should have room for at least three people to stand and carry out procedures. The area should be well lit/illuminated. The surface should be flat and firm.

What are the Requirements for Neonatal Resuscitation?

The cardinal requirement is anticipation and the preparedness for neonatal resuscitation

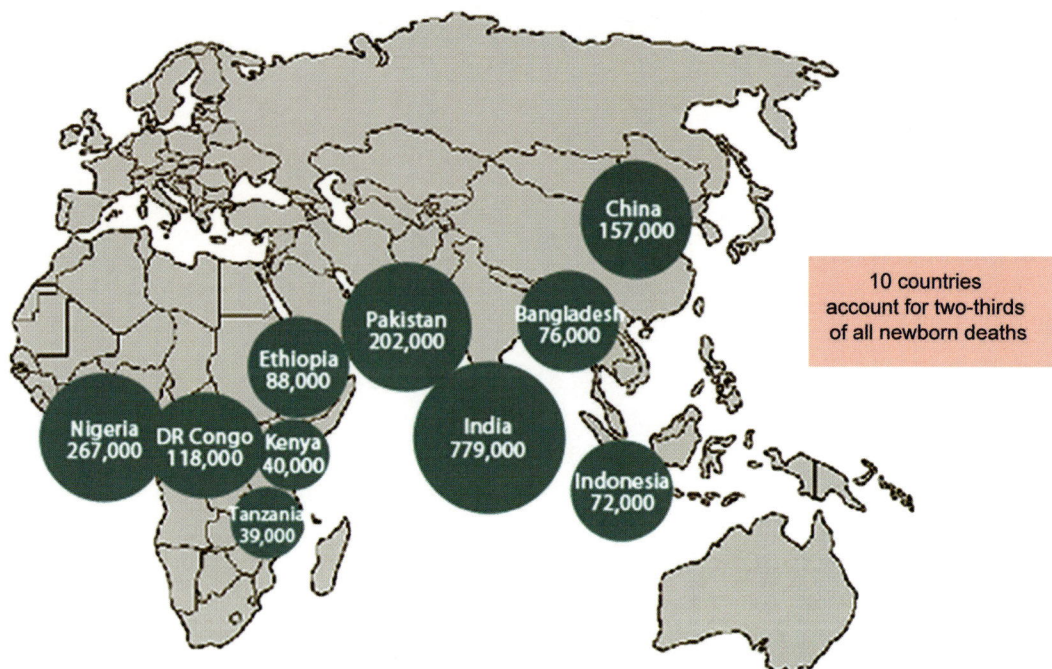

Fig. 28.1: India, the world leader in neonatal births and deaths

at every birth—equipment for neonatal resuscitation should always be available, checked regularly and be ready for use ALL the time.

1. A skilled birth attendant with a helper; emergency contact numbers of pediatrician/neonatologist/newborn ambulance services/NICUs to be kept at hand.

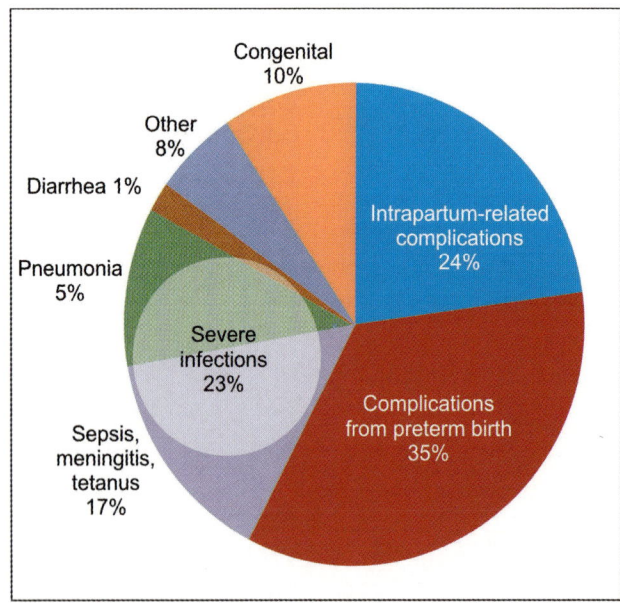

Fig. 28.2: Causes of newborn deaths

2. *Place for resuscitation:* A dedicated newborn corner should be marked in every maternity service where resuscitation would be carried out. In the newborn corner, a source of heat should be available to keep the baby warm—ideally a radiant warmer, or a 200 watt lamp or plastic wrap with a cap. In resource poor setting, skin-to-skin contact with the mother could be the only option.

3. A trolley where all the equipment required for resuscitation is maintained should always be kept ready.

The Equipment Required

Also, do not forget to keep a sterile cord clamp, umbilical cord cutting scissor and an identity band for the baby.

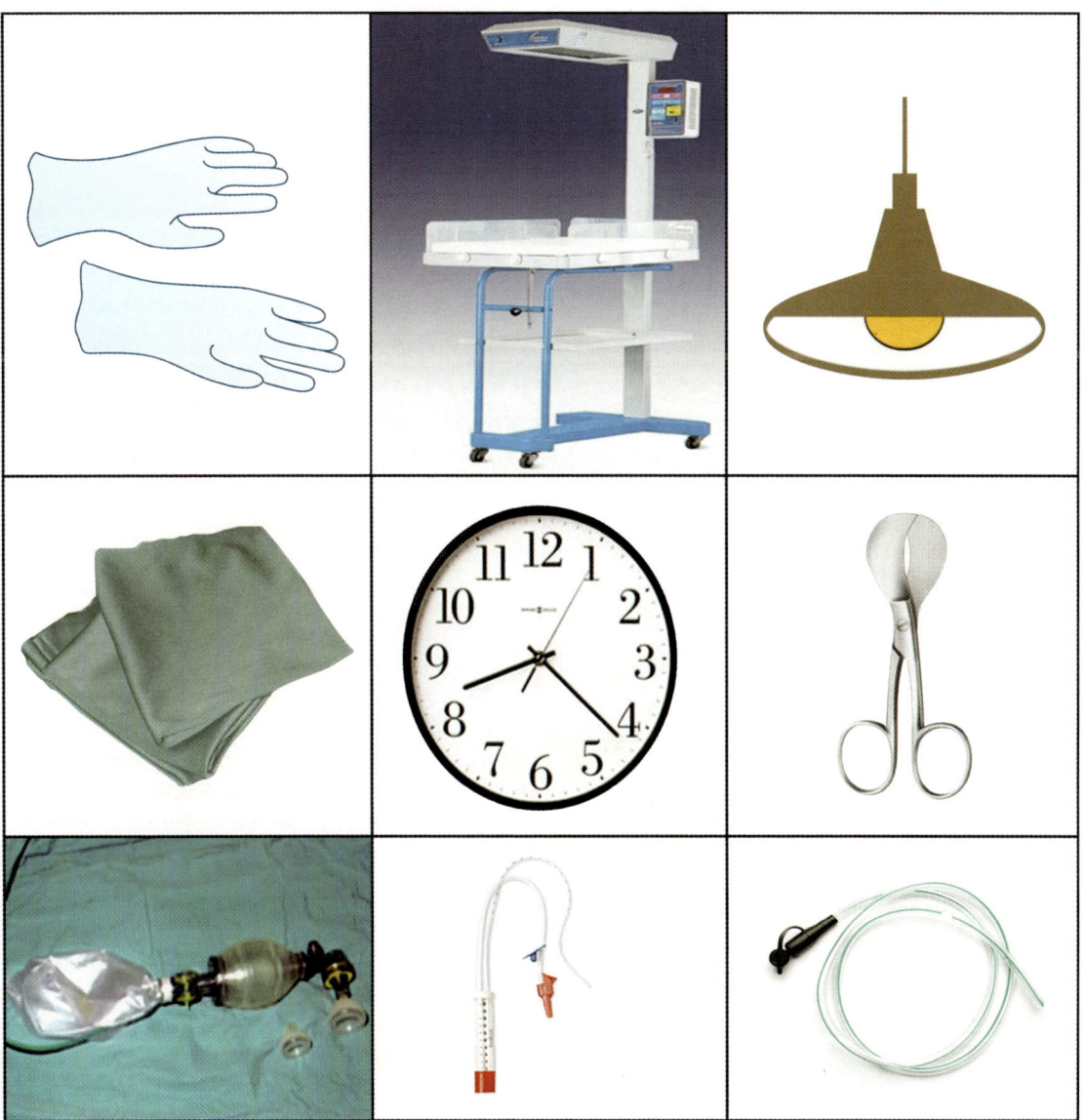

Fig. 28.3a: Minimum equipment required

EQUIPMENT LIST FOR NEONATAL RESUSCITATION IN DELIVERY ROOM

Temperature → Airway → Breathing → Circulation/clean hands → Dextrose

Temperature
- Designated resuscitation area
- Radiant warmer or 200 w bulbs (50 cm)
- Warmed linen (3) and shoulder roll

Airway
- Shoulder roll
- Suction apparatus—80–100 mm Hg with 10 F (2,14) suction catheter, 5 or 6 F and 8 F for ET suction
- Mucus extractor
- Meconium aspirator

Circulation
- Umbilical catheters 3.5, 5 F
- Three-way stop cock
- Syringes 1, 2, 5, 10 , 20 ml
- Sterile gloves
- Medications: Epinephrine, NS, RL
- ECG and SpO$_2$ monitor with ECG leads

Dextrose
- Glucometer with glucose strips
- Dextrose 10% solution

Clean hands and help
- Gloves
- Watch with seconds hand
- Stethoscope
- Helper

Breathing
- Self-inflating ambu bag (250–750 ml) with mask size 0 and 1
- Oxygen supply—source, tubings, reservoir
- T-piece resuscitator
- Laryngoscope with straight blades 0, 1 and 00 (optional) and extra set of batteries
- Endotracheal tube with inner diameter of 2.5, 3, 3.5, 4
- Endotracheal tube stylet (optional)
- Scissors and adhesive tape for fixing endotracheal tube
- 8F feeding tube

And of course
- Calm, cool head
- Systematic approach

Fig. 28.3b: Ideal list of equipment to be kept ready

Steps of Neonatal Resuscitation

Neonatal resuscitation is one of the most important steps in reducing neonatal mortality. The key feature in resuscitating a newborn is to avoid unnecessary delay in initiation of ventilation.

Remember, initial 60 seconds (First Golden Minute) of focused and systematic approach can help a newborn breathe and live!! Once the baby is born:

1. Confirm that it cries and breathes well; and

2. Ask for the presence or absence of meconium in liquor amnii.

If the baby is breathing and crying, look at the tone, color and activity. Chart the APGAR Score.

If the baby is fine, all she requires is routine care: Dry the baby with a pre-warmed sterile cloth—especially the head and the body, care should be taken not to wipe off the amniotic fluid from the palms. Smell of the amniotic fluid is similar to smell of some chemicals on the breast. This helps the baby in identifying the breast during Breast Crawl. Change the wet linen.

Do not be in a hurry to cut the cord-delayed cord cutting/clamping (1–3 minutes) is advocated.

Keep the baby on the mother's abdomen—most babies will crawl to the breast and take the first feed in the first 60 minutes. The mother and the baby are covered by a warm blanket.

All routines like weighing the baby, giving injection vitamin K, anthropologic measurements, wrapping the baby, etc. should be delayed till after the first feed. Cleaning the vernix from the body and baby bath in the first 24 hours should be discouraged.

Primary vs Secondary Apnea

It is not unusual to have a baby who cries (may be weak) immediately after birth and then stops breathing and becomes limp. The birth attendant panics and continues stimulation, may be even violently at times. No amount of stimulation or violent slaps help. The baby remains limp and just does not breathe. This baby is in *secondary apnea.*

> If a baby does not begin breathing immediately after being stimulated, he or she is likely in secondary apnea and will require positive-pressure ventilation. Continued stimulation will not help.

Studies have shown that cessation of respiratory efforts is the first sign that a newborn has had some perinatal compromise (Neonatal Resuscitation Textbook, AAP/AHA, 6th edition).

Perinatal stress results in an initial period of rapid breathing followed by a period of primary apnea (no breathing or gasping). During this period of primary apnea, stimulation, such as drying the newborn or slapping the feet, will cause resumption of breathing.

However, if cardiorespiratory compromise continues during primary apnea, the baby will

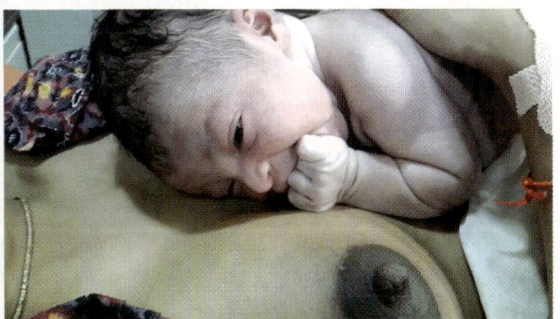

Fig. 28.4: Breast crawl amniotic fluid on hand

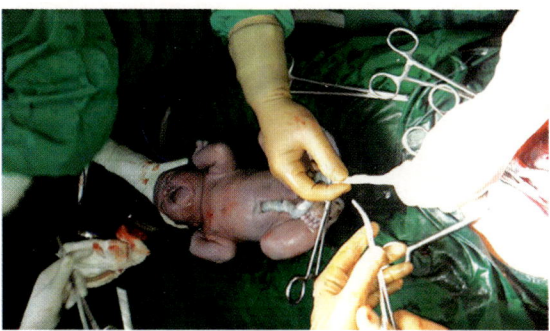

Fig. 28.5: Delayed cord clamping

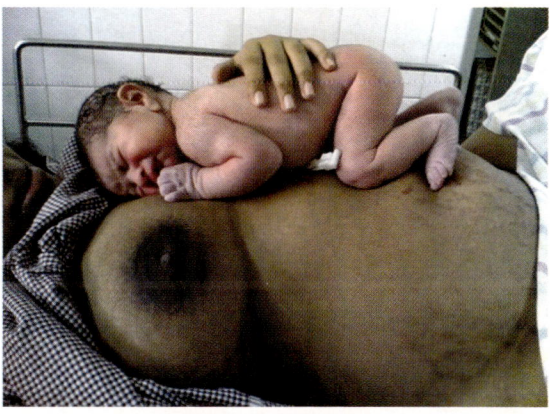

Fig. 28.6: Breast crawl

have an additional brief period of gasping breaths and then will enter a period of secondary apnea.

Heart rate begins to fall at about the same time as baby enters primary apnea.

Blood pressure is usually maintained until the onset of secondary apnea.

During the phase of secondary apnea, no amount of stimulation will restart the baby's breathing. Assisted ventilation must be provided at the earliest to reverse the process. Most of the time the baby will present to us somewhere in the middle of the sequence described above.

The compromising event will have started sometime before and may be during labor; and therefore at the time of birth, it is difficult to determine how long the baby's respiration and circulation have been compromised.

Physical examination of the baby will not allow/help us to distinguish between primary apnea and secondary apnea.

However, the respiratory response to stimulation may help us to distinguish/ estimate how recently the event began. If the baby is breathing as soon as she is stimulated, she was in primary apnea; if she does not breathe right away, she is in secondary apnea and respiratory support must be initiated without wasting time as no amount of stimulation will restart respiration.

The NRP algorithm does not change in either scenarios, however, the difference in physiology and hence diagnosis may indicate the difficulty likely to be encountered in neonatal resuscitation and the prognostication.

So, if the baby does not cry or breathe well immediately after birth, then you need to follow the resuscitation algorithm as under.

Neonatal Resuscitation Algorithm in Practice

Questions to be asked now, at the time of delivery:

1. Is the liquor clear or meconium stained? Yes/ No
2. Is the baby crying or breathing? Yes/No

90% of the newborns will not require any resuscitation after birth, especially if the

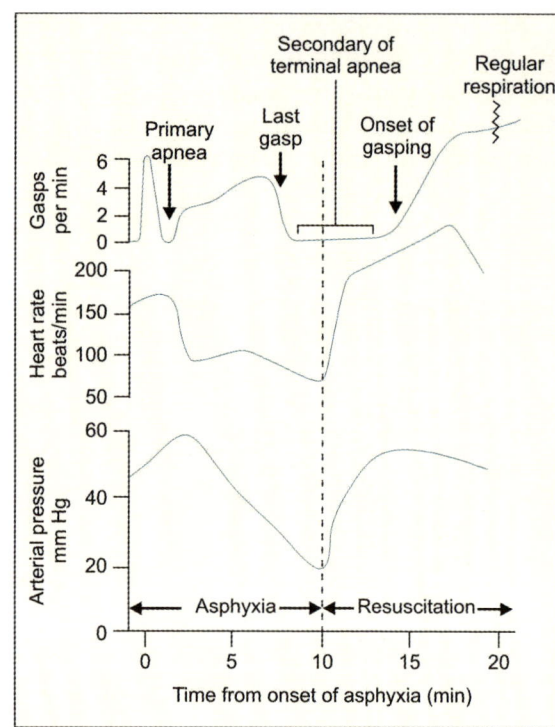

Fig. 28.8: Sequence of physiological events in animal models involving total asphyxia

answer to these questions is 'yes', and will be given to the mother for routine care.

According to the earlier guidelines, questions asked were: Is the baby term? Is the baby breathing or crying? Is the tone good? The focus has shifted from airway and breathing to circulation.

Earlier guidelines asked, apart from crying/ breathing, whether the baby is term and what is the color, tone. Current guidelines ask for presence/absence of meconium in liquor and it is very much possible that the next set of guidelines will ask only one question: Is the baby crying and breathing?

Over the time, the focus seems to have shifted from airway and breathing to circulation as the cardinal event that needs to be preserved primarily.

Resuscitation protocols change according to the gestational age and tone is considered a prognostic factor.

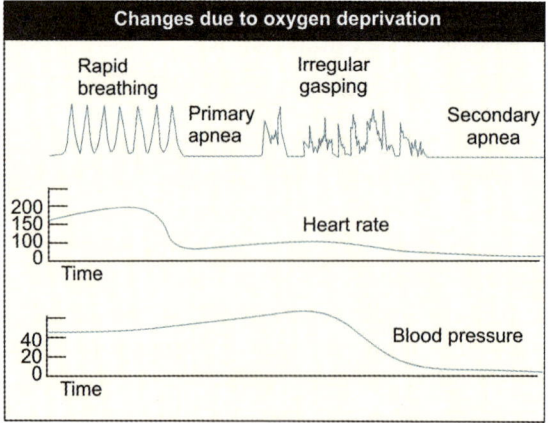

Fig. 28.7: Primary apnea vs secondary apnea

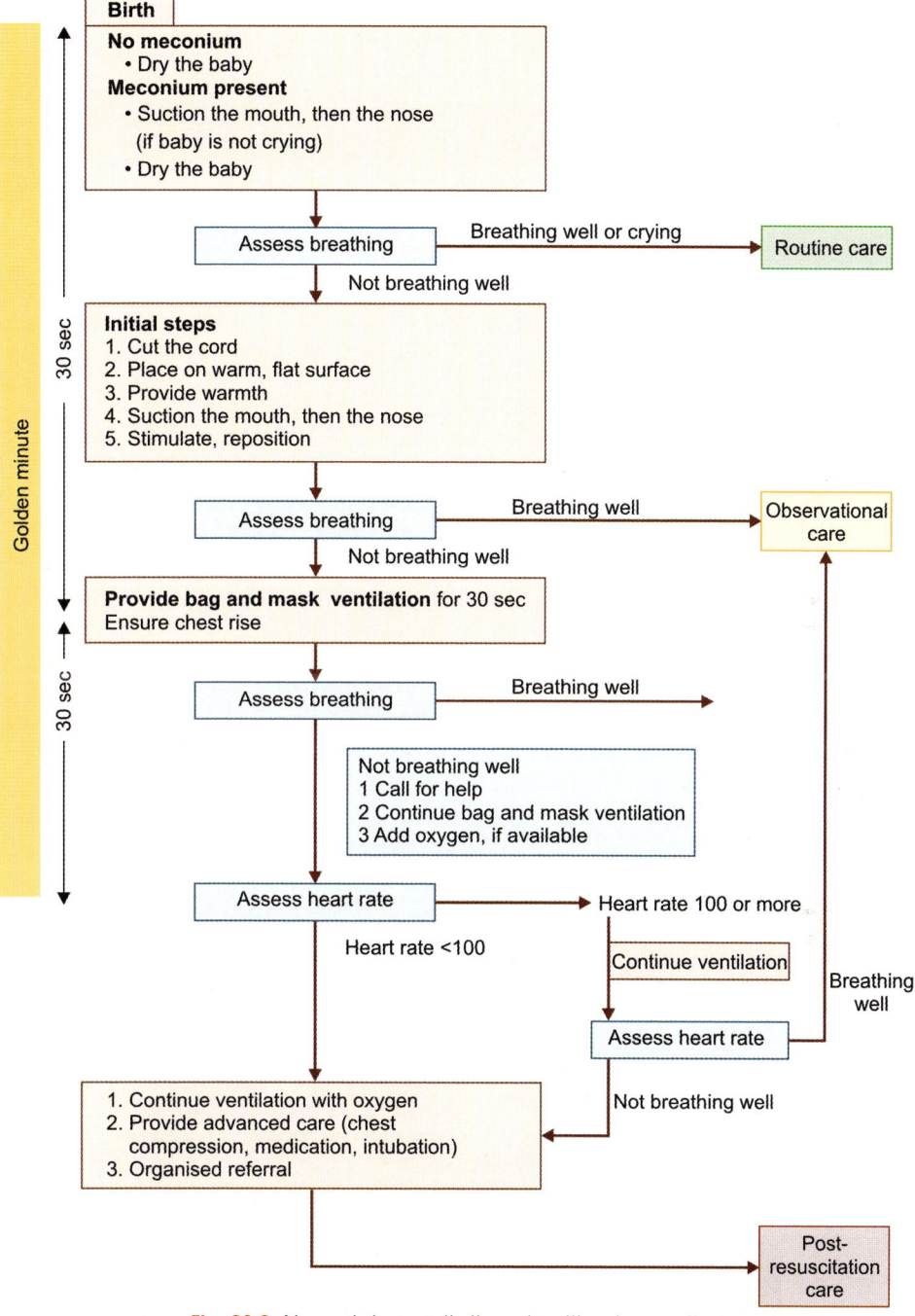

Fig. 28.9: Neonatal resuscitation algorithm in practice

The change in guideline, from the previous one, includes:

- Just two primary questions, instead of three

- Thermoregulation
- Use of ECG to measure heart rate instead of palpating/auscultating or the use of pulse oximetry

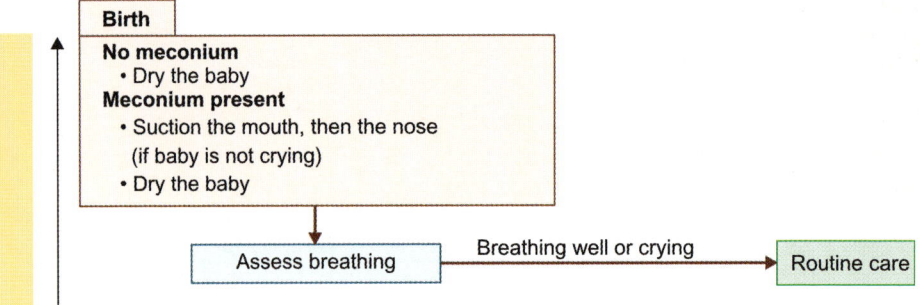

Fig. 28.10: NRP algorithm: Routine care

- Effective ventilation which results in chest rise and use of endotracheal intubation before chest compression
- Use of room air to initiate resuscitation/ventilation; and the use of O_2 blender instead of pure O_2. Also, use of normal saline to correct hypovolemia.

Routine care includes:

1. Provide warmth and maintain normal temperature

   → Skin to skin contact, best achieved by the "*Breast Crawl*"

2. *Clear airway only if necessary:* Check position, clear secretions: Oropharyngeal suction only if required

3. *Dry the baby:* Preferably with a pre-warmed sterile towel: Head, body and limbs-care should be taken to avoid cleaning the amniotic fluid from the baby's hands. Change the wet linen.

4. *Evaluate:* Heart rate, respiration, color, tone, activity. Chart the APGAR score.

5. *Avoid early cord clamping:* Cord to be cut/clamped after 1–3 minutes of birth

6. Baby stays with mother for breast crawl/routine care

If a baby does not cry/breathe immediately after birth

- Tie the umbilical cord
- Tell the mother and
- Transfer the baby to the newborn corner (TTT)

Steps of Neonatal Resuscitation

The initial steps include: Position, suction, stimulation, re-position (PSSR)

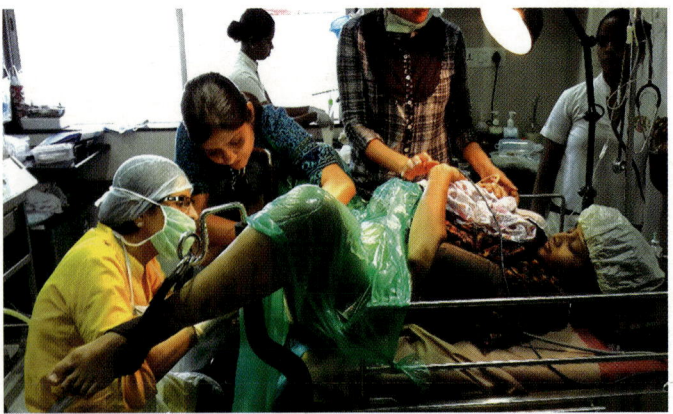

Fig. 28.11: Breast crawl after delivery ...and during cesarean section

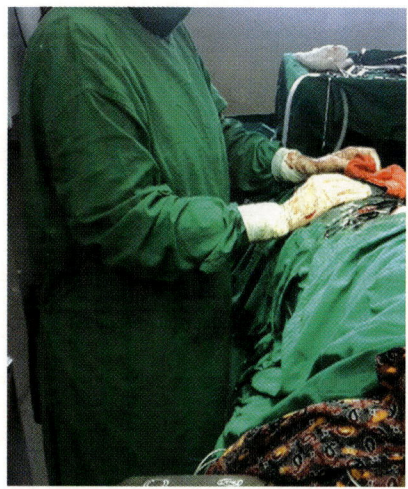

Fig. 28.12: Suction and administration of oxygen is not routinely required in newborns

Position: Sniffing position of slight extension with the help of the shoulder roll in order to align the airway.

Avoid hyperextension or flexion as they cause obstruction to the airway.

Suction: To clear the airway → oropharyngeal suction, mouth before nose.

Appropriate sized suction catheter to be inserted in the mouth up to 5 cm and nose up to 3 cm.

Stimulation: Flicking or gently slapping the sole of the feet. Rubbing the back

Do not harm the baby by following steps:
- Slapping the back
- Squeezing the rib cage
- Forcing thighs into the abdomen
- Dilating anal sphincter
- Hot or cold compresses or baths
- Shaking

Re-position: May be required, because initial steps of suction/stimulation may have changed the sniffing position.

a. *Assess:* Is the baby breathing or crying?
   If Yes—assess the quality of respiration →
   If good → observational care
   If No—bag and mask ventilation
b. *Bag and mask ventilation*
   Round mask is commonly used.
   Size 0 for preterms and
   Size 1 for term babies

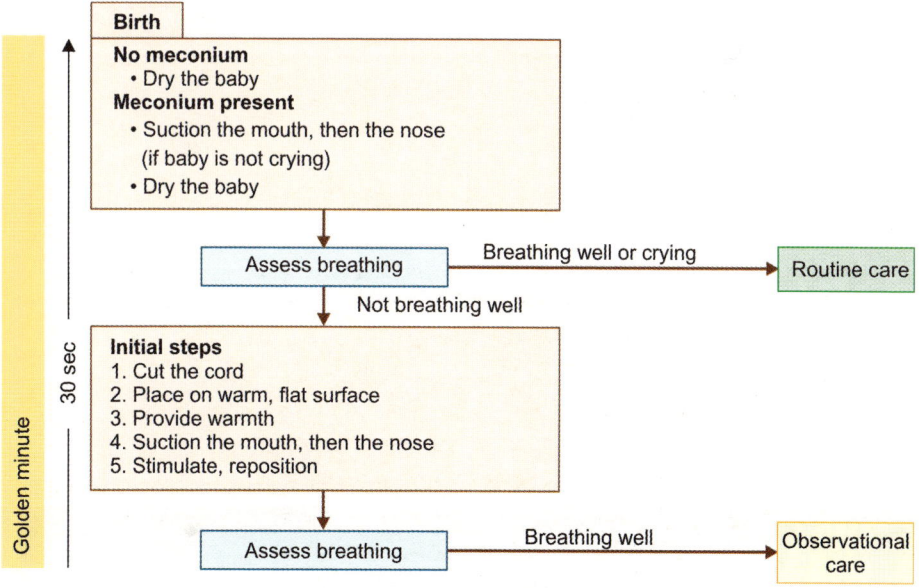

Fig. 28.13: NRP initial steps

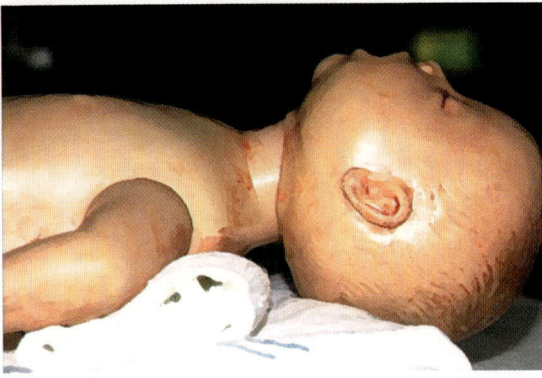

Fig. 28.14: Sniffing position for resuscitation

The correct sized mask is applied appropriately to get a proper seal; and bag and mask ventilation is initiated in the following rhythm:

Breathe……..2………..3

Breathe……..2………..3

Breathe……..2………..3

This provides ventilation at a rate of 40–60 breaths per minute

Re-assessment is done after 3 breaths → look for chest rise

If bag and mask ventilation fails, i.e. no chest rise is seen after a few breaths—ventilation corrective measures are to be instituted (MR SOPA) in the following order; 2 at a time

M—Mask reapplication

R—Reposition

S—Suction

O—Open mouth

P—Pressure increased

A—Alternate airway

Bag and mask ventilation is to be continued for 30 seconds → assessment → if breathing/respiratory efforts present → wean off ventilation gradually → assess quality of respiration.

If no respiratory efforts → continue bag and mask ventilation → ask for help → assess heart rate through umbilical cord pulsations for 6 seconds × 10/auscultation with stethoscope/attach oxygen, reservoir to ambu bag and pulse oximeter and/or ECG leads.

If bag and mask ventilation is continued beyond 2 minutes →

Insert orogastric tube—at this point

i. Measure the length

ii. Insert the tube through the mouth

iii. Gently aspirate stomach contents

iv. Leave the end of the tube open

v. Tape the tube to the cheek of the baby

Remember: Increase in heart rate is the most sensitive indicator of a successful response to each step practiced.

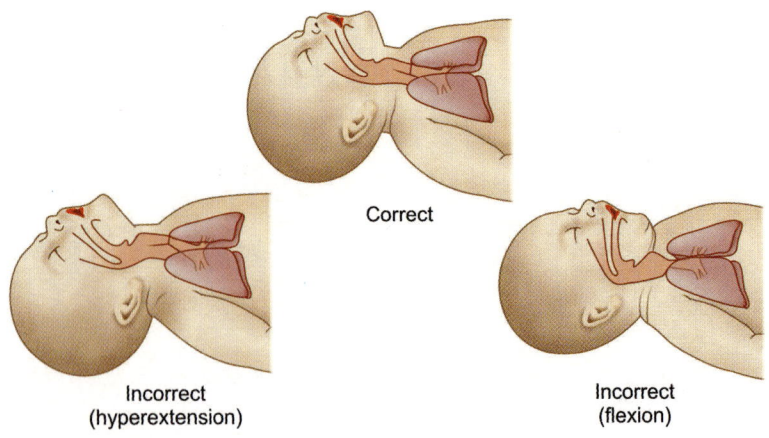

Correct

Incorrect
(hyperextension)

Incorrect
(flexion)

Fig. 28.15: Diagrammatic representation of the correct position for resuscitation

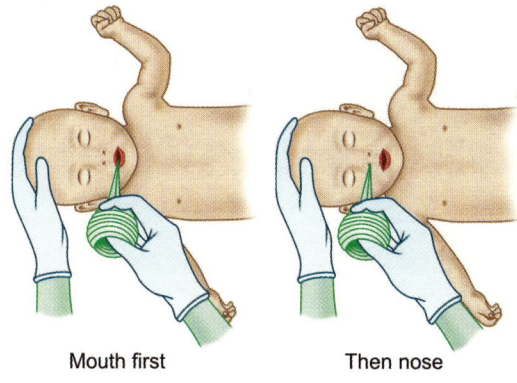

Mouth first   Then nose

Fig. 28.16: Suction, mouth before nose

If the baby is not improving—ask,
- Is the chest movement adequate?
- Is the heart rate good?
- Has the tone/color improved?

At this point consider
a. Adding oxygen to bag and mask ventilation

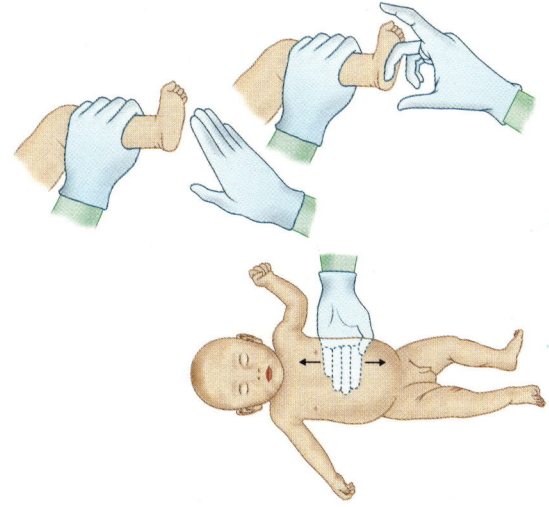

Fig. 28.17: NRP stimulation

b. Endotracheal intubation
c. Chest compression
d. Drugs

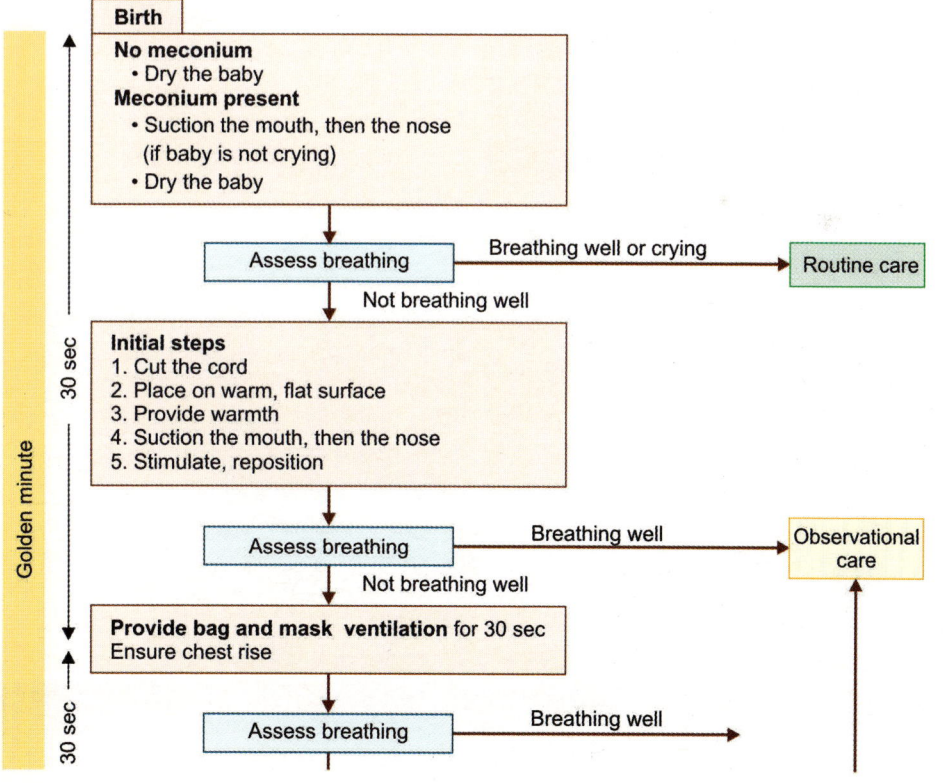

Fig. 28.18: Neonatal resuscitation algorithm in practice

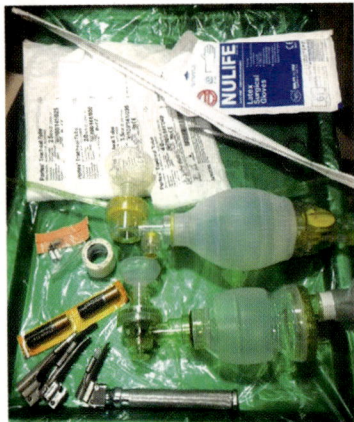

Fig. 28.19: Ambu bag

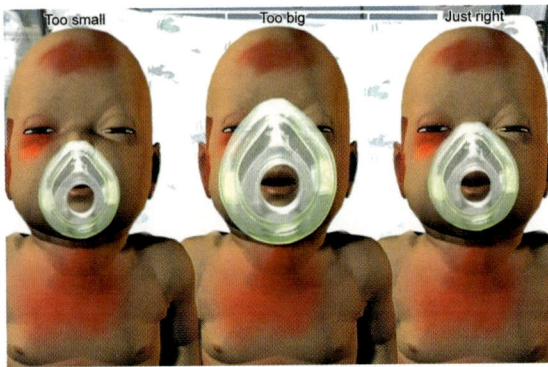

Fig. 28.20: Face mask

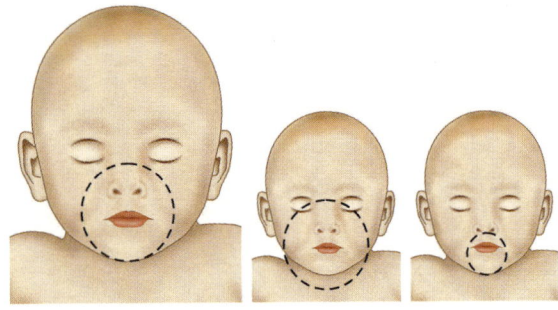

Fig. 28.21: Face mask application

Endotracheal intubation

To be considered under the following indications:

1. Meconium stained liquor, non-vigorous baby—for suction and/or ventilation
2. Failure of bag and mask ventilation, in spite of corrective steps
3. Prolonged bag and mask ventilation
4. Chest compressions required (heart rate < 100 bpm), for better coordination of IPPV and chest compressions
5. Special indications like congenital diaphragmatic hernia, extreme prematurity, planned surfactant administration.

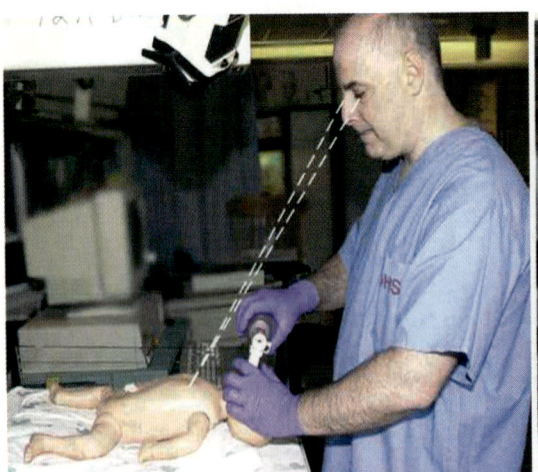
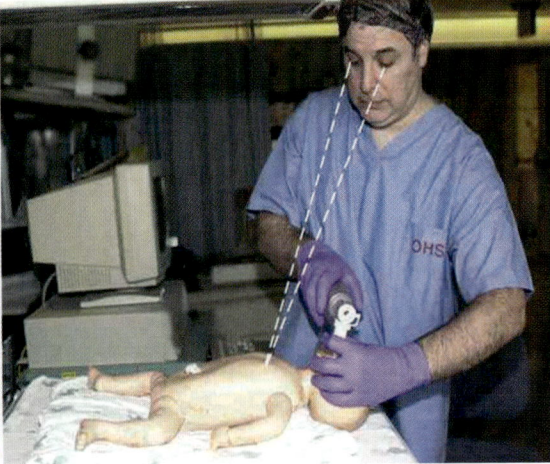

Fig. 28.22: Position for resuscitation

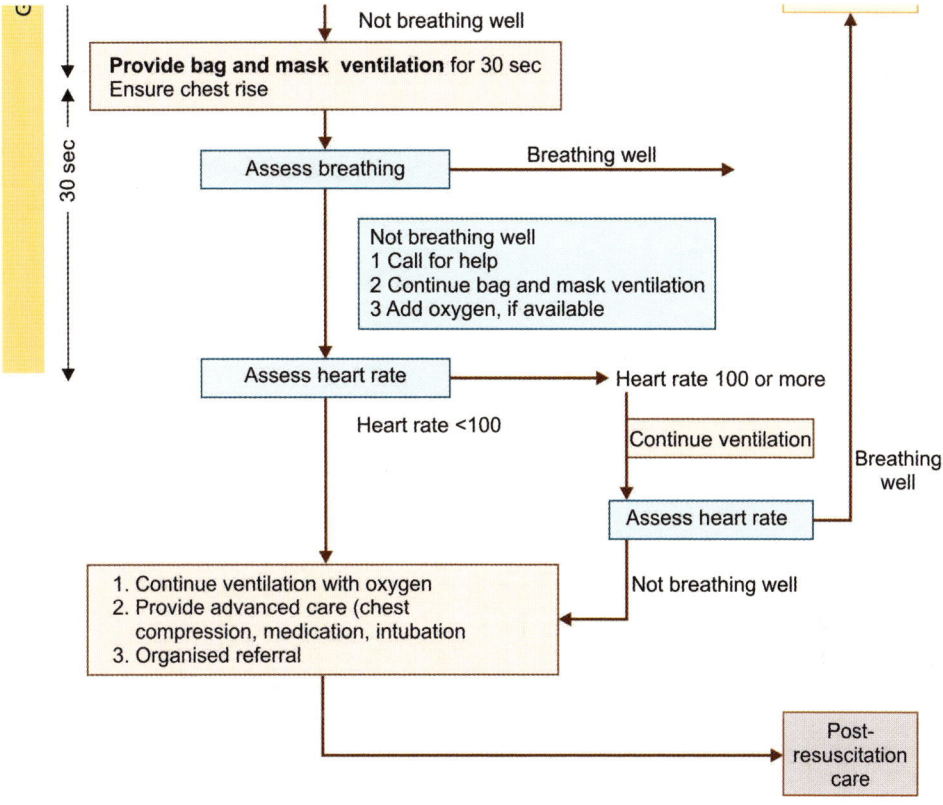

Fig. 2823: NRP agorithm—advanced care

Table 28.1	Selecting the tube size	
Weight (gm)	Gestational age (weeks)	Tube size (mm) inside diameter
Below 1000	Below 28	2.5
1000–2000	28–34	3.0
2000–3000	34–38	3.5
Above 3000	Above 38	3.5–4.0

Tube placement formula—up to (6 + wt of baby) in cm, e.g. for a 2 kg baby up to 6 + 2 = 8 cm.

Chest compression: To be considered if heart rate < 100 bpm despite 30 seconds of effective PPV (bag and mask).

Two techniques: 2-thumb technique/2-finger technique (2-thumb technique is preferred).

Fingers should remain always in contact, encircle 1/3rd of AP diameter

Coordination with bag and mask.

In the following rhythm: One... AND... Two... AND ... Three ... (3 chest compressions) AND ...

Squeeze ... (IPPV)

STOP: When HR > 100 bpm

Medications

a. Epinephrine
   – HR <100/min after ventilation and chest compressions
   – *Dose:* 1:10,000 Adrenaline can be given through IV or ET, 0.1–0.3 ml/kg or 0.3 to 1 ml/kg respectively
   – Normal saline/RL (dilute 1 ml of Adrenaline in 9 ml of NS to get 1:10,000 Adrenaline)

b. Normal saline
   – Tachycardia, low pulse volume, blood loss
   – 10 ml/kg IV over 5 to 10 minutes

Neonatal resuscitation—intubation

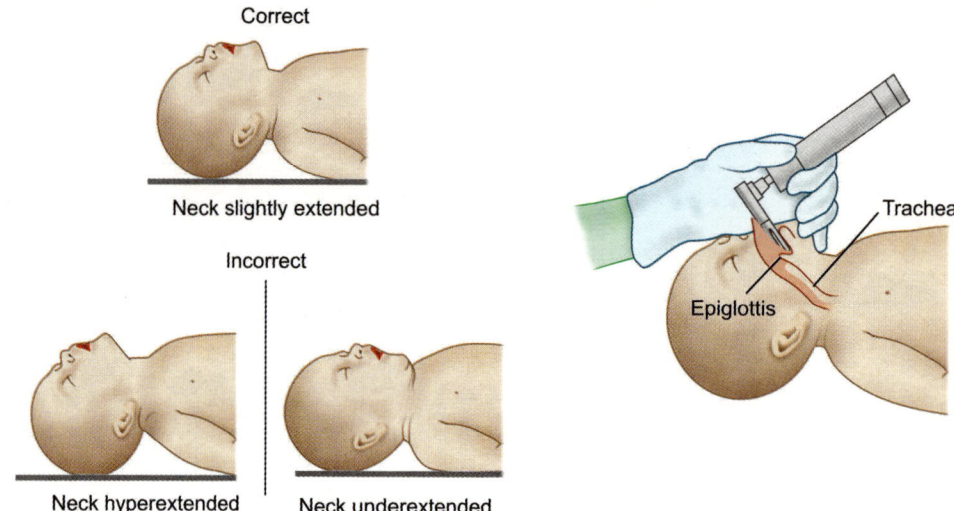

Only experience and well practiced clinicians should attempt to intubate

Fig. 28.24: Intubation

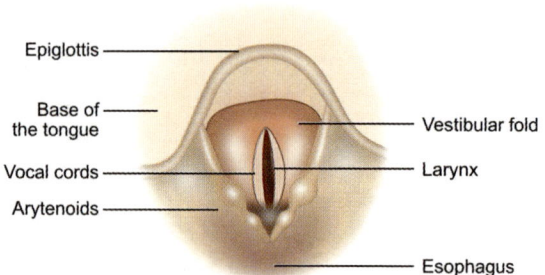

Fig. 28.25: Laryngoscopic view

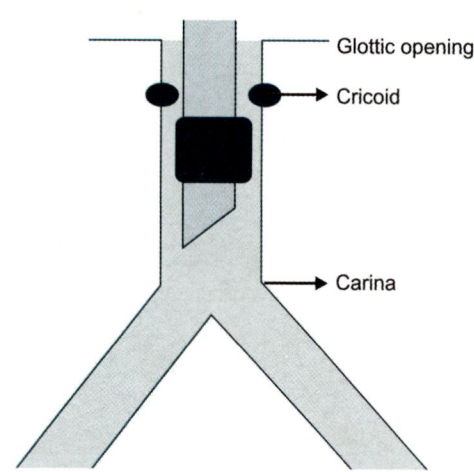

Fig. 28.26: Placement of the ET

c. Naloxone
  – 0.1 mg/kg, IV, respiratory depression
  – Opiods to mother within 4 hours of delivery

Parasympathetic stimulation hence bradycardia

d. Bicarb soda: Say no to bicarbonate!!

Current Recommendations

I. AAP-NRP 2015 recommended addition of ECG leads to pulse oximeter monitoring during newborn resuscitation.

II. Pulse oximetry is useful for helping use oxygen rationally, whereas ECG monitoring is more effective in picking up cardiac activity

III. Use of air-oxygen blenders in labor room and delivery room CPAP with neopuff neonatal resuscitator is also suggested to improve neonatal outcomes.

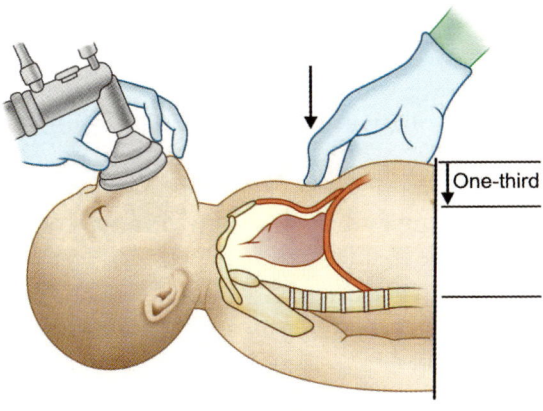

One-third

Fig. 28.27

When is Non-initiation of Resuscitation Appropriate?

- Confirmed gestational age <23 weeks or a birth weight <400 gm
- Anencephaly
- Confirmed lethal genetic disorder or malformation
- When available data support an unacceptably high likelihood of death or severe disability

When do You Give up Resuscitation?

In a newly born baby with no detectable heart rate, it is appropriate to consider stopping resuscitation if the heart rate remains undetectable for 10 minutes (Class IIb, LOE) in spite of resuscitation.

The decision to continue resuscitation beyond 10 minutes with no heart rate should take into consideration factors such as the presumed etiology of arrest, gestation of the baby, the presence or absence of complications, potential role of hypothermia, and the parents' previously expressed feelings about acceptable risk of morbidity.

How do You Monitor a Baby who required Resuscitation at Birth? For how Long?

All babies who require resuscitation should be monitored and followed up.

Baby who required just a little stimulation to initiate breathing or responded to just the initial steps of resuscitation, will need just the routine care, and should be managed like all normal babies and discharged in 3/4 days.

If the baby required bag and mask ventilation for >30 sec, observational care in the maternity service is required; but if the baby required bag and mask for >2 min or extensive resuscitation, it is wiser to shift the baby to a NICU for post-resuscitation care and monitoring.

This baby (full term) may need NICU care for up to 7 days. If the baby does not throw convulsions or show any other sequel or neurological change following resuscitation, he/she could be discharged from the NICU once lactation is established and the baby shows weight gain on two consecutive days.

Transport to NICU

A baby who responds to the initial Steps or required minimum bag-and-mask ventilation for <30 seconds requires no special care and is kept with the mother.

However, a baby who requires bag-and-mask for >30 seconds, endotracheal intubation, chest compressions or drugs needs to be closely monitored and may be shifted to an NICU.

Transport from the maternity service to the NICU is crucial and should be done preferably, in a transport incubator whenever available. Care should be taken to prevent hypothermia, hypoxia, hypovolemia and hypoglycemia.

See the chapter on neonatal transport for detailed discussion on the subject.

Follow-up Protocol

It is recommended that a baby who received significant resuscitation at birth and who goes on to show signs of encephalopathy should be assessed by Sarnat Staging between 24 and 48 hours from birth.

These babies may be seen two days after discharge to confirm established lactation and absence of jaundice; one week later to check weight gain; one, three, six months later for neurological assessment and checking developmental milestones.

Depending on the degree of resuscitation required, following tests may be offered:

1. OAE (screen for hearing) at discharge/ BERA if NICU care
2. Neurological assessment at 1 month, 3 months, 6 months and 1 year
3. EEG at the earliest before stopping anticonvulsants
4. MRI at 3 months to define the degree of neurological damage

Prognosis of a Baby who Required Resuscitation

Long/Short Term

Most babies will do well, if provided with adequate resuscitation care at birth.

More than 50% of babies who require resuscitation, who do not throw convulsions and recover fast, do not show any short/ long term sequelae.

Prognosis depends on stage of the HIE and Sarnat and Sarnat score. Sarnat Staging Scale may be of great use.

Mild HIE (Stage 1), according to the Sarnat scale, usually has a normal outcome; whereas in severe HIE (Stage 3) the mortality rate is as high as 75% and 80% of survivors have neurological sequelae.

Breastfeeding

All babies should be breastfed. Babies who require only the initial steps are given to the mother and breastfed like normal newborns, but babies who required extensive resuscitation may be fed breast milk through the RT in the transition from IV nutrition to exclusive breastfeeding.

Medico-legal Issues

A baby who does not cry at birth is a serious and unexpected situation, not only for the parents/relatives but also for the doctor. An expected moment of joy turns into despair which sometimes may lead to accusation of negligence which may culminate into a possible law suit later.

Since it is difficult to predict which baby will not cry and require resuscitation, the search for cause may be inconclusive. An obvious cause may be found only in very few situations especially where the fetus is compromised (severe IUGR, severe oligohydramnios, fetal distress ....), either in the antenatal or intra-natal period.

A few precautions/measures to prevent such situations and counter such allegations would be:

1. *Antenatal counselling:* Discuss the possibility that 1 in 10 chance that a baby may not cry and require resuscitation, without any obvious cause, and may need to be transferred to a NICU. This should be done in the antenatal period, much before the time of delivery, and the session should preferably involve the husband and the relations.
2. *Good antenatal care:* To pick up high risk features, e.g. IUGR, oligohydramnios, PIH, GDM and address them.
3. Institutional delivery, preferably, or at least a supervised delivery.
4. *Proper intra-natal care (supervised labor):* Use of a partogram to document the progress of labor is a great tool, extremely useful not only during labor, but also later as a medicolegal document for standard of care.
5. Every birth to be supervised, preferably by a person trained in neonatal resuscitation- a doctor, nurse, ANM or a trained birth attendant.

Presence of a neonatologist, though welcome, may not be possible in all deliveries, especially in our country.

6. *Preparation for neonatal resuscitation:* Equipment to be checked and kept ready for every birth.
7. Proper documentation of the steps of resuscitation taken/required.
8. *Arrangement for transfer to NICU:* Contact numbers of the pediatrician/neonatologist and the NICU to be kept handy, intimation, personnel, role of transport incubator, time taken for transport.
9. NICU back up

The most important concept to prevent a medico-legal situation is: The obstetrician and the neonatologist to work together as a team, presenting a single front. Counter accusations and pushing the blame from one to another is the worst event/case scenario and will definitely precipitate into an unwarranted law suit. On the contrary, the team of obstetrician-neonatologist should jointly counsel the parents/relatives of the baby, give them a clear idea of the situation offering a possible explanation wherever possible, presenting a clear picture and giving a realistic prognosis.

CONCLUSION

Neonatal resuscitation remains one of the most important measures to prevent neonatal morbidity and mortality. This can be achieved by institutionalising every delivery where every birth attendant is trained in neonatal resuscitation; improving neonatal transport and creating centers capable of looking after sick newborns close to the place of delivery.

In a country like India, full of contrasts and disparities, a single set of guidelines for neonatal resuscitation will not work: each region will have to have its own guidelines, incorporating the basic principles of neonatal resuscitation.

Care of the sick and extremely premature newborns requires elaborate infrastructure and the onus lies with not only with the care provider but also the Administration/Government. In our country, where 70% health care is provided by the private sector, the disparity between the two sections is huge. Dedicated Herculean efforts will have to be put in backed by sensible health policies to develop centres of excellence if we wish to save every baby.

FIGURE SOURCE

Figure 1—UNICEF, WHO, The World Bank, 2013
Figure 2—Liu L et al, 2014
Figures 10, 11, 14, 19, 21, 24—IAP-NNF NRP Textbook
Figures 8, 9, 16, 17, 18, 22, 23, 25, 26, 27, 28—Table 1 AAP/AHA NRP Textbook, 6th Edition
Figure 3—NSSK Booklet

SUGGESTED READING

1. UNICEF, WHO, Levels and Trends in Child Mortality, 2013.
2. Liu, et al. Statistical Report, Lancet 2014 and WHO World Health Status, 2007, India.
3. Lancet Neonatal Survival Series, 2007.
4. UK Resuscitation Council guidelines on newborn life support.
5. Sarnat H, Sarnat M. Neonatal encephalopathy following fetal distress. Arch Neurol 1976;33: 695-705.
6. Gardiner M, Eisen S, Murphy C. Training in paediatrics: the essential curriculum. Oxford University Press; Oxford 2009.
7. Newborn Life Support: Third Edition. Resuscitation Council (UK), London 2011.
8. AAP Neonatal Resuscitation Textbook, 6th Edition, 2012.
9. Volpe Neurology Textbook, 6th edition.
10. Neonatal Resuscitation Textbook, AAP/AHA, 6th edition.

Care of a Premature Newborn

Vaidehi Dande, Sanjay B Prabhu

Premature newborns are the most fragile of human beings and need to be managed with utmost care. A newborn with gestational age less than 37 completed weeks is defined as being premature.

Gestational age from first day of last menstrual period (day 1/week 0/7) till day 259 or 36 6/7 weeks gestation is preterm.

Late preterm is 34 0/7 till 36 6/7 weeks or day 239–259 gestation.

Early term is 37 0/7 weeks till 38 6/7 weeks or day 260–274 of gestation.

Over the years the cut-off for period of viability has reduced and due to advanced neonatal care there has been a significant increase in number of surviving premature babies. Premature newborns have special needs and problems and require special care right from birth.

Following birth, age of preterm babies is expressed as chronological age and corrected age. Chronological age is the time elapsed after birth and is expressed in days, weeks, months, and/or years. Corrected age represents the age of the child from the expected date of delivery. The corrected age, and not the chronological age, age should be used till up to 3 years of age for babies who were born preterm.

DELIVERY ROOM CARE

The first 60 minutes after birth is called as the 'First Golden Hour' as events and actions during the first hour after birth have great influence on immediate as well as long-term outcome.

The American Academy of Pediatrics—Neonatal Resuscitation Programme 7th edition published in 2015 and modification of its 6th edition by Indian Academy of Pediatrics/National Neonatology Forum have made the following recommendations for care of premature newborn at birth.

1. Delivery room temperature should be maintained at 26°C. A neonatologist or pediatrician trained in newborn resuscitation should be present in the delivery room and all equipment required for newborn resuscitation should be ready.
2. Immediately after delivery, the baby should be wrapped in sterile food grade plastic without drying, to prevent insensible water loss and hypothermia.
3. Use of Continuous Positive Airway Pressure (CPAP) in the delivery room immediately after birth is highly desirable. This helps in better inflation of lungs and prevents the collapse of lungs. Therefore, the work of breathing is reduced. Use of a T-piece resuscitator for delivering CPAP and positive pressure ventilation in delivery room is highly desirable.

4. Use of oxygen blender and pulse oximeter is also desirable. The pulse oximeter probe should be attached to right upper limb. The pulse oximetry reading should be used to titrate oxygen delivery and hyperoxia should be avoided.

5. If baby is hemodynamically stable, cord clamping can be delayed by 1–3 minutes and skin-to-skin contact with the mother can be initiated.

6. Minimum use of oxygen during resuscitation is recommended. IAP/NNF recommends room air resuscitation if gestational age >32 weeks and lowest possible FiO_2 for gestational age <32 weeks to maintain SpO_2 in target range.

Premature babies with gestational age less than 35 weeks and birth weight less than 1.8 kg are best managed in neonatal intensive care unit (NICU). Babies requiring NICU care should be shifted from delivery room to NICU in a transport incubator maintaining ambient air temperature of 33–34°C and relative humidity of 70%. Strict asepsis should be maintained and these babies should be handled minimally (Table 29.1).

CARE IN NEONATAL INTENSIVE CARE UNIT

The first 48 hours after birth are the most critical period in the premature baby's life. This is also a period of decision making for the caregivers and the parents. If it is clearly evident from the clinical status of the baby (e.g. severe birth asphyxia, profound shock, Grade IV intraventricular hemorrhage, severe sepsis and DIC, etc.) that death is imminent and chances of intact survival are grim, an informed decision for withdrawal of care can be made. A newborn with gestational age of >23 weeks gestation and birth weight >500 gm is considered to be viable. However, chances of intact survival are meagre between gestational age of 23–28 weeks at birth.

The importance of thermoregulation in management of premature newborn cannot be over-emphasized. Following transfer to NICU, premature babies should be kept in incubator or under radiant warmer to maintain body temperature in thermo-neutral range (36.6–37.2°C). Their positioning should facilitate flexed and midline position of extremities. Use of shoulder roll, swaddling (wrapping babies in a cloth to prevent free movements of limbs) and nesting are some of the methods by which this can be achieved.

Table 29.1	Initial management of premature newborns in NICU
1. Thermoregulation	Keep under radiant warmer/incubator. Maintain skin temperature between 36.5 and 37.5°C
2. Position	Head in midline and in neutral position by keeping a shoulder roll
3. Ventilation	Gentle ventilation, use of CPAP with PEEP of 6–8 cm of H_2O, volume ventilation
4. Fluids	Secure an umbilical or peripheral venous line. Start with electrolyte free fluid. Initial volume depends on the birth weight. Fluid boluses should be avoided.
5. Nutrition	Dextrose 5%/10% to keep glucose infusion rate of 6 mg/kg/min and amino acid preparation @ 3 gm/kg. Keep blood sugar levels between 80–120 mg/dl
6. Feeding	Start gavage feeding with EBM @10–20 ml/kg/day
7. Skin care	Minimum handling should be practiced. Central line manipulation should be minimized.
8. Surfactant instillation	If signs of RDS especially if gestational age <34 weeks and grunting. Use INSURE(INtubation SURfactant Extubation)
9. Caffeine Citrate	In babies with birth weight <1250 gm and gestational age <32 weeks
10. Monitoring	Continuous monitoring of HR, RR, temperature and SpO_2. Dextrostix, BP and urine output monitoring every 4 hourly

Respiratory support includes delivery room CPAP with PEEP of 6–8 cm of H_2O and use of blended air–oxygen mixture to maintain SpO_2 in minute specific target range (Table 29.2). However, in order to prevent oxygen toxicity, SpO_2 should be maintained in the range of 90–93%.'Early rescue' surfactant instillation therapy (administration of surfactant within 2 hours of birth as soon as signs of RDS set in) preferably by INSURE (intubate, surfactant instillation, extubate) should be practiced. This method avoids mechanical ventilation and its complications. However, extremely preterm (<28 weeks) and extremely low birth weight babies (< 800 gm) may be given prophylactic surfactant immediately following birth. Caffeine citrate stimulates breathing and should be started on day 1 for babies with gestational age <32 weeks and birth weight <1250 gm. Caffeine should be continued till 34–40 weeks of corrected gestational age. Vitamin A prophylaxis (5000 IU/D intramuscular on alternate days) to prevent chronic lung disease is started in babies who are on CPAP or mechanical ventilation.

Cardiovascular support should be provided by ensuring good circulatory volume. Heart rate, capillary refill time, blood pressure and urine output together indicate cardiovascular stability. These parameters should be monitored continuously. Fluid volume is titrated based on the cardiovascular status and fluids boluses are avoided. Giving fluid boluses too frequently may open the ductus arteriosus resulting in a hemodynamically significant PDA. Ionotropic support with dobutamine and vasopressor like dopamine and adrenaline may be used when indicated.

Total parenteral nutrition should be started from Day 1 of life to meet the calorie and protein requirements of the premature newborn. Dextrose of varying strength provides the carbohydrates, amino acid preparation provides the proteins and intra-lipids provide the calories as fats. Apart from this, calcium gluconate should be administered from Day 1 onwards and sodium and potassium after Day 3. In addition, if prolonged parenteral alimentation is expected, multi-vitamin preparation and trace elements should also be added. Fluid volume should be adjusted as per the day of life (Table 29.3).

Feeding of premature babies—initiation of feeding in premature babies largely depends on the gestational age. Very premature babies (<28 weeks) are started on parenteral nutrition and trophic feeding (minimal enteral nutrition) on day 1. Milk feeds are advanced as tolerated. Gavage feeding through an orogastric tube is the preferred method of feeding newborns less than 33 weeks of gestation. After 33 weeks, feeding with a spoon can be attempted. However, non-nutritive sucking (NNS), i.e. sucking on empty breast should be encouraged as it accelerates the maturation of the sucking reflex and has been observed to shorten the transition time from gavage to breastfeeding. NNS helps in initiation and maintenance of successful breastfeeding, during hospital stay and after discharge. Absolute contraindications to feeding are necrotizing enterocolitis, intestinal obstruction and shock. Feeding can be withheld for the initial 48–72 hours in babies who are having severe intrauterine growth

Table 29.2	Minute specific SpO_2 targets after birth
1 min	60–65%
2 min	65–75%
3 min	70–75%
4 min	75–80%
5 min	80–85%
Beyond 10 min	>95%

Table 29.3	Guidelines for fluid therapy in premature babies					
	Day 1	Day 2	Day 3	Day 4	Day 5	Day 6	Day 7
<1 kg	100	120	140	150	160	170	180
1–1.5 kg	80	100	120	140	150	150	150

Fluid volume in ml/kg/day

retardation or abnormal antenatal Doppler suggestive of severe placental insufficiency. For successful breastfeeding, sucking, swallowing and breathing should be well coordinated. This co-ordination is achieved after 33–34 weeks of gestation. Hence, direct breastfeeding should not be attempted before 33–34 weeks of gestation.

Premature babies with gestational age between 34 and 37 weeks; commonly called late preterm babies; and those with birth weight more than 1.8 kg can be roomed-in with mother after initial stabilization. However, they should be monitored every 4 hourly for initial 3–4 days for adequacy of feeding and identification of common problems encountered in these babies such as hypothermia, hyperbilirubinemia and hypoglycemia (Table 29.4).

Skin care of these tiny babies is of utmost importance. Their skin is fragile, non-keratinized and easily abraded. It acts as a portal of entry for micro-organisms leading to infections. Abraded skin is also a source of fluid loss. Hence maintenance of skin integrity should be ensured by minimum and gentle handling and use of semipermeable coverings like Tegaderm.

Prevention and treatment of infections is a major challenge in the management of premature babies. Premature babies are at high risk of early onset sepsis as well as late onset sepsis including fungal sepsis. Strict handwashing and use of hand rub like chlorhexidine before handling should be ensured. Inline suction systems for patients on mechanical ventilation prevents ventilator associated pneumonia. All the activities like insertion of intravenous cannula, collection of blood, administration of drugs, sponging and taking weight should be clustered together. Equipment used should be disposable and single use as far as possible and parenteral nutrition solutions should be prepared under laminar flow. Longer duration of mechanical ventilation, use of long lines and parenteral nutrition are risk factors for late onset sepsis. Unnecessary use of antibiotics should be avoided as this practice favors the emergence of resistant species. Whenever indicated, antibiotics with narrowest spectrum should be used for shortest duration warranted. Treatment of proven infection is usually 14–21 days.

Developmentally supportive care (DSC) forms an integral part of management of premature babies. It should be designed in a way which treats every neonate as a person and ensures conducive environment for development of this vulnerable population. Such practices enable infants to experience soothing auditory, visual and sensory-motor inputs without causing undue stress and without disrupting their sleep pattern and autonomic and motor function. Keeping sound levels in NICU below 50 dB, cycled lighting to simulate day–night pattern, regulation of the environmental temperature by air conditioning are some of the methods to provide comforting surrounding to a growing premature neonate. Positioning of baby by swaddling, nesting or by facilitated tuck, kangaroo mother care, massage therapy and non-nutritive sucking helps in maintaining autonomic stability and better recovery. DSC is a team responsibility should

Table 29.4	Feeding volume and frequency			
Birth weight (g) volume	Starting volume (ml/kg/d)	Increment each day (ml/kg/d)	Maximum volume (ml/kg/d)	Frequency of feeds
<1200	10–20	20	180	2 hourly
1200–1600	30	30	180	2 hourly
>1600	60	30	150	3 hourly

involve doctors, nurses, physiotherapist and the parents. Providing early and ongoing developmentally supportive care goes a long way to improve neurological outcome and family adaptation of babies born before term.

PREMATURITY RELATED PROBLEMS

1. *hsPDA i.e. hemodynamically significant Patent ductus arteriosus:* Symptomatic PDA or PDA size > 1.5 mm is considered to be hemo-dynamically significant. Medical closure of PDA with intravenous indomethacin or ibuprofen and more recently paracetamol should be attempted. Surgical ligation of PDA should be done if medical management fails.

2. *Intraventricular hemorrhage:* The chances of intraventricular hemorrhage are maximum in the first week of life and in most premature of babies. The grade of hemorrhage may vary from grade I to grade IV. Posthemorrhagic hydrocephalus and periventricular leukomalacia are long-term sequelae of IVH which adversely affect the neurological outcome.

3. *Broncho-pulmonary dysplasia (BPD) and chronic lung disease (CLD):* Requirement of oxygen beyond 28 days of life defines BPD/CLD. Prolonged mechanical ventilation, aggressive ventilation, delayed enteral nutrition and sepsis increase the risk of BPD while use of antenatal steroids, non-invasive ventilation and early enteral nutrition reduces the risk of BPD/CLD.

4. *Necrotising enterocolitis (NEC):* It is the most common gastrointestinal emergency in a preterm neonate. NEC is caused due to injury to gastro-intestinal mucosa and is characterized by feeding intolerance, bilious RT aspirates and abdominal distension. Advanced NEC results in intra-abdominal abscess and intestinal perforation and has a very poor prognosis. Delayed initiation of feeds, rapid advancement and use of formula feeds instead of human breast milk increases the risk of NEC.

5. *Osteopenia of prematurity:* Deficiency of phosphorus and calcium are main causes but vitamin D deficiency may also be causative. Symptoms include chronic respiratory insufficiency, pathological fractures, signs of rickets like craniotabes, widely open fontanelles and beading and decreased linear growth. Prolonged parenteral nutrition, use of unsupplemen-ted milk, excessive fluid restriction and use of furosemide all increase the risk of osteopenia. Osteopenia is treated by supplementing the deficient nutrient by using dietary supplements or special formula feeds.

6. *Retinopathy of prematurity:* Excessive use of oxygen and low birth weight are proven risk factors for ROP. An ROP screen should be routinely performed at 28 days of life in all preterm neonates less than 34 weeks gestation and less than 1.75 kg birth weight (and larger babies 34–36 weeks and up to 2 kg if they have risk factors) for early diagnosis and treatment of the condition. Use of laser therapy is the mainstay of treatment. In unstable babies, intravitreal injection of anti-VEGF can be attempted.

DISCHARGE PLANNING

Screening Tests

Newborn metabolic screening as per institutional protocol-suggested are for congenital hypothyroidism, G6PD deficiency, congenital adrenal hyperplasia and cystic fibrosis.

Hearing screen by otoacoustic emission/auditory brain response as per institutional protocol.

Eye screen-red reflex and retinopathy of prematurity screen as mentioned above.

Cardiac screen with pulse oximetry to detect cyanotic congenital heart disease.

Screening for developmental dysplasia of the hip-clinical and ultrasound.

Thermoregulation: Babies should be able to maintain temperature without warmer/incubator.

Nutrition: Preferably child should be discharged on full breast feeds with weight gain 10–15 gm/kg/day documented over 3 consecutive days. Breast milk should be fortified by human milk fortifier (HMF) till 40 weeks corrected age. Some babies may require preterm formula for achieving adequate weight gain. Competent oral feeding very essential.

Parental Involvement and Confidence

Anthropometry and growth monitoring is an integral part of discharge. Growth should be plotted on Fenton's growth charts till 40 weeks of gestation and WHO growth charts beyond 40 weeks. Corrected age and not the chronological age should be taken into consideration while plotting anthropometric parameters.

Apnea resolution: 3–7 apnea free days are required prior to discharge. Apnea of prematurity spontaneously resolves by 37–40 weeks gestational age.

Neurodevelopmental monitoring especially of babies who are extremely preterm or those which had a stormy course/intraventricular hemorrhage. Physiotherapy and an early intervention should be started by a team involving a developmental specialist.

Supplements

Calcium and Phosphorus till term. Fortification of human milk with human milk fortifier (HMF) meets the RDA of calcium and phosphorus for these babies, hence additional supplementation should be avoided.

Vitamin D drops 800 IU/day till 1 year of age.

Multivitamin drops and zinc till 1 year of age is recommended by some.

Iron supplements at dose of 3–4 mg/kg/day should be started at by 2 weeks and after establishment of full enteral feeds and should be continued till 1 year of age.

SUGGESTED READING

1. Age terminology during the perinatal period DOI: 10.1542/peds.2004-1915 Pediatrics 2004; 114; 1362.
2. Averys Diseases of the Newborn, 9th edition. Christine Gleason and Sandra Juul. 2012.
3. Manual of neonatal care, 7th edition. John P. Cloherty, Eric C. Eichenwald, Ann R. Stark Lippincott Williams & Wilkins, 2012.
4. National Neonatology Forum—Evidence based clinical practice guidelines, 2010. ON Bhakoo, Praveen Kumar, Naveen Jain, Rhishikesh Thakre, Srinivas Murki, S Venkataseshan. NNF publications.
5. NRP—Neonatal Resuscitation Textbook (6th Edition). Edited by American Academy of Pediatrics and American Heart Association. 2011.

Transporting the Neonate

Shweta Chawla, Prashant Dixit

'Life has very simple ways of teaching us the relevant things; we just need to keep our mind open.'

INTRODUCTION

As a medical graduate, training to be a pediatrician working in the NICU, I once received four critically ill newborns. Two out of the four died and the other two barely survived. When I reached home and vented the frustrations to my mother, all she said was: I wish you had received them in good condition.

The penny dropped! That was true! I remembered the glasses which broke while I carried them from one room to another as a 7-year-old; and the most expensive mangoes I carried in my bag as a teenager which were squashed by the time I reached home….

Transporting the sick newborn in the best of condition to the NICU was the answer to my dilemma of receiving very critical babies.

We are aware of the significance of the Golden Minute and the Golden Hour for Neonatal Resuscitation, but bringing the sick, fragile newborn to the NICU in the optimal condition is the key to the quality of care provided to the sick newborn and an ideal beginning for the best results.

What is Neonatal Transport?

Neonatal transport is…

Taking the right newborn at the right time, by the right personnel, to the right place, by the right form of transport, and receiving the right care throughout.

Treatment of the sick neonate in specialized units (NICU) has been associated with decrease in mortality and morbidity. Since neonatal intensive care facilities may not always be attached to maternity services, a significant number of neonates reach the facility in suboptimal condition, increasing the morbidity and mortality.

In the early 1960s, neonatal transport was first used to make intensive care accessible to those neonates who needed it but delivered outside the facility. Subsequently, organised neonatal transport systems developed and became an integral part of newborn care, especially in the developed world.

Before the advent of specialised Neonatal Transport Services, 'in utero' transfer was the safest method to transport an extremely preterm baby before delivery. "*In utero transfer*", till date, still remains one of the best methods to transfer a preterm/sick baby before delivery but perinatal illnesses, asphyxia and congenital malformations cannot always be anticipated, thus requiring a continued need to transfer babies after delivery.

Why is the Transport Important?

Labor has been described as the most difficult journey in life. Newborns are essentially fragile, more so if they have been through a difficult delivery.

Even when a normal newborn may take time to adjust to the hostile cold and dry extra uterine environment, a compromised baby will definitely be at a much greater disadvantage to begin an independent life. These babies are often critically ill, and the outcome is partly dependent on the effectiveness of the transport system even before specialized care is given.

Facilities for neonatal transport in India are dismal. Most neonates are transported without any pretransfer stabilization or care during transfer.

Any available vehicle is used, which often takes long hours and the referral place is not identified. There is an acute shortage of neonatal beds and majority of the sick neonates in need of urgent admission are dumped in pediatric facilities with inadequate infrastructure. Often, these neonates are shunted from one health facility to another.

Many of the babies thus transported are cold, blue and hypoglycemic and 75% of the babies transferred this way have serious clinical implications.

Which Newborn may Require Transport?

Commonly, babies require NICU care under the following conditions:

a. Preterms (born at less than term gestational age < 36 weeks)
b. Low birth weight (typically less than 2000 gm)
c. Birth problems (like perinatal depression, infant of diabetic mother, respiratory distress, meconium related distress)
d. Apnea requiring bag and mask ventilation
e. Cyanosis persisting despite oxygen therapy
f. Sepsis with signs of systemic involvement
g. Neonatal seizures
h. Jaundice with potential for exchange transfusion
i. Active bleeding from any site
j. Hypoglycemia or IDM
k. Congenital heart disease (antenatally diagnosed or suspected)
l. Suspected metabolic disorder
m. Severe electrolyte abnormalities

A newborn to be transported though sick, should be fit enough to withstand transport. An unstable neonate not likely to make it to the NICU should be managed at the same place in the best possible way, rather than attempting heroic transfers and losing the baby on the way

The Golden Hour is very crucial in determining the newborn's morbidity and mortality

n. Surgical emergencies (like tracheo-esophageal fistula, congenital diaphragmatic hernia, imperforate anus, genetic malformations, etc)
o. Neonate requiring special diagnostic and/or therapeutic service

Since all the problems which affect the newborn adversely (hypothermia, inability to establish adequate respiration or circulation, hypoglycemia, hypoxia, hypovolemia, tachypnea, etc.) can be accentuated during the transport; safe transportation of these highly fragile and vulnerable newborns in the best of condition from the place of birth to a NICU is a challenge to intact survival. Procedures/protocols/algorithms for neonatal resuscitation should continue during the transport, maintaining temperature all the time.

Transferring a newborn improperly is like adding insult to injury, and could be the deciding factor of its fate. It can increase mortality as well as morbidity.

What is the Right Time to Initiate Newborn Transport?

An ideal condition would be a planned transport, wherein the maternity service keeps the Neonatal team ready to attend the delivery and transport the baby to the NICU if required at the earliest after initial stabilization. This may however, not be possible in all cases; and therein lies the challenge of stabilizing the newborn at the maternity home, arranging for transport in the shortest possible time, and transferring the baby with the trained personnel, in a stable condition to the NICU for achieving optimal outcomes.

Who should Transport? The Neonatal Emergency Transport System (NETS)

Most of the developed countries have an advanced neonatal care system incorporating transporting the needy babies from maternity services or level I and level II neonatal care to tertiary centers or level III/ IV care services. These involve the use of Neonatal Emergency Transport System for achieving the same.

Types of Vehicle with their Features

Though neonatal care has grown by leaps and bounds in our country in the last decade or so, a disparity exists between the affluent and the rural India. In terms of newborn transport,

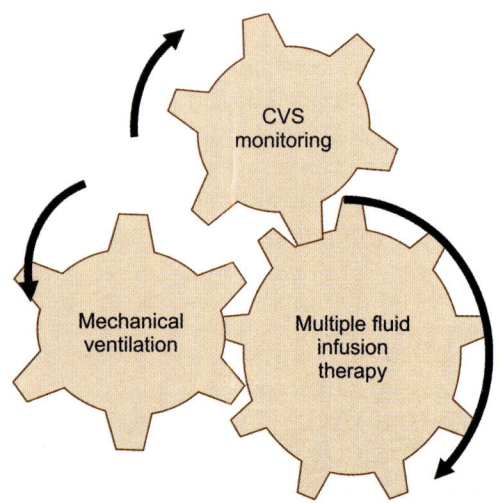

Fig. 30.1: Constituents of an ideal transport system

Table 30.1	Features of vehicles used for neonatal transport (choosing the correct vehicle)		
	Ground ambulance	*Rotor-wing aircraft*	*Fixed-wing aircraft*
Departure times	Excellent	Excellent	Poor to fair
Arrival times	Fair to poor	Excellent	Good
Out-of-hospital time	Poor	Excellent	Fair to excellent
Patient accessibility	Good	Poor	Fair
Weather issues	Excellent	Poor	Fair to good
Cost	Low	High	High

however, we are still in an age of infancy and even most cities do not have good neonatal transport services (Fig. 30.2).

An ideal neonatal transport system should provide for cardiorespiratory monitoring, multiple fluid Infusion and mechanical ventilation.

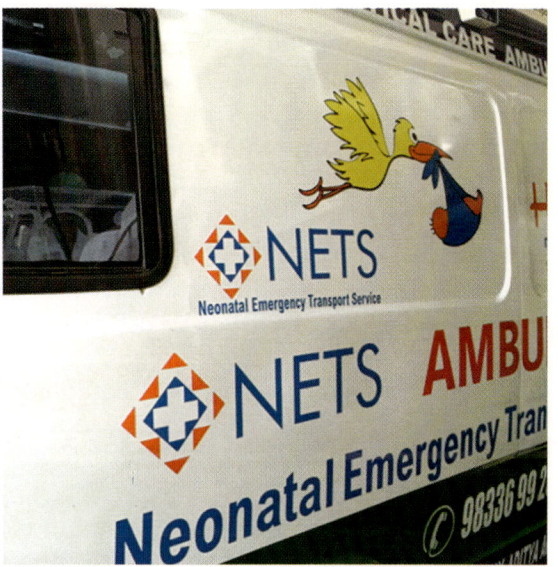

Fig. 30.2: Neonatal emergency transport system

Ours is a country of paradoxes: We have 'Multispeciality Hospitals' (Hospitals) in the metros on one hand and places where there is no doctor on the other.

A birth may occur in a maternity service/ unit in the cities or at home or a PHC in villages. A newborn may need to be transferred to a specialised level III/IV NICU in the metros or a SCNU/PHC/district hospital in the rural India.

Since the terrain and the infrastructure differ in different geographical areas, a single method of transport may not work. However, the basic principles governing and ideal newborn transport remain consistent.

Ideal NETS will consist of:

a. A Full-sized Ambulance

With enough room for a transport incubator, emergency trolley and 3–4 people to work freely in the ambulance. The ambulance needs to have an inverter back up, oxygen cylinders, suction facilities, a small scrub area and fire extinguishers as recommended. It should be registered with the RTO as an ambulance vehicle, with all documents and road worthiness determined periodically. Speed and stability (lateral roll and front-back impact on braking) of the vehicle is very important. The quality of new gen chassis in the vehicles has improved performance in terms of acceleration, cruising speed, braking, etc. However, unfortunately most of the commonly used vehicles for transport in India are highly inadequate for smooth travel.

Speed of the vehicle should not be more than 15–20 km/hr over the posted speed limit.

Power backup: All the equipment used in transport should have a battery backup and should be kept fully charged in anticipation of transport at all times. An AC 240 V power source can be provided in the ambulance by 2 methods-inverter or a dedicated generator. Sufficient adapters, to make a quick changeover to available mode of power supply should be available.

Gas supplies: Ensure that the oxygen cylinders are filled prior to onset of journey and will last the transport. Spare cylinders and equipment to change the cylinders are desirable. Large cylinders are preferable as the small ones may last only for a couple of hours, usually. The personnel accompanying the ambulance should be well versed in the technique of changing the cylinders.

b. A Trained Ambulance Driver

Who knows the traffic rules and conditions adequately and is able to think smartly on the road. His role is to minimize time on the road, and at the same time not endangering the life and limb of passengers or pedestrians. A communication window in the ambulance, and understanding with the driver is essential for the working of the team.

c. A Helper/Ward Assistant

Who can help the driver as well as assist in mobilizing the transport incubator. Both the driver and the helper should have knowledge of the working of incubator, inverter system, oxygen, suction machines and fire extinguishers.

d. Medical Personnel

Ideally a pediatrician/neonatologist/trained birth attendant should be attending the delivery and be there to transport the baby to the NICU. He/she will lead the team. Having two or more helping hands/staff nurses/paramedics trained in newborn care is desirable. Having transport protocols, management algorithms, documentation helps and pre- and post-transport briefing should be in place.

The team members should be recruited carefully, taking into account their motivation levels, fitness and physical strengths, their knowledge and efficiency. They must be trained and oriented regularly regarding their job.

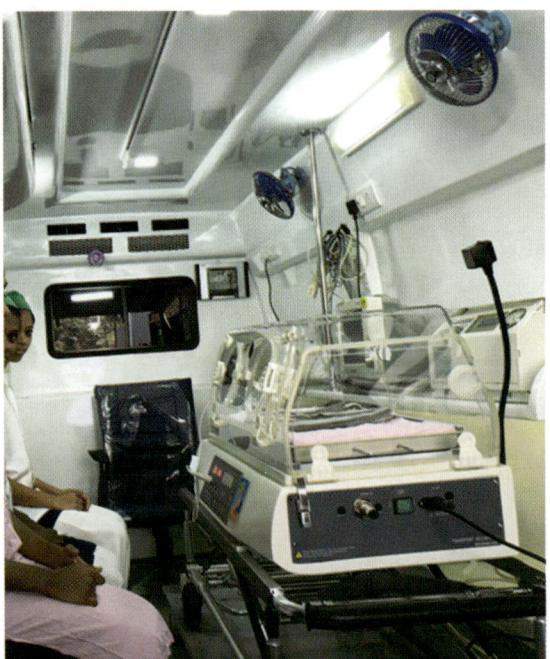

Fig. 30.3: Layout of ambulance interiors with personnel

The neonatal transport team should be capable of handling the airways, intravenous access and invasive procedures like umbilical line insertions, chest tube insertions and intraosseous access.

Team competence in neonatal airway management is imperative. The team should be capable of:
1. Recognizing impending respiratory failure,
2. Performing effective bag-valve-mask ventilation,
3. Performing atraumatic intubation with appropriate endotracheal tubes,
4. Instillation of artificial surfactant,
5. Managing fluid and electrolytes, and
6. Management of ventilator settings.

Nearly all ill neonates require peripheral or central intravascular access during transport. The team must have the necessary equipment and skills for routinely and reliably securing intravenous (IV) access in these tiny and challenging patients.

Ideally, staff competency also includes training in other unusual invasive procedures such as percutaneous needle aspiration of the chest, chest tube insertion, umbilical catheter insertion, and intraosseous vascular access.

e. Transport Incubator

A CE approved transport incubator, with pre warmed environment and baby towels is essential to prevent hypothermia. It is desirable to have an off-side mounting incubator as compared to transverse mounting.

The transport incubator should be cleaned regularly and kept sterile with use of appropriate solutions.

In resource poor settings, the baby may be wrapped in a warm clean cloth, taking care to cover the head and keeping the mouth and nose free. A warm water bag may be used to provide warmth, care taken not to keep the hot water bag in direct contact with the skin to prevent burns. In extreme premies, plastic wraps/plastic bags may be used in addition.

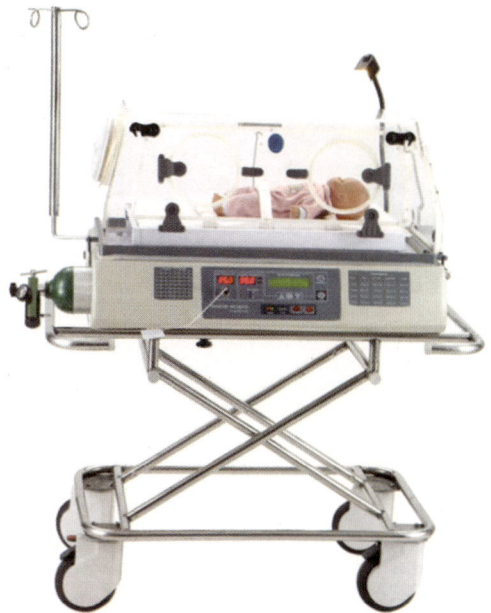

Fig. 30.4: Transport incubator

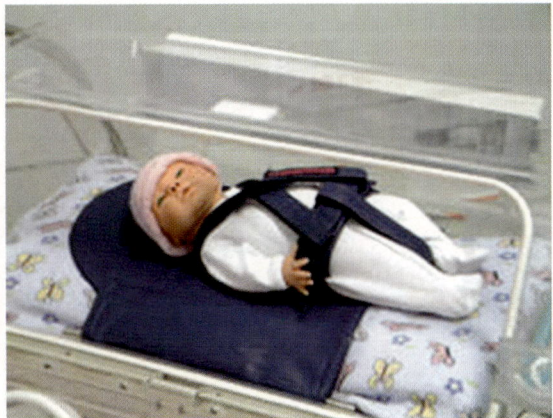

Fig. 30.5: Newborn restrained with safety harness during transport

In extremely resource poor setting, where nothing is available, mother's body could be the only source of heat and the baby may be transferred in skin-to-skin contact. In the absence of transport ambulance, closed vehicles like a car or a tempo is desirable rather than transporting the newborn on a scooter or a three-wheeler. The air current in open modes of transport may exacerbate the risk of hypothermia and infections.

f. Emergency Medications/Kit

The list of equipment/medications in the emergency kit should include the following:

- *Temperature*: Incubator on wheels which can be pushed right into the OT and labor room, sterile linen, can be warmed in the incubator while on the way for retrieval and polythene wrap for extremely premise/ELBW babies.
- *Airways:* Shoulder roll, portable suction machine, suction catheters, mouth suction (De Lees mucus extractor), meconium aspirator.
- *Breathing:* Self-inflating ambu bag (250 ml for preterm, 500 ml for term), masks (sizes 0 and 1), oxygen supply, T-piece resuscitator, laryngoscope (blades 00, 0, 1), endotracheal tubes (2.5, 3, 3.5, 4), ET stylet, scissors, adhesive tapes, splint, feeding tube (8F).
- Circulation—IV cannulas (24, 26 G), Umbilical catheters—3.5, 5 F, three ways, syringes—1, 2, 5, 10, 20 ml, IV sets, micro-drip sets, sterile gloves, ECG and SpO_2 monitors.

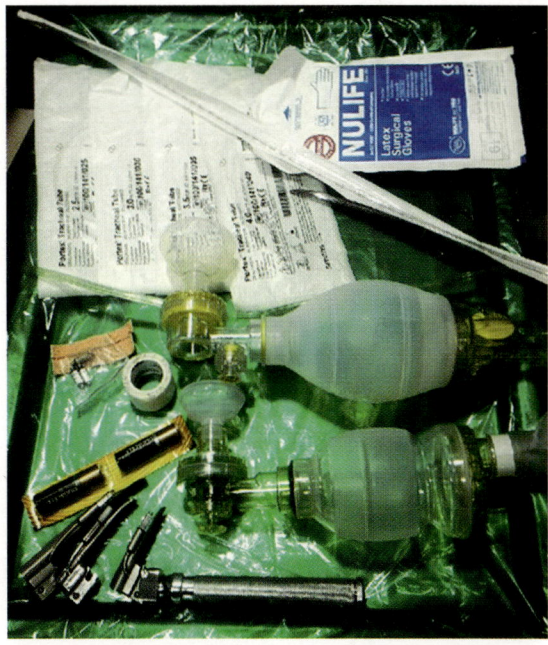

Fig. 30.6: Equipment to establish respiration

- C—*also clean hands, help:* Watch, stethoscope, helping assistant, calm cool head
- *Drugs:* Epinephrine, NS, RL, Dextrose 10%, calcium gluconate, midazolam, phenobarbitone, surfactant
- Glucometer with disposable strips
- Neonatal T-piece resuscitator, transport ventilators, blenders, oxygen hood

The right place (neonatal intensive care unit): Needless to say, having NICU facilities within the maternity care complex is desirable, but difficult to achieve in our country. In such a scenario, establishing a good rapport and understanding, with a NICU facility offering the requisite neonatal care and services is essential. The newborn case requiring NICU care should be discussed with the neonatologist in charge, and relatives counseled adequately and appropriately.

The Right Care Protocols?

The first and foremost aim of ideal neonatal transport is working towards 'intact survival'.

The services of a specialized neonatal transport team have been shown to be associated with reductions in hypothermia and acidosis, as well as with reduced mortality and morbidity in sick and critical newborns.

A systematic approach of maintaining temperature, airway, breathing, circulation and sugars (T A B C D) not only during neonatal resuscitation but also during the neonatal transport will ensure not only survival but also the quality of life later. Assessment of the baby and depending on facilities available check for temperature, airway, breathing, circulation and sugar.

   i. *Temperature:* Maintenance of thermoneutral environment for the baby is very crucial, especially for the tiniest of the sick newborns. Transport incubators, dry warm linen, plastic wraps and appropriate measures go a long way in maintaining the perfect temperature during transfer.

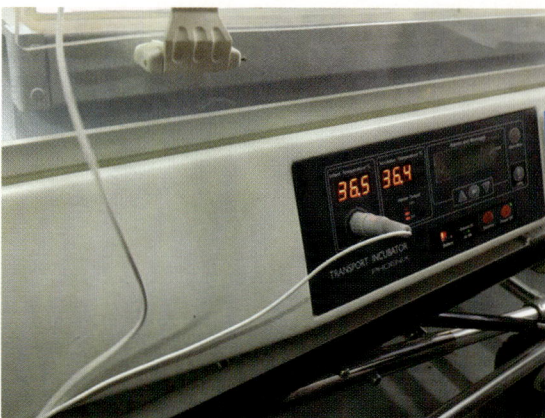

Fig. 30.7: Temperature maintenance—transport incubator

Newborns are born wet and very prone to heat loss not only through evaporation of amniotic fluid but also through conduction, convection and radiation.

Temperature maintenance is the weakest link in neonatal transport.

It is well known that hypothermia initiates a vicious cycle in these newborns, leading to acidosis, hypoxia, anaerobic metabolism, persistent pulmonary hypertension, stress and hypoxemia.

Maintaining temperature in these newborns is essentially the first step to good neonatal care.

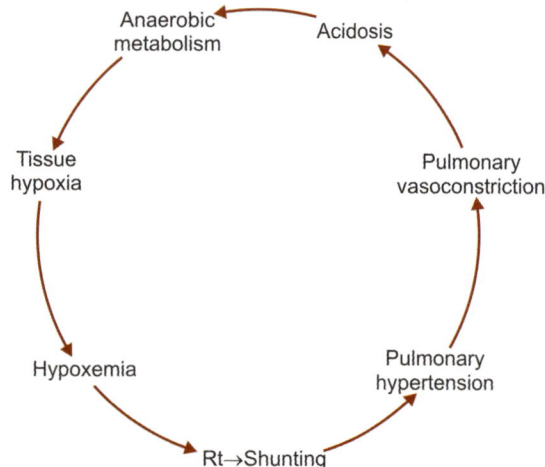

Fig. 30.8: Vicious cycle of hypothermia

Fig. 30.9: ELBW/extremely premise in a plastic wrap

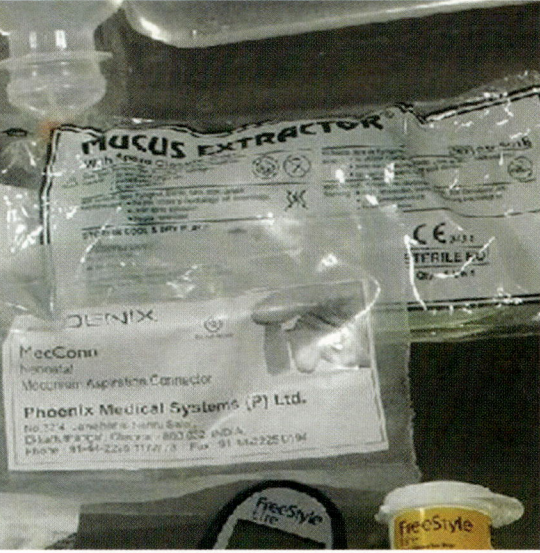

Fig. 30.10: Mucus extractor, meconium aspirator

ii. *Airway:* Maintaining the patency of the airway in order to facilitate smooth, unobstructed airflow can be achieved by use of shoulder rolls, suction of orpharyngeal, nasopharyngeal passages and positioning. The basic principles are to prevent obstruction to the airway and to facilitate airflow into the neonatal lungs.

Suction (as and when required) and position of neck (placement of shoulder roll) is the key to maintaining a patent airway.

iii. *Breathing:* Establishing and ensuring adequate respiratory efforts, and assisting them wherever required with oxygen, nasal CPAP and/or ventilation is crucial to maintain the supply of oxygen to meet the needs of various organs. Delivery room CPAP, T-piece resuscitator and use of blenders are the recent modalities with strong evidences in improving neonatal outcomes.

iv. *Circulation:* Check CRT, heart rate, blood pressure, urine output for monitoring and establishing adequate circulation. Well felt peripheral pulses and normotensive state is crucial for maintaining the ideal pH status of the delicate newborn. IV

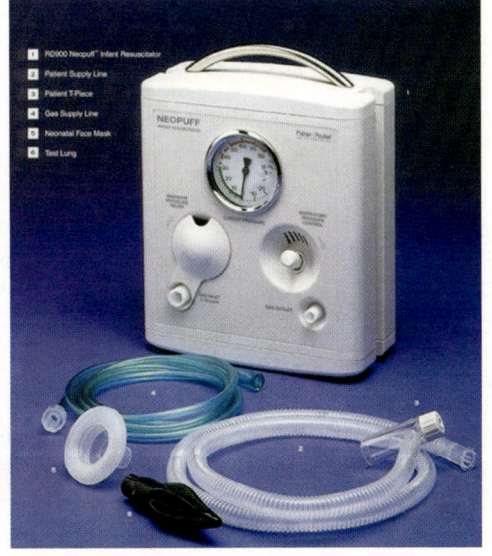

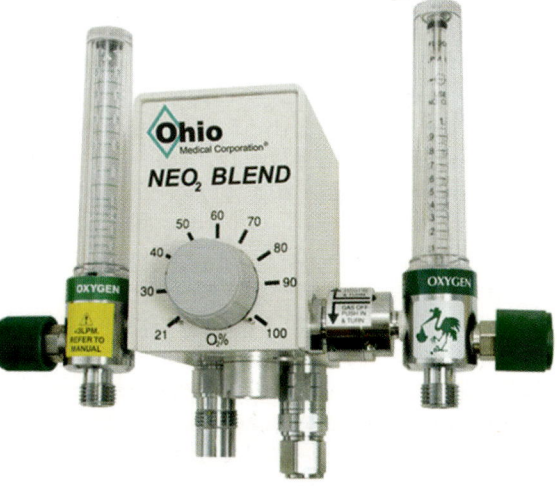

Fig. 30.11: T-piece resuscitator and air-oxygen blender

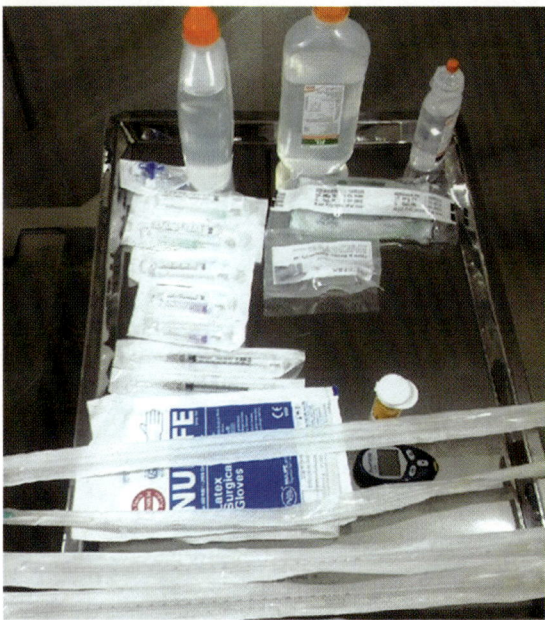

Fig. 30.12: Circulation establishment is crucial for intact survival

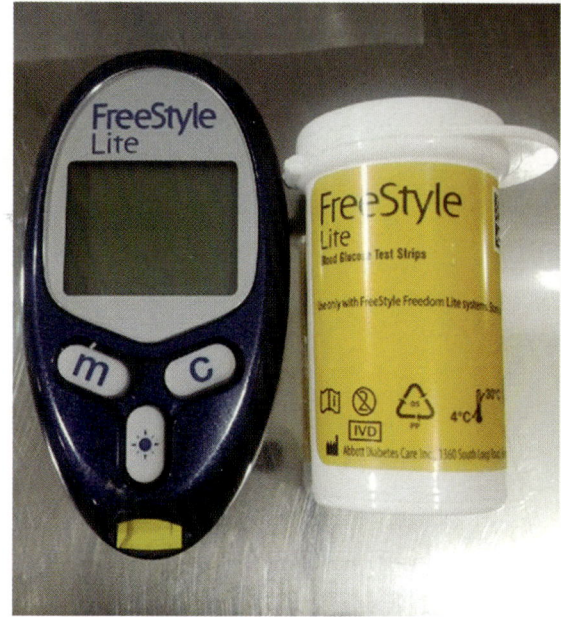

Fig. 30.13: Monitoring glucose levels of at-risk newborns

fluids resuscitation and inotropes help towards this end.

Antenatal history and perinatal status of the newborn may guide towards early use of volume infusions. Inotropes may be used if indicated.

v. *Dextrose:* Last but not the least, ensuring adequate blood glucose levels to provide substrate for brain metabolism is very important from a long-term perspective. IV dextrose or early initiation of feeds helps promote a euglycemic state. Adequate GDR (glucose delivery rate) needs to be maintained, in order to achieve appropriate blood glucose levels in the newborn. If blood glucose < 40 mg/dl, give 2 ml/kg of 10% dextrose through IV line.

Maintaining and developing algorithms and protocols to manage neonatal emergencies and training the transport team to follow them will lead to minimization of errors and improved outcomes.

Documentation of all the treatment done during transport in a chronological order is mandatory.

Transport in Special Conditions

Specific conditions like respiratory distress syndrome (HMD, MAS, congenital diaphragmatic hernia, etc.) need to be managed as per the primary cause. If the baby requires ventilation, it should be provided. It is advisable to support ventilation during transport in babies with apneas and perinatal asphyxia.

What if the Baby Deteriorates during Transport?

The answer to this is not backed by sufficient evidence and is largely dependent on the level of skills of transport team in resuscitation, space and equipment available in the ambulance, and the distance to be travelled.

Intubation before Transport

In babies where the respiratory/hemodynamic status is unstable, it is preferable to intubate, and secure airway and breathing before initiating transport. Common indications are:

i. Congenital diaphragmatic hernia

ii. Extremely low birth weight babies or extremely premature neonates

iii. Respiratory distress with increasing oxygen requirement (FiO_2, more than 40%)

iv. Congenital cardiac disease on prostaglandin E infusion, because of risk of apnea

v. Recurrent apnea or seizures

vi. Limited space and skills to perform any resuscitation in the ambulance

If neonatal transport ventilator is not available, T-piece resuscitator or Neopuff or bag and tube ventilation can be provided.

Early surfactant administration can be done in specific cases as indicated.

Medicolegal Issues

Most medicolegal issues arise because of improper communication. Proper communication between the obstetrician, the pediatrician/neonatologist and the relatives is the key to prevent medicolegal situations. Communication should be continuous, at different times and places/situations: Antenatal, during labor, immediately after birth, at the time of transfer and in the NICU. Parties involved would be the obstetrician, the neonatologist and the relatives. A detailed account on the subject may be found in the chapter on "Communication in the NICU".

i. *Antenatal counselling:* Important to discuss the possibility of complications at the time of birth: 1 in 10 babies may require resuscitation, and possibility of neonatal transfer. The place where the baby will be transferred may also be decided before birth, if problems anticipated. In severely compromised babies, e.g. severe IUGR or extreme prematurity, 'in utero transfer' may be the best option.

ii. Minimum time wasted in arranging transport.

iii. Transport vehicle to be properly equipped.

iv. Baby to be accompanied by a staff trained in resuscitation or preferably by a doctor.

v. Complete notes (antenatal, perinatal and resuscitation) to be carried along and notes of vital parameters during transport to be maintained.

vi. Baby to be handed over to the doctor on duty in the NICU/higher center, condition of the baby to be documented.

vii. Contemporaneous communication with the neonatologist maintained till the time of discharge.

Because the outcome of a neonate with major medical or surgical problems (including extreme prematurity) remains worse than that for an inborn infant, primary emphasis should always remain on prenatal diagnosis and subsequent planned transfers whenever possible. A system to transport a sick or critically ill newborn should always be available and ready at all times. Herein lays the organizational challenge.

Despite advanced training and technology, mothers (*in utero* transfers) usually make the best transport incubators; however, is not always possible practically.

SUGGESTED READING

1. NNF Neonatal Transport Guidelines.
2. Manual for transport of High Risk Newborn, PCNA, Segal S, 1972.
3. National guidelines, Health & Welfare, Canada, 1988.
4. Guidelines for Perinatal care, AAP, ACOG, 2002.
5. Comprehensive Neonatal Care in India, Mathur NB, Journal of Neonatology, 2006.
6. Optimising neonatal transport, Fenton AC, Arch Dis Child Fetal Neonatal, 2004.
7. NNF module, Neonatal referral and transportation, Saluja S, Mathur NB, 2005.
8. Transporting the sick neonate, Cornette L, Current Pediatric, 2004.
9. Audit of Neonatal Intensive Care Transport, Leslie AJ, Acta Pediatrica, 1997.

Neonatal Screening and Challenges of Implementation

Saurabh Dani

History

The idea of newborn screening (NBS) was initiated way back in 1960 by the pioneering work of Dr Robert Guthrie in the USA with the screening of Phenylketonuria (PKU). Today NBS is recognized internationally as an essential, preventive public health program for early identification of disorders in newborns that can affect their long-term health.

Newborn screening has expanded from single disorder screening in 1960 to multiple disorders screening as of today. With the introduction of system of collection and transportation of blood samples on filter-paper wide scale screening was made possible.

With the introduction of Tandem Mass Spectrometry (TMS) in 1990 there was a move from one test per disorder to one test for multiple disorders. Today in several parts of the world screening is performed universally for as many as 50–60 treatable disorders.

Disorders

NBS can include several disorders ranging from one to sixty. The list of disorders that should be screened in any population depends on the principles of the Wilson and Jugners classical screening criteria[1] which has now been updated by WHO[2] as laid down in Table 31.1. Most countries always start with

Table 31.1 Screening criteria[1,2]
• The screening programme should respond to a recognized need.
• The objectives of screening should be defined at the outset.
• There should be a defined target population.
• There should be scientific evidence of screening programme effectiveness.
• The programme should integrate education, testing, clinical services and programme management.
• There should be quality assurance, with mechanisms to minimize potential risks of screening.
• The programme should ensure informed choice, confidentiality and respect for autonomy.
• The programme should promote equity and access to screening for the entire target population.
• Programme evaluation should be planned from the outset.
• The overall benefits of screening should outweigh the harm.

a small list and expand as time goes by. As per guidelines issued by the National Neonatal Forum of India (NNF India)[3] every baby in India should be screened for congenital hypothyroidism (CH), congenital adrenal hyperplasia (CAH) and G6PD deficiency. It also states that in urban or resource rich settings the babies should be screened for the expanded panel of 50–60 treatable disorders. This list of treatable

Recommended Uniform Screening Panel
Core Conditions
(as of March 2015)[5]

ACMG Code	Core Condition	Metabolic disorder			Endocrine disorder	Hemoglobin disorder	Other disorder
		Organic acid condition	Fatty acid oxidation disorders	Amino acid disorders			
PROP	Propionic acidemia	X					
MUT	Methylmalonic acidemia (methylmalonyl-CoA mutase)	X					
Cbl A,B	Methylmalonic acidemia (cobalamin disorders)	X					
IVA	Isovaleric acidemia	X					
3-MCC	3-Methylcrotonyl-CoA carboxylase deficiency	X					
HMG	3-Hydroxy-3-methyglutaric aciduria	X					
MCD	Holocarboxylase synthase deficiency	X					
βKT	β-Ketothiolase deficiency	X					
GA1	Glutaric acidemia type I	X					
CUD	Carnitine uptake defect/carnitine transport defect		X				
MCAD	Medium-chain acyl-CoA dehydrogenase deficiency		X				
VLCAD	Very long-chain acyl-CoA dehydrogenase deficiency		X				
LCHAD	Long-chain L-3 hydroxyacyl-CoA dehydrogenase deficiency		X				
TFP	Trifunctional protein deficiency		X				
ASA	Argininosuccinic aciduria			X			
CIT	Citrullinemia, type I			X			
MSUD	Maple syrup urine disease			X			
HCY	Homocystinuria			X			
PKU	Classic phenylketonuria			X			
TYR I	Tyrosinemia, type I			X			
CH	Primary congenital hypothyroidism				X		
CAH	Congenital adrenal hyperplasia				X		
Hb SS	S,S disease (sickle cell anemia)					X	
Hb S/βTh	S, βeta-thalassemia					X	
Hb S/C	S,C disease					X	
BIOT	Biotinidase deficiency						X
CCHD	Critical congenital heart disease						X
CF	Cystic fibrosis						X
GALT	Classic galactosemia						X
GSD II	Glycogen Storage Disease Type II (Pompe)						X
HEAR	Hearing loss						X
SCID	Severe combined immunodeficiencies						X

disorders was the effort of American College on Medical Genetics.[4] All disorders were grouped in three categories based on the screening criteria. Only disorders that fulfilled the screening criteria were included in the list, i.e. 32 core and 26 secondary panel. Any disorders that were not treatable or did not fulfill the screening criteria were excluded.

Since India does not have any universal screening program/guideline; healthcare provider can choose what to offer to the parents. It would be a good practice though to include a minimum of three disorders as advised by NNF to a maximum of 58 as in the current guideline by the US Department of Health and Human Services.[5] Anything outside the list of 58 disorders would be inappropriate for now.

Method

The process of screening newborns has been refined over the years to perfection. The salient features are: Collection, transportation, technology and reporting.

Recommended Uniform Screening Panel
Secondary Conditions
(as of March 2015)

ACMG Code	Secondary Condition	Metabolic Disorder			Hemoglobin Disorder	Other Disorder
		Organic acid condition	Fatty acid oxidation disorders	Amino acid disorders		
Cbl C,D	Methylmalonic acidemia with homocystinuria	X				
MAL	Malonic acidemia	X				
IBG	Isobutyrylglycinuria	X				
2MBG	2-Methylbutyrylglycinuria	X				
3MGA	3-Methylglutaconic aciduria	X				
2M3HBA	2-Methyl-3-hydroxybutyric aciduria	X				
SCAD	Short-chain acyl-CoA dehydrogenase deficiency		X			
M/SCHAD	Medium/short-chain L-3-hydroxyacyl-CoA dehydrogenase deficiency		X			
GA2	Glutaric acidemia type II		X			
MCAT	Medium-chain ketoacyl-CoA thiolase deficiency		X			
DE RED	2,4 Dienoyl-CoA reductase deficiency		X			
CPT IA	Carnitine palmitoyltransferase type I deficiency		X			
CPT II	Carnitine palmitoyltransferase type II deficiency		X			
CACT	Carnitine acylcarnitine translocase deficiency		X			
ARG	Argininemia			X		
CIT II	Citrullinemia, type II			X		
MET	Hypermethioninemia			X		
H-PHE	Benign hyperphenylalaninemia			X		
BIOPT (BS)	Biopterin defect in cofactor biosynthesis			X		
BIOPT (REG)	Biopterin defect in cofactor regeneration			X		
TYR II	Tyrosinemia, type II			X		
TYR III	Tyrosinemia, type III			X		
Var Hb	Various other hemoglobinopathies				X	
GALE	Galactoepimerase deficiency					X
GALK	Galactokinase deficiency					X
	T-cell related lymphocyte deficiencies					X

- *Collection:* The sample is collected by heel prick using an auto-lancet from the sides as shown by the shaded area in Fig. 31.1. Blood should not be collected within 24 hr of birth. The sample is directly collected on

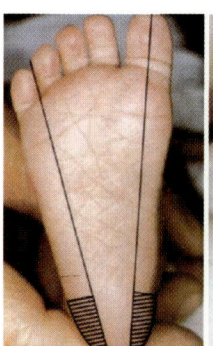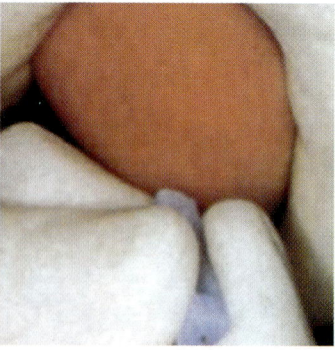

Fig. 31.1: Sample collection procedure

a CLSI (clinical and laboratory standards institute) approved filterpaper.[6] Umbilical cord blood should not be used for screening as it has limited value and not recommended for comprehensive screening programs.

- *Transportation:* Once the sample has been collected allow it to dry for 2–4 hours. Once dried it can be shipped to the newborn screening laboratory by post or courier.

- *Technology:* NBS tests are performed using different technologies for different disorders. For CH. CAH. G6PD, galactosemia, biotinidase deficiency and cystic fibrosis biochemical enzyme assay is preferred. For hemoglobin variants HPLC (high-performance liquid chromatography) is preferred. For amino acid disorders, organic acid disorders and fatty oxidation disorders tandem mass spectrometry (TMS) is preferred.

- *Reporting:* The results of NBS will mention the following possibilities: Normal, presumptive positive/deficient or request repeat. Repeat may be requested by the screening lab to reanalyze those samples which are not conclusive or sample is insufficient. Presumptive positive/deficient means that the baby has a high likelihood of a certain disorder and more definitive tests should be done to confirm or disprove the same.

Incidence

Individually these disorders may be rare but collectively they are not uncommon. In several studies over the years we consistently observe that the incidence in India,[7,8] is much higher as compared to international statistics.[9–12] One probable reason could be more number of marriages in India happen within the same caste, unlike the western counterparts.

Disorder	World	India
CH	1:3,000	1:900
CAH	1:7,000	1:2,000
ACMG Panel	1:1,350	1:650 ~ 1:1,100

Counseling

Counseling plays a very crucial role for NBS due to the absence of any nationwide NBS program and also since it is still not a very popular advice. Antenatal care giver or an Obstetrician has to play this very important role of counseling the parents. NHS[13] of United Kingdom and ACOG[14] (American College of Obstetricians and Gynecologists) recommend that obstetricians should counsel parents about NBS during the antenatal period. Even in countries like USA and UK where NBS is mandatory, obstetricians play a very crucial role. In a study done on focus groups of parents it was suggested that information on NBS should be incorporated into the prenatal care.[15]

Implementation

Newborn screening is not just a diagnostic test. It is a screening program. Like any screening test we need to identify the target population, which in this case is every newborn. There are no absolute contraindications to NBS. Relative contraindication is severe prematurity. But in that situation also NBS may be performed after birth and repeated again after the baby attains maturity. Since there are no government guidelines on NBS it will be the discretion of the individual hospital/doctor to make it mandatory or optional. If mandatory which panel? The choice of panel would depend on the financial status of the parents. It would be wise to make a small basic panel mandatory for all and give option to parents to opt-in for a complete ACMG panel.

It would save a lot of time if all parents are counseled together somewhere in the early third trimester. Counseling aids could be used in the form of PPT, videos, etc. There are plenty of instruments available for patient education in various languages since NBS is a popular public health program in many countries:

Ontario NBS program (Hindi)
https://youtu.be/WFWl6HqBvIM
Save Babies Through Screening Foundation (English)
https://youtu.be/G_qjpjO3gFE

Choosing a NBS Laboratory is very important. Choose a laboratory which participates regularly in CDC's NSQAP. NSQAP (newborn screening quality assurance program) is a voluntary, non-regulatory program of Centers for Disease Control and Prevention, Atlanta to help laboratories maintain and enhance the quality of test results. The program provides services to newborn screening laboratories in 67 countries. NSQAP has been the only comprehensive source of essential quality assurance services for dried-blood spot testing for more than 33 years. An accreditation like CAP or NABL could be a desirable addition.

It is generally very easy to read and understand a NBS report. The report clearly mentions normal, positive/deficient or repeat. For a normal report nothing needs to be done. For a positive report a confirmatory test may be needed to confirm the screening test result. For repeat you would be asked to send a second sample to the laboratory. Generally no expertise is required to decipher the report. In case of a normal report an obstetrician does not seek help of any other specialist.

Challenges

Whenever a new screening program is to be introduced, we would encounter several challenges till the program reaches its goal. It would be no different in newborn screening program as well. The various challenges we encounter today are:

- Lack of awareness in policy makers and public at large.
- Lack of Government Advisory or National Newborn Screening Policy.
- Federal Structure of India wherein "Health" is a State subject and all health policies are independently enacted by the various State governments and UT's.
- Accepting that though individually these diseases are rare but collectively they are not uncommon.
- Poor understanding of the cost–benefit analysis of newborn screening program.
- Segmented focus on non-communicable diseases.
- Pregnancy caregivers not geared to counsel for NBS.
- Roles of obstetricians and pediatricians not defined with respect to counseling for NBS.
- Poor understanding of the advantages associated with universal screening vis-à-vis targeted screening.
- Lack of consensus on which disorders to screen what not to screen.
- Delivery to discharge interval.
- Follow-up or recall rate of newborns.
- Ease of availability of screening laboratories and confirmatory testing facilities.

LABORATORY REPORT FOR First Step™ NEWBORN SCREENING TEST

Acylcarnitine Profile Fatty Acid Oxidation & Organic Acid Disorders	**Within Normal Limits**
Amino Acid Profile Amino Acid Disorders, Urea Cycle Disorders	**Within Normal Limits**
Glucose-6-Phosphate Dehydrogenase (G6PD) Deficiency Glucose-6-Phosphate Dehydrogenase (G6PD)	**Within Normal Limits**
Congenital Adrenal Hyperplasia (CAH) 17-hydroxyprogesterone (17-OHP)	**Within Normal Limits**
Cystic Fibrosis (CF) Immunoreactive Trypsinogen (IRT) *Not valid after 2 months of age	**Within Normal Limits**
Galactosemia (GAL) Total Galactose (TGAL)	**Within Normal Limits**
Congenital Hypothyroidism (CH) Thyroid Stimulating Hormone (TSH)	**Within Normal Limits**
Biotinidase Deficiency (BIOT) Biotinidase	**Within Normal Limits**
Hemoglobinopathies Sickle Cell & Other Hemoglobinopathies *Not valid after 3 months of age	**Within Normal Limits**

QNS Quantity Not Sufficient   **NA** Not Applicable   **Specimen** Dried Blood Spot (DBS)   **Panel** Hb+FS+

Courtesy: NeoGen Labs Pvt. Ltd.

- Ease of availability of screening to manage confirmed positive cases.
- Difficult to procure drugs and specialized diets to manage metabolic disorders.

CONCLUSION

Every baby should be screened for basic disorders. All parents should be educated and counseled during the antenatal period, preferably in the early third trimester. All parents should be offered NBS after delivery. An opt-out form may be used for parents refusing a NBS test. Heel-prick should be done after 24 hrs of birth (not before). A complete ACMG panel should be offered to all affording patients. There is no contraindication of performing a newborn screening test.

REFERENCES

1. Revisiting Wilson and Jungner in the genomic age: a review of screening criteria over the past 40 years, Anne Andermann a, Ingeborg Blancquaert b, Sylvie Beauchamp b, Véronique Déry.
2. Wilson JMG, Jungner G. Principles and practice of screening for disease. Geneva: WHO; 1968.
3. NNF Clinical practice guidelines, October 2010, 289–299.
4. Newborn Screening: toward a uniform screening panel and system, May 2006 Vol. 8 No. 5.
5. US Department of Health & Human Services, http://www.hrsa.gov/advisorycommittees/mchbadvisory/heritabledisorders/recommendedpanel/.
6. Clinical Laboratory Standards Institute. Blood Collection on Filter Paper for Newborn Screening Programs: Approved Standard. 5th ed. LA4-A5, Vol 27 No 20.
7. High incidence of hypothyroidism in newborns in Chennai: ICMR study; http://www.thehindu.com/todays-paper/tp-national/high-incidence-of-hypothyroidism-in-newborns-in-chennai-icmr-study/article4517892.ece.
8. The Goa Newborn Screening Program 3 Year Review 2008 – 2011; http://www.dhsgoa.gov.in/documents/new_born.pdf.
9. Waller DK, Anderson JL, Lorey F, Cunningham GC. Risk factors for congenital hypothyroidism: an investigation of infant's birth weight, ethnicity, and gender in California, 1990–1998. Teratology. 2000;62:36–41.
10. Morreale de Escobar G, Obregon MJ, Escobar del Rey F. Is neuropsychological development related to maternal hypothyroidism or to maternal hypothyroxinemia? J Clin Endocrinol Metab.2000; 85:3975– 3987.
11. Lubani MM, Issa A-RA., Bushnaq R, Al-Saleh QA., Dudin KI, Reavey PC, El-Khalifa MY, Manandhar DS, Abdul Al YK, Ismail EA, Teebi AS. Prevalence of congenital adrenal hyperplasia in Kuwait. Europ. J. Pediat. 149: 391–392, 1990.
12. Virginia Department of Health–Health Practitioner Manual, March 2006.
13. Antenatal and newborn screening timeline; http://cpd.screening.nhs.uk/getdata.php?id=10610.
14. Committee Opinion on Newborn Screening; Number 481, March 2011.
15. Incorporating newborn screening into prenatal care; Campbell ED(1), Ross LF; Am J Obstet Gynecol. 2004 Apr;190(4):876–7.

Retinopathy of Prematurity and Role of Oxygen in Neonatal Resuscitation

Prachi Agashe

INTRODUCTION

Retinopathy of prematurity (ROP) is one of the important causes of preventable childhood blindness.

ROP is a specific proliferative disorder of the immature retinal blood vessels of preterm and low birth weight babies. In its more severe forms, it results in severe visual impairment or blindness, both of which carry a high financial cost not only for the community but also for the individual by affecting the normal motor, language, conceptual, and social development of the child.[1,2] Hence, it is necessary to have a good screening programme to reduce the burden of avoidable blindness arising from ROP.

History

The first clinical correlation of this disease with prematurity was made by Terry in 1942. His observations were based on noticing a retrolental proliferation of the embryonic hyaloid system which was seen as a white reflex in the pupillary area and he termed this as 'retrolental fibroplasia'.

As the pathoanatomy became more appreciated the term 'Retinopathy of Prematurity' was coined.

During the 1940s and early 1950s blindness due to ROP had assumed epidemic proportions due to indiscriminate use of oxygen in the perinatal period. In the late 1950s and 1960s oxygen therapy was curtailed, which however led to increased infant morbidity and mortality. A second epidemic of ROP was seen in the West in the 1980s owing to improved neonatal facilities which allowed the smaller babies to survive.

There has been an alarming increase in the incidence of ROP in developing countries like India owing to improvement in neonatal care facilities allowing smaller babies to survive which is referred to as the 'third epidemic'.

Incidence

In India, the reported incidence of ROP of varying grades is between 38% and 51.9% in low-birth-weight premature infants and about 10% of them can go blind if they do not receive timely treatment.[3] The incidence of ROP requiring timely intervention is inversely proportional to the maturity and the birth weight of the baby. ROP remains prevalent in very low birthweight (VLBW) premature infants, with as many as 12.5% of infants born between 23 and 26 weeks gestation requiring treatment for threshold disease.

Etiopathogenesis

Role of oxygen in causation of ROP:

The normal retinal vessels begin to proliferate at 16 weeks of gestation, they reach the nasal periphery of the retina (ora serrata) at 38 weeks and the process of vascularisation is generally completed at 40 weeks of gestation. Any interruption in this growth process of the retinal vessels leads to retinopathy of prematurity.

In utero, the fetus remains in a hypoxic state with a stable PaO_2 of 22–24 mm Hg while a normal fully mature fetus has a higher PaO_2 of 70–90 mm Hg. When an infant is delivered prematurely, the retinal growth continues to happen in this altered environment creating a risk for development of ROP. The main mediators implicated in the causation of ROP are vascular endothelial growth factor (VEGF) and insulin growth factor-1(IGF-1).

This generally happens in two phases— Phase 1: Hyperoxic phase, Phase 2: Hypoxic phase.

Phase 1: Hyperoxic Phase

The primary effect of supplemental oxygen on the incompletely vascularised retina is vasoconstriction and vaso-obliteration. Production of VEGF may be inhibited by the high levels of supplemental oxygen the infant may receive in the NICU, which causes cessation of normal retinal growth, and vessel constriction with a potential for vaso-obliteration of new immature vessels. This may cause subsequent death of vascular endothelial cells. This leads to delayed retinal growth and partial regression of the existing retinal vessels. This phase lasts from birth to 32 weeks of gestation.

In utero the fetus receives IGF-1 through the placenta. After premature delivery IGF-1 is suppressed by various factors like poor nutrition, sepsis, acidosis. A low level of IGF-1 in turn suppresses VEGF activation, necessary for retinal growth and survival of the retinal endothelial cells.

As the infant matures the non-vascularised retina becomes metabolically active leading to tissue hypoxia. This in turn triggers phase 2.

Phase 2: Hypoxic Phase

Prior to 32 weeks gestation, the retina is very immature with photoreceptors that are not yet fully functional and the retinal metabolic demand is low. As the retina matures, there is an increased metabolic demand and oxygen consumption, creating a relative retinal hypoxia. This hypoxia leads to upregulation of the angiogenic factors causing increase in VEGF and IGF-1 leading to precipitation of pathological new vessel proliferation causing ROP (Fig. 32.1).

Risk Factors

Prematurity and low birth weight are the two most important risk factors in the causation of ROP. The lower the birth weight or smaller the infant the higher are the chances of the baby developing ROP. Other important significant risk factors include respiratory distress syndrome (RDS), RDS requiring administration of surfactant, oxygen administration, anemia requiring multiple blood transfusions, multiple births (twins, triplets), apnea, sepsis, intraventricular hemorrhage, delayed postnatal weight gain (Table 32.1).[4]

International Classification of ROP

The International Classification of ROP[5] describes the location relative to the optic

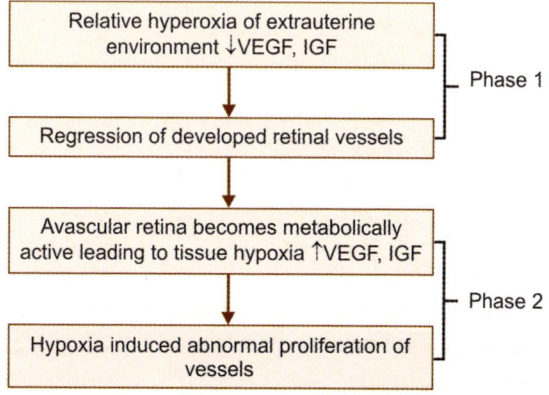

Fig. 32.1: Flowchart depicting the two phases of hyperoxia and hypoxia leading to causation of ROP

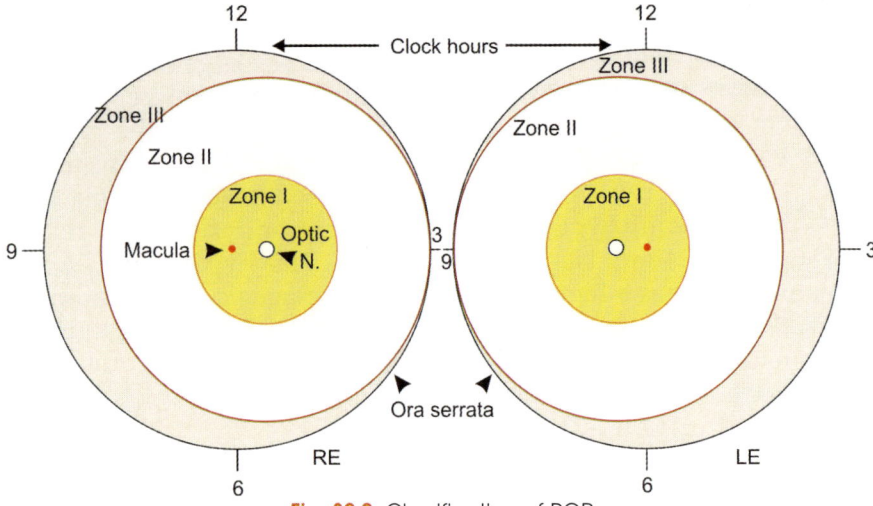

Fig. 32.2: Classification of ROP

Table 32.1	Risk factors causing ROP

- Prematurity
- Low birth weight
- Respiratory distress syndrome (RDS)
- RDS requiring surfactant administration
- Oxygen administration
- Anemia requiring blood transfusions
- Sepsis
- Intraventricular hemorrhage
- Multiple births (twins, triplets, etc.)
- Patent ductus arteriosus
- Delayed postnatal weight gain

nerve, the extent of the developing vasculature, the progressive staging of the disease and presence or absence of plus disease.

Zones

Zone 1 (posterior pole or inner zone) is a posterior circle centered on the disc, and extends twice the distance from the disc to the center of the macula in all directions.

Zone 2: Zone II extends all the way from the zone I perimeter to the ora serrata (the most peripheral part of the retina) on the nasal side.

Zone 3: Zone III is basically a temporal crescent, yet is defined by what is observed nasally (Fig. 32.2).

Stages

Stage 1 is the least involvement and stage 5 is the worst (Fig. 32.3)

Stage 1: Demarcation line seen at the junction of vascular and avascular retina.

Stage 2: Formation of a ridge when the line gains some height and width at the junction of vascular and avascular retina.

Stage 3: Formation of extraretinal tissue (new unwanted vessels) in the vitreous cavity.

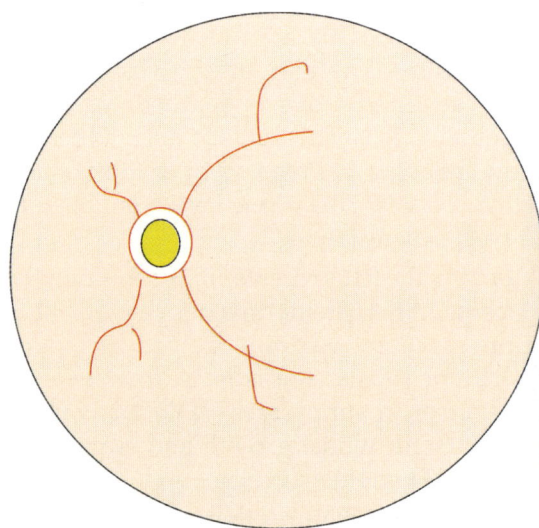

Fig. 32.3a: Stage 0—immature retina (retinal vessels have not reached till the ora serrata)

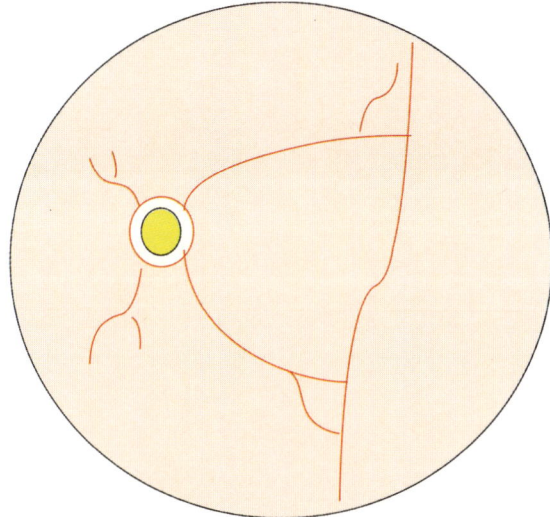

Fig. 32.3b: Stage 1—demarcation line between vascular and avascular retina

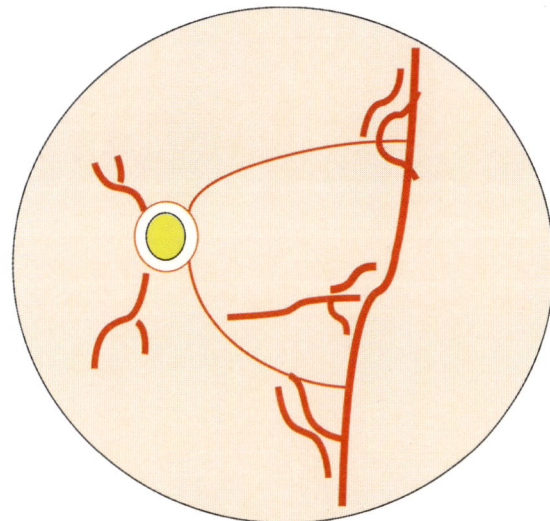

Fig. 32.3d: Stage 3—formation of extraretinal tissue extending in the vitreous

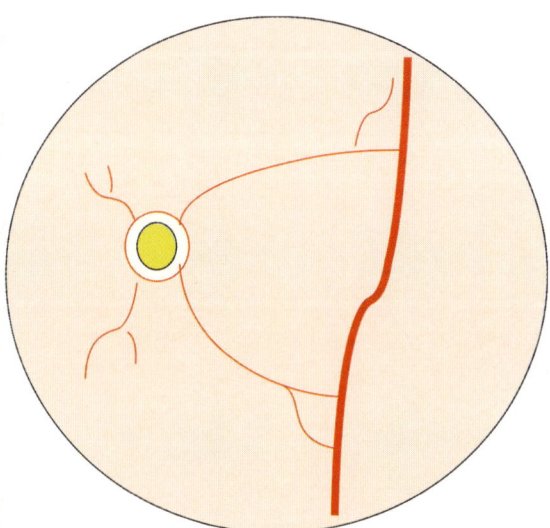

Fig. 32.3c: Stage 2—ridge formation

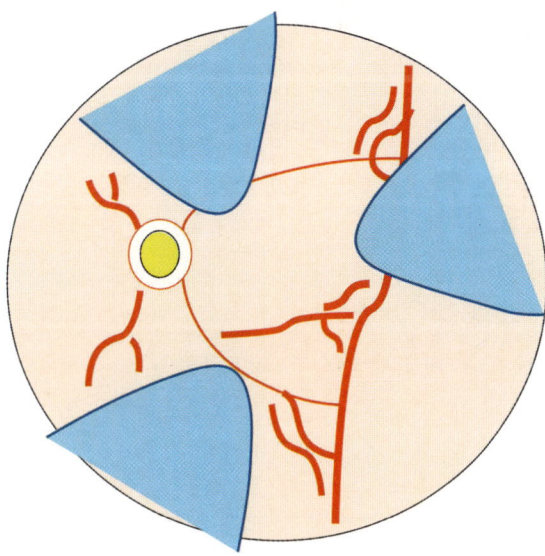

Fig. 32.3e: Stage 4—partial retinal detachment

Stage 4a: Partial retinal detachment sparing the macula.

Stage 4b: Partial retinal detachment involving the macula.

Stage 5: Total retinal detachment

Extent: It is described in the number of clock hours of the retina involved.

Plus disease: Plus disease implies increased venous dilatation and arteriolar tortuosity of the posterior retinal vessels near the optic nerve. It also includes the growth of abnormal blood vessels on the surface of the iris, rigid non-dilating pupil, and vitreous haze. The diagnosis of plus disease is usually made on the appearance of the vessels near the optic

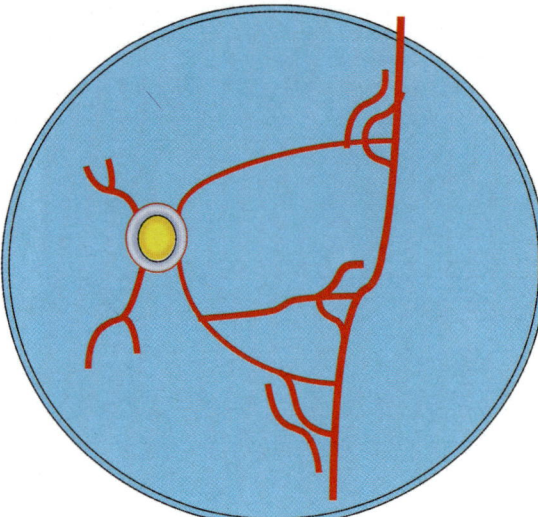

Fig. 32.3f: Stage 5—total retinal detachment

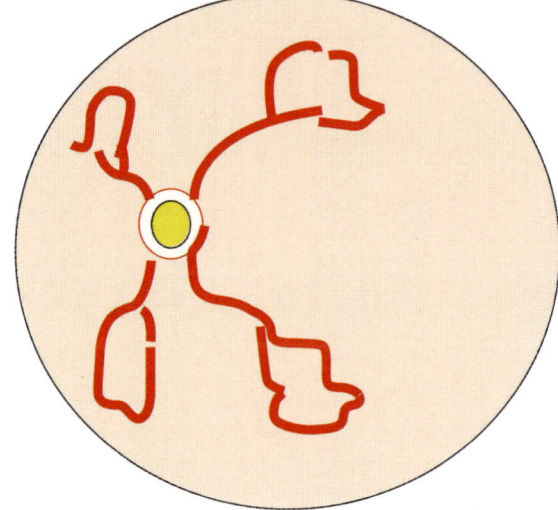

Fig. 32.5: Schematic representation of aggressive posterior ROP showing abnormal looping of retinal vessels

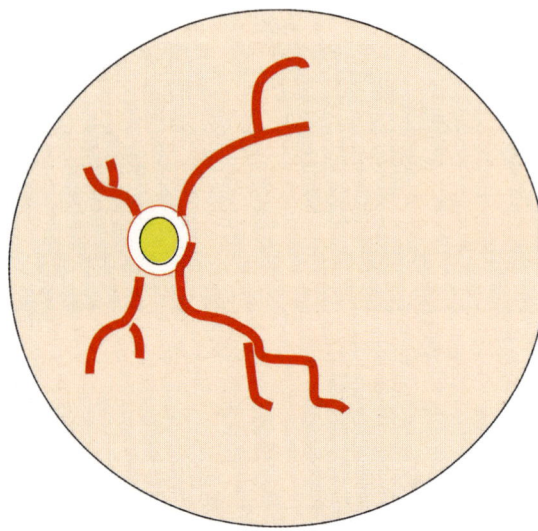

Fig. 32.4: Schematic representation of plus disease showing increased venous dilatation and arterial tortuosity at the posterior pole

nerve, as compared with standard retinal photographs (Fig. 32.4).

Aggressive posterior retinopathy of prematurity: It was initially called 'Rush disease'. It is a term used to denote ROP in Zone 1 and is different as it tends to progress very speedily and may not pass through the stages of ROP in a sequential manner. The prognosis may be poorer as compared to other eyes unless timely intervention is done (Fig. 32.5).

Threshold disease: "Threshold ROP" describes 5 contiguous or 8 cumulative clock hours of stage 3 ROP in zones I or II in the presence of "plus" disease, indicating an increased likelihood of progression to retinal detachment.

Pre-threshold disease: Presence of less than threshold disease in zone 1, or stage 2 plus disease in zone 2, or stage 3 (without plus) disease in zone 2, or stage 3 plus disease with extent less than that for threshold disease.

Management

Management of ROP requires a multidisciplinary approach involving the obstetrician, neonatologist and the treating pediatric ophthalmologist. Though it was initially believed that high levels of oxygen is the only risk factor for causing retinopathy of prematurity; it is evident from the above mentioned risk factors that many of them are modifiable risk factors; if taken care of can reduce the incidence of treatable ROP.

The management of ROP can be divided into screening protocols, definitive treatment and salvage methods.

Screening

Screening forms the basis of identifying babies with risk and is the backbone of preventing any child from becoming blind due to ROP. The screening criteria for babies in India differ from the Western population where babies less than 32 weeks and less than 1.5 kg are screened.

Owing to the finding of bigger babies also developing ROP, the screening criteria for Indian babies is any baby <34 weeks and/or weighing less than 1750 grams.[6] The first screening should be done at 3–4 weeks of life or 31 weeks of gestation whichever is earlier. Screening should commence earlier at 2 weeks of life in smaller babies who are <28 weeks and weigh less than 1.5 kg. These very low birth weight babies may develop aggressive posterior ROP which may progress very quickly to retinal detachment which has to be looked out for, more so in the Indian babies. The first screening should never be delayed beyond day 30 of life.

The screening can be done in the neonatal care units under the supervision of a neonatologist. The babies eyes can be examined even through, the incubators. Once the baby is discharged, the examination can be done in the clinic on an outpatient basis.

It requires the pupils to be dilated with 2.5% phenylephrine drops along with 1% tropicamide an hour before the expected time for screening. Examination is done with the use of an indirect ophthalmoscope and a 20 D lens under topical anesthesia (Fig. 32.6).

In developing countries like India, where majority of people live in remote areas which may not have access to the tertiary-level care, telescreening may bring more children into the screening programme. This model has been successfully used by Vinekar, et al. in Karnataka Internet Assisted Diagnosis of Retinopathy of Prematurity (KIDROP).[7] This involves taking retinal images by a specialised

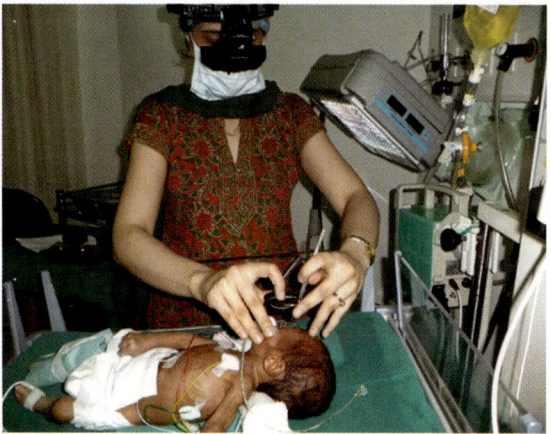

Fig. 32.6: ROP screening in NICU setup

retinal camera (RetCam) and are then assessed by an expert at a central tertiary eye care institute.

These babies need to be followed up weekly to biweekly depending on their presentation till the retina fully matures or till sure signs of regression are seen which is usually around 40 weeks of age.

The importance of screening lies in the fact that no child is born with ROP and these changes occur over a period of time which gives scope for a 'screening window'. Excluding the APROP patients the occurrence of ROP follows a particular pattern based on the corrected gestational age where in changes of ROP often start appearing at around 32 weeks of gestation and usually reach pre-threshold or threshold disease at 36–37 weeks of corrected gestational age. Thus, the crucial period for detection and management of ROP is between 32 and 40 weeks of gestational age (Table 32.2).

Treatment

A landmark study done was an ETROP (early treatment for retinopathy of prematurity) trial[8] conducted in 26 centers in USA which concluded a very significant benefit for eyes with high risk pre-threshold disease. Based on the results two types of ROP were described:

Table 32.2	Screening criteria for ROP

1. Gestational age less than 34–35 weeks
2. Birth weight less than 1750 gm
3. Oxygen exposure more than 30 days
4. Infants who are born before 28 weeks and weigh less than 1500 gm need to be screened at 2 weeks of life as they are at increased risk of developing ROP and particularly to look for development of APROP
5. Other preterm babies <37 weeks weighing less than 2000 gm where other risk factors for development of ROP may be there
   - Respiratory distress syndrome
   - Anemia requiring multiple blood transfusions
   - Aponeic episodes
   - Neonatal sepsis
   - Intraventricular hemorrhage
   - Multiple births (twins, triplets, etc)
   - Pediatrician has index of concern for ROP

Type 1 ROP

- Zone I, any stage ROP with plus disease
- Zone I, stage 3 ROP with or without plus disease
- Zone II, stage 2 or 3 ROP with plus disease

Type 2 ROP

- Zone I, stage 1 or 2 ROP without plus disease
- Zone II, stage 3 ROP without plus disease

Peripheral retinal laser ablation is necessary for all eyes with type 1 ROP and close follow up is necessary for eyes with type 2 ROP.

Modalities of Treatment

- Laser treatment
- Cryotherapy
- Intravitreal anti-VEGF injections
- Surgery

Laser photocoagulation delivered by the indirect ophthalmoscope is the mainstay of ROP treatment. Laser treatment involves the use of diode or argon green laser that is used to ablate the peripheral avascular retina. The reason for ablating the avascular retina is to ablate the source through which excess VEGF

production occurs so as to abolish the stimulus that leads to formation of the abnormal blood vessels. The laser treatment is done anterior to the ridge up to the ora serrata with spots that are near confluent burns. The procedure is usually done under topical anesthesia in the presence of a neonatologist or a pediatric anesthesiologist. In exceptional cases general anesthesia may be required.

Cryotherapy has largely become obsolete owing to higher rate of postoperative ocular and systemic complications after the procedure.

Bevacizumab and Ranibizumab are the two popular anti-VEGF molecules used in the management of ROP. BEAT-ROP study was one of the landmark studies which reported beneficial effect of Bevacizumab in eyes with APROP and Zone I stage 3+ disease.[9] However, some of these eyes may have a late recurrence of the disease and need a longer follow up. Eyes with recurrence may need a second dose of anti-VEGF or may need laser treatment. There is some concern about the systemic absorption of these drugs which may suppress the levels of anti-VEGF in the body which in turn is postulated to be responsible for causing developmental delay in these babies in some studies.

Salvage

Eyes that progress to Stage 4 and stage 5 need complex vitreoretinal surgeries to salvage some useful vision. Early surgeries lead to a favorable outcome in eyes with stage 4a, however, the visual prognosis is extremely guarded in eyes with advanced stages.

Complications

Untreated retinopathy of prematurity can lead to formation of retinal folds and dragging of the retinal vessels called cicatricial ROP. Other complications include early development of cataracts, iris neovascularization, glaucoma, retinal pigmentation, retinal folds, dragging of the retina, lattice-like degeneration, retinal tears, and rhegmatogenous and exudative retinal detachments.

Other Eye Problems associated with Prematurity

Premature babies are at higher risk of developing refractive errors like myopia, astigmatism, amblyopia (lazy eye), nystagmus and squint. This can occur in premature babies with or without ROP. They can also develop delayed visual maturation or sometimes cortical visual impairment; which implies vision problems occurring due to problems in the central nervous system.

These children need to be regularly screened for these ocular conditions every 6 months.

Babies who turn out to be severely visually impaired due to ROP need to be specifically trained and rehabilitated. The use of certain low visual aids may help improve their residual vision.

Role of Gynaecologist/Obstetrician

The awareness about the screening guidelines among the obstetrician may go a long way in educating expecting parents about ROP. Since a bond of trust is created between the treating obstetrician and the expectant parents, sensitisation of the parents by the treating obstetrician helps form a good network between the neonatologist, obstetrician and the ophthalmologist. With proper screening and management strategies we can reduce the burden of blindness arising due to retinopathy of prematurity (Fig. 32.7).

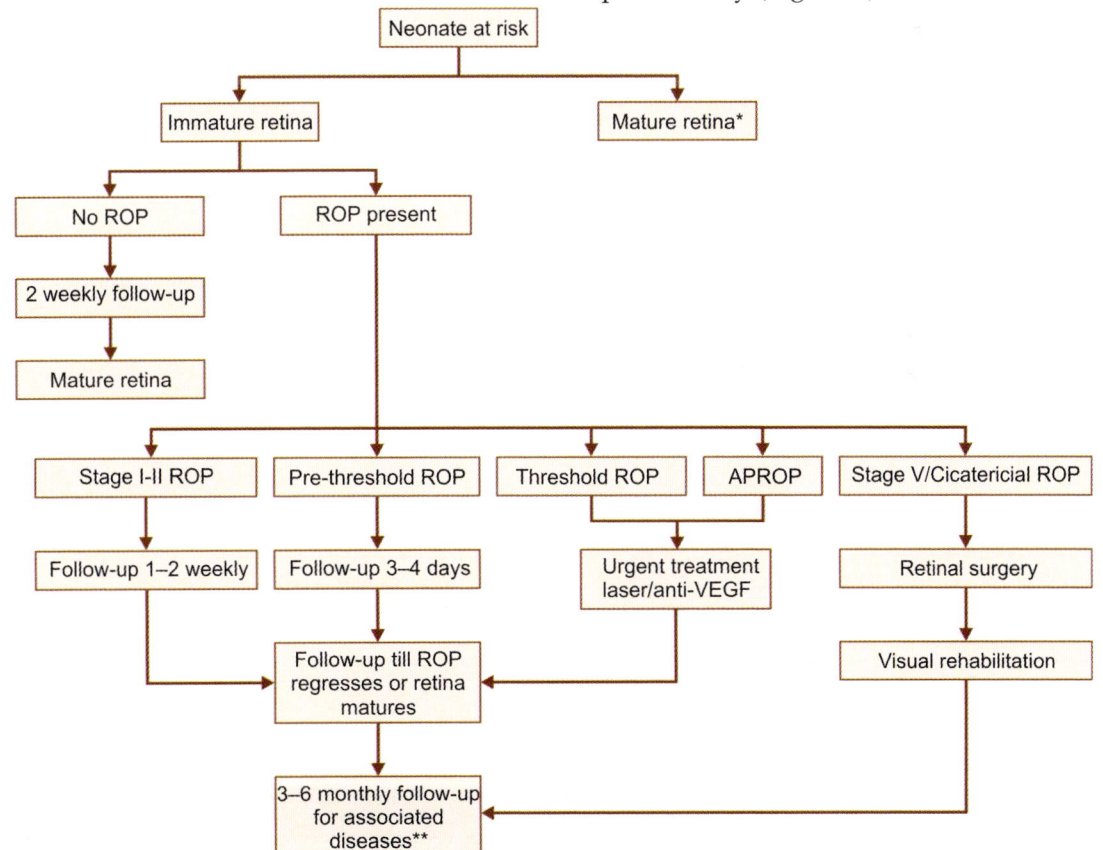

*No risk of ROP 3–6 monthly follow-up for associated diseases
**Associated diseases—Refractive errors, Squint, Nystagmus, Cataract, Glaucoma, Retinal detachment

Fig. 32.7: Flowchart summarising screening and management protocols in ROP

REFERENCES

1. Piccioni A, Lanners J, Goergen E. Early rehabilitation in retinopathy of prematurity children (0–4 years). Progress in retinopathy of prematurity. Proceedings of the international symposium on retinopathy of prematurity, 1997, Taormina, Italy. Amsterdam/New York: Kugler Publications, 1997.

2. Mets MB. Childhood blindness and visual loss: an assessment at two institutions including a "new" cause. Trans Am Ophthalmol Soc 1999;97:653–96.

3. Jalali S, Anand R, Kumar H, Dogra MR, Azad R, Gopal L. Programme planning and screening strategy in retinopathy of prematurity. Indian J Ophthalmol. 2003 Mar;51(1):89–99.

4. Chaudhari S, Patwardhan V, Vaidya U, Kadam S, Kamat A. Retinopathy of prematurity in a tertiary care center--incidence, risk factors and outcome. Indian Pediatr 2009;46:219–24.

5. The Committee for the Classification of Retinopathy of Prematurity: an international classification of retinopathy of prematurity. Arch Ophthalmol 1984;102:1130–1134.

6. Jalali S, Matalia J, Hussain A, Anand R. Modification of screening criteria for retinopathy of prematurity in India and other middle-income countries. Am J Ophthalmol. 2006 May;141(5):966–8.

7. Vinekar A, Gilbert C, Dogra M, Kurian M, Shainesh G, Shetty B, Bauer N. The KIDROP model of combining strategies for providing retinopathy of prematurity screening in underserved areas in India using wide-field imaging, tele-medicine, non-physician graders and smart phone reporting. Indian J Ophthalmol. 2014 Jan;62(1):41–9.

8. Early Treatment For Retinopathy Of Prematurity Cooperative G. Revised indications for the treatment of retinopathy of prematurity: results of the early treatment for retinopathy of prematurity randomized trial. Arch Ophthalmol 2003;121:1684–94.

9. Mintz-Hittner HA, Kennedy KA, Chuang AZ; BEAT-ROP Cooperative Group Efficacy of intravitreal bevacizumab for stage 3+ retinopathy of prematurity. N Engl J Med. 2011 Feb 17;364(7):603–15.

Apnea in Neonates

Swati Manerkar, Jayashree Mondkar

Apnea is the cessation of airflow during respiration. The American Academy of Pediatrics defines apnea in infants as "an episode of cessation of breathing for 20 seconds or longer, or a shorter respiratory pause associated with bradycardia, cyanosis, pallor, and/or marked hypotonia".[1]

Apnea is more common in preterm infants. The incidence increases with decreasing gestational age, being 10% at 34–35 weeks to almost 100% below 28 weeks of gestation.[2,3] Apnea in term babies is almost always pathological.

Apnea in newborns is of 3 types depending on the presence/absence of airflow and inspiratory efforts:[4]
1. *Central (40%):* When respiratory efforts are absent.
2. *Obstructive (10%):* When there is airway obstruction but inspiratory efforts persist.
3. *Mixed (50%):* When airway obstruction precedes/ follows cessation of inspiratory efforts.

Pathogenesis

Although the exact mechanism remains unknown, several have been proposed:[5]
1. Developmental immaturity of the CNS with an immature respiratory drive and prolonged REM sleep periods, especially in preterm infants, during which irregular breathing patterns occur.

2. Blunted central and peripheral chemo-receptor response to hypoxia and hypercarbia.
3. Reflexes invoked by lung inflation, stimulation of posterior pharynx or larynx during suctioning or feeding can cause apnea.
4. Passive neck flexion or nasal obstruction can also cause apnea in preterms.

Apnea should not be confused with periodic breathing which a normal pheno-menon in newborns during which cessation of breathing lasts for less than 20 seconds and does not cause change in heart rate or color. Apnea, especially in term babies, can be an unusual presentation of subtle seizures.

Causes

1. *Apnea of prematurity (AOP):* Onset is from day 2 to 7 of life. Any term baby with apnea or any preterm with apnea on day 1 of life needs investigations to rule out any underlying cause. This is a diagnosis of exclusion and secondary causes should always be ruled out.
2. *Infections:* Meningitis, sepsis.
3. *Cardiovascular:* Patent ductus arteriosus, hypo/hypertension, cardiac failure, hypovolaemia.
4. *Respiratory:* Pneumonia, lesions causing airway obstruction, upper airway collapse, phrenic nerve paralysis,

respiratory distress syndrome, pneumothorax, hypoxia, malformations of chest, pulmonary hemorrhage.

5. *Central nervous system:* Intraventricular hemorrhage, seizures, hypoxic injury, neuromuscular disorders.

6. *Gastrointestinal:* Necrotising enterocolitis (NEC), oral feeding, bowel movement, intestinal perforation, gastroesophageal reflux, abdominal distension.

7. *Hematologic:* Anemia.

8. *Pain:* Acute and chronic pain.

9. *Metabolic:* Hypoglycemia, hypocalcemia, hypo/hypernatremia, hypo/hyperthermia.

10. *Drugs:* Opiates, phenobarbitone, chloral hydrate or general anesthetics administered to the mother.

11. Neck flexion/nasal block.

Monitoring

All infants <35 weeks need routine pulse oximetry monitoring for the first week of life for apnea or till they have been apnea free for 5 days. Above 35 weeks, they need monitoring if unstable. Various apnea monitors are available which detect respiratory movements, however, these are ineffective in monitoring obstructive apnea. AOP usually resolves by 37 weeks postmenstrual age in babies >28 weeks. In preterms <28 weeks, apneic spells may continue till 44 weeks postmenstrual age.

Evaluation of An Infant with Apnea

Apnea of prematurity (AOP) is a diagnosis of exclusion and it is essential to investigate for other causes before labeling apnea as AOP. Check for history of administration of magnesium sulphate or opioids to mother or opioids to baby, birth trauma, perinatal asphyxia or risk for neonatal sepsis. Evaluate for upper airway anomalies and features of systemic illnesses like NEC and patent ductus arteriosus.

Laboratory evaluation: Complete blood count to look for anemia and infection, blood sugar, electrolytes to check for hypo/hypernatremia,

hypocalcemia, hypermagnesemia, blood gas to look for acidosis, sepsis screen and blood culture for infection, screen for toxic substances in urine.

Cranial ultrasound is required to look for bleeds, malformations or cerebral edema.

Management

Prolonged and frequent spells, requiring frequent tactile stimulation or bag and mask ventilation need treatment. Treatment includes general care, drug therapy (methylxanthines) and/or respiratory support.

Acute Management

This includes general care like maintaining the infant's temperature and sugar, ensuring the infant's head and neck are positioned in neutral position to maintain airway alignment and administering oxygen if required to maintain saturations between 91 and 95%. Gentle stimulation may re-establish breathing during mild episodes. Carefully suctioning the mouth and nostrils may be needed to clear any obstruction caused by secretions.

Treatment with Methylxanthines

Caffeine reduces the number of apneic spells and need for mechanical ventilation.[6] Caffeine is used in a loading dose of 20 mg/kg of caffeine citrate (10 mg/kg of base) intravenously or orally over 30 minutes, followed by 5 to 8 mg/kg/day (2.5 to 5 mg/kg caffeine base) once daily maintenance dose. Preterm infants weighing <1250 gm can be started on caffeine soon after birth. Those preterms receiving mechanical ventilation can be started on caffeine before extubation. Caffeine is continued till 34 to 36 weeks postmenstrual age and stopped if baby has not had apneic spells for 5 to 7 days. Infants need monitoring for 5 days after stopping caffeine for any recurrence of apnea. Preterms less than 28 weeks may have recurrent apneic spells till 44 weeks postmenstrual age and therefore need caffeine therapy till 44 weeks postmenstrual age.

In the caffeine for apnea of prematurity (CAP) trial, it was shown that caffeine improves survival without neurodevelopmental disability and decreases the rate of bronchopulmonary dysplasia.[7] The neurodevelopmental outcomes at 5 years were unchanged but caffeine reduced the risk of developmental coordination disorder.[8,9]

Respiratory Support

Infants unresponsive to methylxanthines may require respiratory support with nasal continuous positive pressure ventilation (nCPAP) or high flow therapy (HFT). nCPAP splints the upper airway, improves oxygenation and ventilation by optimising the functional residual capacity and thus reduces the incidence of obstructive or mixed apnea.[10] nCPAP can be provided with short binasal prongs or CPAP mask and initiated at a pressure of 4–6 cm of water. HFT has also been shown to be beneficial in treating AOP.[11] HFT is provided with small nasal prongs with a high flow of heated humidified air oxygen mixture at flow rates of 2–8 liters/min.

Infants failing methylxanthine therapy and CPAP or HFT may require nasal intermittent positive pressure ventilation (NIPPV) or intubation and mechanical ventilation.

Role of transfusion: There is an increased incidence of apnea in infants who develop anemia. Studies have shown that PRBC transfusions in ELBW infants with hematocrit < 25–30% decreases the severity and frequency of apnea and hypoxic episodes.[12]

Secondary causes of apnea need definitive management as per the diagnosis.

Discharge Planning

Control of respiration is mature in most infants, even <28 weekers by 44 weeks postmenstrual age. Infants are discharged once caffeine has been discontinued for 5–7 days and they remain apnea free during this period. Home cardiorespiratory monitoring is usually not required

as the the risk of significant apnea is extremely low after discharge.[13]

REFERENCES

1. Committee on Fetus and Newborn. American Academy of Pediatrics. Apnea, sudden infant death syndrome, and home monitoring. Pediatrics. 2003 Apr. 111(4 Pt 1):914–7.
2. Eichenwald EC, Aina A, Stark AR. Apnea frequently persists beyond term gestation in infants delivered at 24 to 28 weeks. Pediatrics 1997; 100:354.
3. Hofstetter AO, Legnevall L, Herlenius E, Katz-Salamon M. Cardiorespiratory development in extremely preterm infants: vulnerability to infection and persistence of events beyond term-equivalent age. Acta Paediatr 2008; 97:285.
4. Finer NN, Barrington KJ, Hayes BJ, Hugh A. Obstructive, mixed, and central apnea in the neonate: physiologic correlates. J Pediatr 1992; 121:943.
5. Abu-Shaweesh JM, Martin RJ. Neonatal apnea: what's new? Pediatr Pulmonol 2008; 43:937.
6. Henderson-Smart DJ, De Paoli AG. Methylxanthine treatment for apnoea in preterm infants. Cochrane Database Syst Rev 2010; CD000140.
7. Schmidt B, Roberts RS, Davis P, et al. Long-term effects of caffeine therapy for apnea of prematurity. N Engl J Med 2007; 357:1893.
8. Schmidt B, Anderson PJ, Doyle LW, et al. Survival without disability to age 5 years after neonatal caffeine therapy for apnea of prematurity. JAMA 2012; 307:275.
9. Doyle LW, Schmidt B, Anderson PJ, et al. Reduction in developmental coordination disorder with neonatal caffeine therapy. J Pediatr 2014; 165:356.
10. Finer NN, Barrington KJ, Hayes BJ, Hugh A. Obstructive, mixed, and central apnea in the neonate: physiologic correlates. J Pediatr 1992; 121:943.
11. Sreenan C, Lemke RP, Hudson-Mason A, Osiovich H. High-flow nasal cannulae in the management of apnea of prematurity: a comparison with conventional nasal continuous positive airway pressure. Pediatrics 2001; 107:1081.
12. Zagol K, Lake DE, Vergales B, et al. Anemia, apnea of prematurity, and blood transfusions. J Pediatr 2012; 161:417.
13. Lorch SA, Srinivasan L, Escobar GJ. Epidemiology of apnea and bradycardia resolution in premature infants. Pediatrics 2011; 128:e366.

Transient Tachypnea of Newborn

Sushma Malik, Vinaya Singh

INTRODUCTION

- Transient tachypnea of newborn (TTN) is a benign, physiological, self-limited process affecting 0.5 to 4% of late preterm and term babies.

- It results from pulmonary edema secondary to delayed clearance of fetal alveolar fluid.[1]

- TTN is characterized by tachypnea with mild respiratory distress along with retractions or expiratory grunting and reduced oxygen saturation. Usually hypoxemia is corrected by supplemental oxygen with FiO_2 less than 0.04.[2, 3]

- TTN however continues to pose a diagnostic dilemma as its initial presentation closely mimics other causes of respiratory distress including pneumonia/sepsis and meconium aspiration.[1]

Pathophysiology

The term transient tachypnea of newborn (TTN) was first described by Avery and coworkers in 1966. TTN results from delayed clearance of fetal lung fluids. It is also termed "wet lungs".[2] In the fetal life the lungs, are mainly in secretory mode, which provides the fetal lung fluid required for normal lung growth and development *in utero*.[4] The lungs switch from this secretory mode, to an absorptive mode, to accommodate the transition to breathing air at birth. Initial theories of lung fluid clearance focused on the role of the mechanical force of thoracic compression during birth canal squeeze and "starling forces", however these have been proved to account for only a fraction of the fluid absorbed.[1] Any delay or disruption in the clearance of the fetal lung fluid results in the transient pulmonary edema and the neonate then presents as TTN. Recent studies have revealed that alveolar fluid transition is facilitated by:

- Lung fluid clearance by amiloride-sensitive sodium channels

- Adrenergic stimulation near birth leading to clearing of interstitial lung fluid into pulmonary capillaries and lung lymphatics.[5,6]

- Fetal alveolar fluid is continuously secreted during pregnancy through an epithelial chloride secretion mechanism. Near birth the balance of fluid movement in alveolus switches from chloride secretion to sodium absorption, causing resorption of intra-alveolar fluid. Activated endothelial sodium channels (ENaC) at the apical surface of lung type II epithelial cells transport sodium and water from the alveolar space into the type II cells. Sodium is then actively moved from the type II cells into interstitium by NaKATPase, causing passive movement of water which is then resorbed into the pulmonary circulation and lymphatics.[7] Retention of lung fluid in the peribronchiolar lymphatics and bronchovascular spaces causes compression and bronchiolar collapse. This further lead to air trapping and hyperinflation. All these factors eventually lead to reduced lung compliance, accounting for clinical features of TTN.[2]

Risk Factors

- Premature birth, precipitous birth, cesarean delivery with or without labor increase risk of TTN.[2]
- Factors like delayed cord clamping or cord milking, promotes placento-fetal transfusion and raised central venous pressure of infant. This raised CVP causes impaired clearance of fluid by thoracic duct or pulmonary lymphatics leading to TTN.[2]
- Babies of asthmatic mother have an altered sensitivity to catecholamines which may lead to delayed clearance of lung fluid.
- Other factors like male gender[1,7,8] maternal diabetes, macrosomia, multiple gestation, prolonged labor, excessive maternal sedation have been less consistently associated with TTN.[3]
- Rarely genetic polymorphisms in beta-adrenergic receptors in type II alveolar cells may be associated with TTN.[1]

Clinical Manifestations

Affected newborns usually present within 6 hours of birth.[2,3] In mild cases clinical signs usually persist for 12 to 24 hours, but can last up to 48 to 72 hours in severe cases. Common clinical features include

- Tachypnea, mild cyanosis and subcostal retractions. Nasal flaring and expiratory grunting are suggestive of baby's effort to compensate for reduced lung compliance. Cyanosis usually responds to supplemental oxygen at less than 0.40 FiO_2 and respiratory failure and mechanical ventilation are rare.
- In preterm babies TTN and RDS can be seen simultaneously. Fetal lung fluid retention

in TTN causes inadequate lung expansion, leading to further decrease in lung compliance.

- Hyperinflation of chest leads to barrel-shaped thorax which further pushes liver and spleen downward in abdominal cavity, making them palpable.
- On auscultation of chest air entry is usually good with occasional crackles may be heard.

Investigations

There is no reliable diagnostic test for TTN, hence diagnosis remains one of exclusion, and vigilance for other, more severe disorders is imperative.[1, 2, 3, 8]

1. History and Physical Examination

Careful history evaluation may be helpful in identifying risk factors like meconium, infections, prematurity. Cardiac and neurological abnormal findings on physical examination helps to conduct more targeted investigations.

2. Chest Radiography

- Chest X-ray in TTN shows characteristic prominent perihilar streaking called "sunburst" pattern which is suggestive of fluid accumulation in periarterial lymphatics.
- Alveolar edema is reflected by coarse, fluffy densities.
- Hyperinflation with wide intercostal spaces, cardiomegaly, and mild pleural effusion and intralobar fissural fluid collection can also be seen.

Follow-up radiographs will show clearance to resolution in a peripheral to central and upper to lower lung zone pattern. Rapid resolution of radiographic findings in TTN within 48–72 hours helps to rule out other close differential diagnosis like pneumonia, HMD and meconium aspiration.[8]

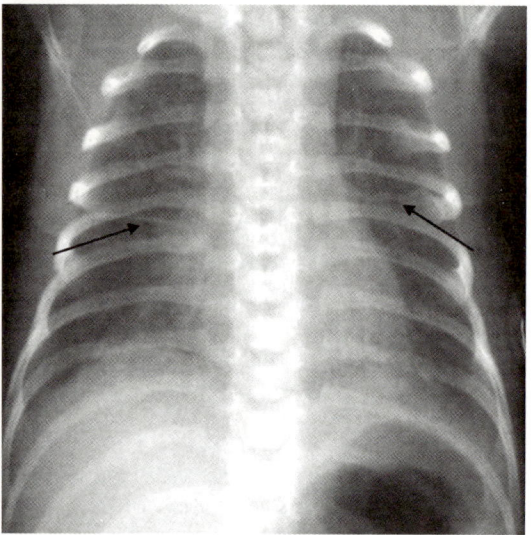

Fig. 34.1: Chest radiograph of an infant who has transient tachypnea of the newborn. Arrows point at fluids in the interlobar fissures. Note the increased pulmonary interstitial markings in both lung fields.

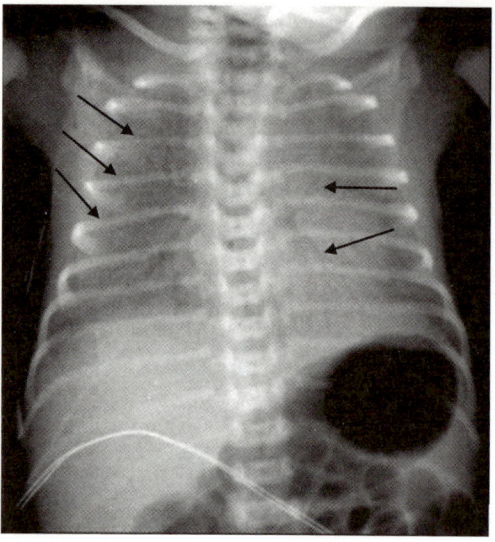

Fig. 34.2: Chest radiograph of a preterm infant who has hyaline membrane disease. Arrows on the right lung field point at ground glass appearance. Arrows on the left demonstrate air bronchogram.

3. *Ultrasonography*

Ultrasonography of chest is also useful in diagnosis and helps in prediction of the need for respiratory support.[1]

4. *Lab Parameters*

Complete blood counts and blood culture helps to rule out infectious etiology of respiratory distress. ABG in TTN often shows respiratory acidosis and hypoxemia.

Differential Diagnosis

It includes all the causes of respiratory distress in neonate in first 6–8 hours of life. Examples include pneumonia, sepsis, congenital heart disease, RDS, MAS, pulmonary hypertension, CNS abnormalities.[2,3]

- *Pneumonia:* Exposure to pathogens *in utero*, delivery and post-delivery lead to pneumonia. Chest radiographs show granular coarse, interstitial patterns, progressive or persistent opacities, hyperinflation, and pleural effusion suggests pneumonia.

- *RDS:* Also known as hyaline membrane disease, occurs in preterm babies due to deficiency of lung surfactant. Chest radiograph suggestive of 'white' out lung due to hypoaeration.
- *MAS:* Fetal hypoxia leads to intrauterine passage of meconium, there may be intra-uterine gasping or postpartum aspiration. It will lead to atelectasis, air trapping and pulmonary hypertension. Typical chest radiograph is characterized by patchy infiltrates, coarse streaking, increased anteroposterior diameter and flat diaphragm.
- *Congenital heart disease:* Echocardiography helps to rule out congenital heart disease in the neonate.
- *Congenital malformations* like congenital diaphragmatic hernia and cystic adeno-matiod malformations can mimic TTN but classical radiological findings help to exclude these conditions.[2]

Treatment

- Treatment is supportive with supplemental oxygen.

- In severe cases use of continuous positive pressure ventilation is useful as it helps in lung recruitment. Need for invasive mechanical ventilation is very rare.
- Empircal anitibiotics may be initiated while baby is undergoing septic screening and can be continued for 48 hours till blood cultures negative.
- If tachypnea persists gavage feeds or intravenous fluids should be started.
- Furosemide has no role in treatment of TTN.
- Antenatal steroids are known to enhance the lung fluid clearance from alveolar spaces, however, use of exogenous steroids after birth is not beneficial.[1]

Complications

- TTN is self-limited disease. Some complications because of supportive therapy like CPAP may lead to air leak (pneumothorax).
- Admission into the neonatal intensive care unit may lead to delayed establishment of breastfeeding and prolonged hospitalization.
- Some studies have suggested that infants of TTN may have increased risk for development of asthma in later life.[3]

Prognosis

Prognosis is excellent without any significant residual effects.

Acknowledgment: The authors wish to acknowledge and thank the Dean of our institute, Dr Ramesh Bharmal, for giving permission to publish this article.

REFERENCES

1. Crowley MA. Neonatal respiratory disorder-Transient tachypnea of the newborn. In Neonatal and Perinatal Medicine, Diseases of the fetus and infant, Eds Martin RJ, Fanaroff AA, Walsh MC, 10th edn, Elsevier, Philadelphia, 1127–1128.
2. Kienstra KA. Transient tachypnea of the newborn. In Manual of Newborn Care. Eds Cloherty JP Eichenwald EC, Hansen AR, Stark AR, 7th edn, 2012, Wolters Kluwer, New Delhi, 403–405.
3. Ambalavanan N, Cario W, Transient tachypnea of the newborn. In Nelson textbook of pediatrics, Eds Kliegman RM, Stanton BF, St.Geme JW, Schor NF, Behrman RE, 19th edn, 2011, Elsevier, New Delhi, 590.
4. Olver RE, Strang LB. Ion fluxes across the pulmonary epithelium and secretion of lung liquid in foetal lamb. J physiol 241:327, 1974.
5. Baines DL, Folkesson HG, Norlin A, Bingle CD, Yuan HT, Olver RE. The influence of mode of delivery, hormonal status and postnatal O_2 environment on epithelial sodium channel (ENaC) expression in guinea pig lung. J Physiol 2000, 522, 147–157.
6. Barker PM, Brown MJ, Ramsden CA, Strang LB, Walters DV. Synergistic action of triiodothyronine and hydrocortisone on epinephrine-induced reabsorption of fetal lung liquid. Pediatr Res,1990, 27, 588–591.
7. Ramchandrappa A, Jain L. Late preterm, In Neonatal and Perinatal Medicine-Diseases of the fetus and infant, Eds Martin RJ, Fanaroff AA, Walsh MC, 10th edn, Elsevier, Philadelphia, 583.
8. Hellinger JC, Mendelson N, Seshia MK, McDonald MG. Radiologic imaging of neonate. In Avery's Neonatology-Pathophysiology of the Newborn. Eds, McDonald MG, Seshia MK, 7th edn, 2016, Wolters Kluwer, New Delhi, 70.

Role of Surfactant

Bhupendra Avasthi, Anuja Rege

Surfactant therapy is one of the few treatments that decrease overall mortality in preterm newborns with RDS and many other clinical situations, and has significantly changed clinical practice in neonatology. This chapter mainly focuses on confusing points faced by clinicians today like when to give surfactant, dose and route, frequency, etc. with practical tips and general pointers regarding innovative techniques of surfactant use.

A systematic PubMed search up to January 2013 was undertaken to identify manuscripts addressing the following three specific questions:

1. Which infants should we treat with exogenous surfactant therapy?
2. When should preterm infants with RDS be treated with exogenous surfactant?
3. How should preterm infants with RDS be treated with exogenous surfactant?

Surfactant

- Is a complex mixture of phospholipids (PL) and proteins (SP)
- Synthesized and secreted by Type II alveolar cells (pneumocytes)
- Synthesized between 24 and 34 weeks of gestation
- It is made up of 70 to 80% phospholipids, approximately 10% protein and 10% neutral lipids, mainly cholesterol.

- The primary surface-active material is dipalmitoylphosphatidylcholine (DPPC)
- Surfactant proteins are SP-A, SP-B, SP-C and SP-D.
- It reduces surface tension at the air–liquid interface of the alveolus and prevents its collapse during end-exhalation
- It participates in innate host defense against inhaled pathogens.
- Defective pulmonary surfactant metabolism results in respiratory distress with resulting morbidity and mortality.
- Pulmonary surfactant is on the WHO Model List of Essential Medicines, the most important medications needed in a basic health system.
- Metabolism of surfactant is slower in newborns, especially preterm, than in adults.

Mechanism

During expiration (breathing out) the lungs have a tendency to collapse, if they are allowed to do so then a much greater inspiratory effort is required to open them with the next breath. Surfactant prevents this by reducing surface tension throughout the lung. Surfactant forms a very thin film which covers the surface of the alveolar cells; the components of surfactant work together to reduce surface tension and therefore reduce the tendency of the alveoli to collapse during expiration. The lungs are less stiff (improved pulmonary compliance) and therefore reduced effort is needed to expand the lungs making breathing easier. The natural production of surfactant increases at approximately week 30 to 32 and babies born after the end of the 32nd week usually have sufficient surfactant to breath normally.

Defects in Surfactant

Defective surfactant metabolism leads to both morbidity and mortality in preterm and term neonates. In general, defects in surfactant metabolism occur due to accelerated break-down of the surfactant complex by oxidation, proteolytic degradation, and inhibition. Some inherited surfactant gene defects have also been implicated.

Surfactant Replacement Therapy

Setting: Administered by a trained personnel in NICU (doctor and nurse).

Pre-requisites: Suction prior to surfactant is crucial (after surfactant avoid suction for 6 hr).

Indications
Prophylactic administration (within 15 min of delivery). Recommended in:
1. Gestational age less than 28 weeks
2. Gestational age between 28 and 30 weeks. with no ANC steroidal cover
3. Long transport time from delivery setup to NICU

Early rescue surfactant (within 15 min to 2 hr)
Two new RCTs have demonstrated that routine early surfactant administration within 2 hours of life:
- Reduces the need for mechanical ventilation in the first week of life among preterm infants with RDS on nasal CPAP, born between 28 and 32 weeks GA.
- Decreases intra-ventricular hemorrhage (≥ grade III) and pneumothorax rates but does not have any effect on BPD when compared to delayed surfactant administration.

Late rescue surfactant: Both animal and human studies have demonstrated that early administration of surfactant is more effective than late rescue surfactant treatment because of better surfactant distribution and avoidance of ventilator-induced lung injury.

Administration equipment (for systematic administration of surfactant)
- Syringe containing the ordered dose of surfactant, warmed to room temperature
- 5-Fr feeding tube or catheter, or endotracheal tube connector with delivery port

Contraindications

Relative contraindications to surfactant administration are
- The presence of congenital anomalies incompatible with life beyond the neonatal period
- Respiratory distress in infants with laboratory evidence of lung maturity

InSurE Strategy

- A strategy in which surfactant is administered during brief intubation followed by immediate extubation and nasal respiratory support
- The InSurE strategy is associated with a significantly lower incidence of mechanical ventilation and a trend towards a decrease in BPD and air-leak syndromes.
- More recently, Bhandari et al. have demonstrated that the InSurE strategy is associated with a significantly lower incidence of BPD or death (20% vs. 52%; p = 0.03)

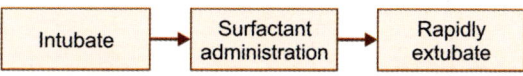

Adverse effects
Short-term risks
- Hypoxemia
- Bradycardia due to hypoxia

- Plugging of endotracheal tube (ETT) by surfactant
- Reflux of surfactant into the ETT
- Relative risk of pulmonary hemorrhage
- Over distension and hyperventilation with low partial pressure of carbon dioxide (PCO_2)

To date, there is no evidence that there are any immunological changes of clinical concern due to bovine or porcine sources of proteins contained in natural surfactants.

Frequency

Surfactant re-dosing should not be needed more often than every 12 hours, unless surfactant is being inactivated by an infectious process, meconium, or blood. Dosing intervals shorter than 12 hours recommended by some manufacturers are not based on human pharmacokinetic data.

Animal Derived Versus Synthetic Surfactant

Animal derived surfactant have a few advantages over first generation synthetic surfactants like lower mortality rate (RR 0.86; 95% CI 0.76–0.98; number needed to harm of 40) and fewer pneumothoraces (RR 0.63; 95% CI 0.53–0.75; NNTB 22). Animal-derived surfactants contain variable amounts of surfactant protein B (SP-B). SP-B enhances the rate of adsorption of phospholipids at the air–water interface, is involved in the formation of tubular myelin, and has anti-inflammatory properties. However, it is unclear whether significant differences in clinical outcomes exist among the available animal-derived products.

In India we have only animal derived surfactant available.

Synthetic surfactants like Exosurf or Surfaxin are used in other countries.

Candisurf (Goat lung surfactant)
- Under trial in India
- Multicentric trial at 18 centers

Newer Advances

- MIST (minimally invasive surfactant therapy)/NIST (non-invasive surfactant therapy)
- Avoids endotracheal tube related risks
- No need for positive pressure mechanical ventilation
- *LMA (laryngeal mask):* The Cochrane experts concluded that there was some evidence that selective surfactant administration through an LMA to preterm infants >1200 g with established RDS reduced oxygen requirement in the short term, although the LMA technique needed to be further evaluated in clinical trials, including those dedicated to investigating the size of the LMA used according to gestational age.
- *IPI (intrapartum pharyngeal instillation):* Evidence from animal and observational human studies suggests that the pharyngeal instillation of surfactant before the first breath is potentially safe, feasible and effective. Well-designed trials are still needed to confirm this.
- Feeding and vascular catheter
- Aerosolized surfactant

In conclusion, surfactant administration via an endotracheal tube remains the gold standard for surfactant administration in intubated infants. However, CPAP, as the

Trade name	Origin	Dose	Company	Vials (ml)
Survanta	Bovine	4 ml/kg	Abbvie Inc, North Chicago, IL	
Curosurf	Minced porcine lung extract	2.5 ml/kg or 1.5 ml/kg	Chiesi Farmaceutici, Italy	
Infasurf	Bovine calf lung lavage	3 ml/kg	ONY Inc., Amherst, NY	

primary respiratory support technique in infants with RDS, has been shown to be at least as effective as mechanical ventilation, and newly emerging techniques for surfactant administration in non-ventilated infants are currently under investigation. The optimization of a less or non-invasive method of surfactant administration will be one of the most important challenges in the field of surfactant therapy for RDS in the coming years.

Common Pathologies Related to Surfactant Deficiency

1. Respiratory Distress Syndrome (RDS)

- One of the most common causes of morbidity in preterm neonates.
- Slight male predominance.
- Apnea, cyanosis, grunting, inspiratory stridor, nasal flaring, poor feeding, and tachypnea shortly after birth.
- Intercostal or subcostal retractions.
- There is increasing evidence to suggest that CPAP immediately after birth is a reasonable alternative to systematic intubation for surfactant administration to preterm infants.
- Radiological findings include a diffuse reticulogranular "ground glass" appearance (resulting from alveolar atelectasis) with superimposed air bronchograms.
- Prenatal corticosteroids and postnatal surfactant replacement therapy significantly reduce the incidence, severity and mortality associated with RDS, and surfactant therapy has become the standard of care in management of preterm infants with RDS.

2. Meconium Aspiration Syndrome (MAS)

- Important cause of morbidity and mortality from respiratory distress in the perinatal period
- Approximately 25,000 neonates in the United States each year.
- Meconium staining of the amniotic fluid or fetus is an indication of fetal distress.
- Fetal respiration is associated with movement of fluid from the airways out into the amniotic fluid. However, in the presence of fetal distress, gasping may be initiated in utero leading to aspiration of amniotic fluid and its contents, which includes meconium, into the large airways.
- Acute lung injury is characterized by airway obstruction, pneumonitis, pulmonary hypertension, ventilation/perfusion mismatch, acidosis and hypoxemia.
- It has been shown that meconium destroys the fibrillary structure of surfactant and decreases its surface adsorption rate.
- MAS is associated with an inflammatory response characterized by the presence of elevated cell count and pro-inflammatory cytokines IL-1, IL-6, and IL-8 as early as in the first 6 hours and significantly decreased by 96 hours of life.
- Phospholipase-A2, (PLA2) present in meconium, has been found to inhibit the activity of surfactant *in vitro* in a dose-dependent manner, through the competitive displacement of surfactant from the alveolar film.
- Exogenous surfactant replacement either as bolus therapy or with a diluted surfactant lung lavage have been shown to reverse the

Table 35.1	International guidelines for RDS treatment		
Country	*Year*	*Gestational age*	*Prophylactic surfactant use*
UK	1999	<29 weeks	Systematic
		<32 weeks	If need for intubation at birth
Canada	2005	< 26 weeks	Systematic
		26–27 weeks	If no antenatal steroids
US	2008	<28 weeks	Systematic

hypoxemia and reduce pneumothoraxes caused by meconium aspiration, decrease requirement for extracorporeal membrane oxygenation (ECMO), decrease duration of oxygen therapy and mechanical ventilation, and reduce the duration of hospital stays.

- A comparison of various surfactant treatment regimens in MAS did not find the superiority of one form of therapy over another, and may be related to the heterogeneous nature of this form of lung injury.

To Summarise

- An understanding of the complex metabolic process involving phospholipids and surfactant proteins is the key in the management of respiratory failure secondary to defects in surfactant metabolism.

- The combined use of prenatal corticosteroids and postnatal surfactant replacement therapy can be credited with a dramatic improvement in the outcome of patients with RDS.

- Early rescue surfactant treatment (<2 hours of age) in infants with RDS decreases the risk of mortality, air leak, and chronic lung disease in preterm infants (LOE 1).

- Early initiation of CPAP with subsequent selective surfactant administration in extremely preterm infants results in lower rates of BPD/death when compared with treatment with prophylactic surfactant therapy (LOE 1) (to search for more studies as contradictory sentence).

- Surfactant replacement has not been shown to affect the incidence of neurologic, developmental, behavioral, medical, or educational outcomes in preterm infants (LOE 2).

- Surfactant treatment improves oxygenation and reduces the need for ECMO without an increase in morbidity in neonates with meconium aspiration syndrome (LOE 2).

- Surfactant treatment of infants with congenital diaphragmatic hernia does not improve clinical outcomes (LOE 2).

- Lung transplantation has been successful in treating infants with inherited SP-B deficiency and has also afforded the opportunity to investigate surfactant composition and function.

- Gene therapy could overcome the limitations of surfactant replacement therapy in inherited defects of surfactant metabolism.

SUGGESTED READING

1. Andersson S, Kheiter A, Merritt TA. Oxidative inactivation of surfactants. Lung. 1999;177:179–189. [HYPERLINK "https://www.ncbi.nlm.nih.gov/pubmed/10192765" PubMed].

2. Manalo E, Merritt TA, Kheiter A, Amirkhanian J, Cochrane C. Comparative effects of some serum components and proteolytic products of fibrinogen on surface tension-lowering abilities of beractant and a synthetic peptide containing surfactant KL4. Pediatr Res. 1996;39:947–952. [HYPERLINK "https://www.ncbi.nlm.nih.gov/pubmed/8725253" PubMed].

3. Emmanuel Lopez1, Géraldine Gascoin2, Cyril Flamant, Mona Merhi, Pierre Tourneux, and Olivier Baud6* for the French Young Neonatologist Club: Exogenous surfactant therapy in 2013: what is next? who, when and how should we treat newborn infants in the future?, Lopez et al. BMC Pediatrics 2013, 13:165.

4. Bhandari V, Gavino RG, Nedrelow JH, Pallela P, Salvador A, Ehrenkranz RA, Brodsky NL: A randomized controlled trial of synchronized nasal intermittent positive pressure ventilation in RDS. J Perinatol 2007, 27:697–703.

5. Richard A. Polin, MD, FAAP, Waldemar A. Carlo, MD, FAAP, and COMMITTEE ON FETUS AND NEWBORN : Surfactant Replacement Therapy for Preterm and Term Neonates With Respiratory Distress, American Academy of Paediatrics 156-163.

6. Soll RF, Blanco F. Natural surfactant extract versus synthetic surfactant for neonatal respiratory distress syndrome. Cochrane Database Syst Rev. 2001;(2):CD000144.

Postmature Infant

Deep Parekh, Pallavi Untwal

INTRODUCTION

The adjectives post-term, prolonged, postdates, and postmature are often loosely used interchangeably to describe pregnancies that have exceeded a duration considered to be the upper limit of normal.

The definition of postmature pregnancy as one that persists for 42 weeks or more from the onset of a menstrual period assumes that the last menses was followed by ovulation 2 weeks later.

That said, some pregnancies may not actually be post-term, but rather are the result of an error in gestational age estimation because of faulty menstrual date recall or delayed ovulation.

Thus, there are two categories of pregnancies that reach 42 completed weeks: (1) those truly 40 weeks past conception, and (2) those of less-advanced gestation but with inaccurately estimated gestational age. These variations in the menstrual cycle may partially explain why a relatively small proportion of fetuses delivered post-term have evidence of postmaturity syndrome.

Definition

Postmaturity is reserved for the relatively uncommon specific clinical fetal syndrome in which the infant has recognizable clinical features indicating a pathologically prolonged pregnancy (Williams, 24 ed, pg 862).

Pathology

Most postmature infants are not technically growth restricted because their birthweight seldom falls below the 10th percentile for gestational age. On the other hand, severe growth restriction—which logically must have preceded completion of 42 weeks—may be present. Attributed the postmaturity syndrome to placental senescence, although he did not find placental degeneration histologically. Still, the concept that postmaturity is due to placental insufficiency has persisted despite an absence of morphological or significant quantitative findings (Larsen, 1995; Rushton, 1991).

There are findings that placental apoptosis—programmed cell death—was significantly increased at 41 to 42 completed weeks compared with that at 36 to 39 weeks (Smith, 1999). Several proapoptotic genes such as kisspeptin were shown to be upregulated in post-term placental explants compared with the same genes in term placental explants (Torricelli, 2012). The clinical significance of such apoptosis is currently unclear due to placental aging in post-term pregnancies. Another scenario is that the post-term fetus may continue to gain weight and thus be unusually large at birth. This at least suggests that placental function is not severely compromised. Indeed, continued fetal growth is the normal albeit at a slower rate.

Who is at Risk of Postmaturity?

Postmaturity is more likely to happen when a mother has had a post-term pregnancy before. After one post-term pregnancy, the risk of a second post-term birth increases by 2 to 3 times. Other risk factors include:
- First pregnancy
- Male baby
- Older mother
- Mother or father personal history of postmaturity
- White mother

Problems of Postmature Infant

- Meconium aspiration
- Malnutrition, hypoglycemia
- Hypothermia
- Asphyxia, death

Can Postmaturity be Prevented?

- Knowing your due date is the best way to know if your baby may be post-term.
- An ultrasound test early in pregnancy can help your healthcare provider figure out your baby's age by checking the baby's size. Ultrasound is also a good way to check the placenta for signs of aging.

Management of Postmature Infant

How is postmaturity in the newborn diagnosed? By checking following:
- Baby's physical appearance
- The length of pregnancy
- How old baby appears to be

Features of Postmature Infant

- Wrinkled, patchy, peeling skin (Figs 36.1 and 36.2).
- A long, thin body suggesting wasting and advanced maturity.
- The infant is open-eyed, unusually alert, appears old and worried.
- Skin wrinkling can be particularly prominent on the palms and soles. The nails are typically long.

Nursing Interventions with the Post-term Infant

Delivery → NRP → Admission to nursery/NICU.

Nursing Interventions at Delivery and Admission to Nursery/NICU

- Place ID bands on
- Obtain footprint
- Allow mother to see newborn
- Transport to nursery or NICU
- Obtain weight and length
- Complete physical and gestational age assessment
- Administer vitamin K

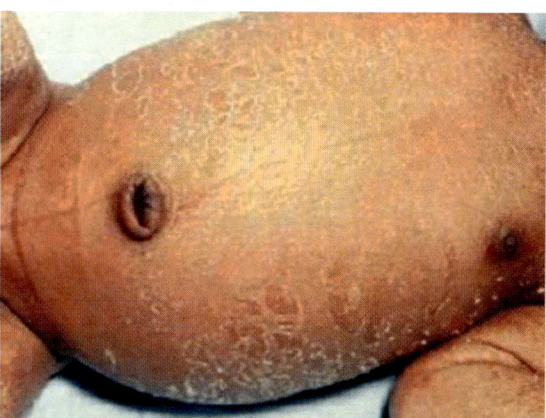

Fig. 36.1: Patchy and peeling skin in postmature infant

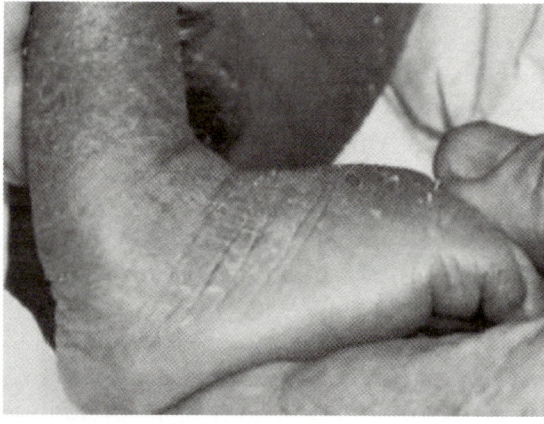

Fig. 36.2: Dry and wrinkled skin in postmature infant

Special care of the post-term baby may include:

- Checking for breathing problems caused by baby's breathing in fluid containing the first stools (meconium)
- Blood tests for low blood sugar

Treatment

- Treatment of complications
- Prognosis and treatment depend on complications. Neonates with meconium aspiration may have chronic respiratory insufficiency and secondary pulmonary hypertension, if untreated.

Complications

Postmature infants have higher morbidity and mortality than term infants. During labor, postmature infants are prone to develop

- Asphyxia—may result from cord compression secondary to oligohydramnios (Fig. 36.3).
- Meconium aspiration syndrome—may be unusually severe because amniotic fluid volume is decreased and thus the aspirated meconium is less dilute.
- Hypoglycemia—neonatal hypoglycemia is caused by insufficient glycogen stores at birth. Because anaerobic metabolism rapidly uses the remaining glycogen stores, hypoglycemia is exaggerated if perinatal asphyxia has occurred.

Hypoglycemia Management

Symptoms of hypoglycemia: Hypothermia, temperature instability, poor suck or refusal to eat, vomiting, cyanosis, jitteriness, lethargy, hypotonicity, apnea, irregular respirations, high-pitched or weak cry, seizures.

Blood Sugar Guidelines

Avoid enteral feedings (PO or NG) due to increased risk of aspiration with respiratory rate >60 breaths per minute and impaired bowel blood flow which may lead to necrotizing enterocolitis.

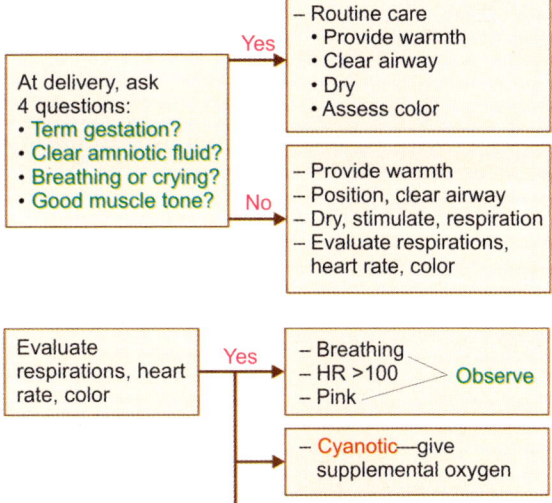

Fig. 36.3: Asphyxia—neonatal resuscitation

↓

Establish IV access quickly to normalize the blood sugar.

↓

Blood sugar screening—repeat every 15–30 minutes until > 50 mg/dl (2.8 mmol/L) on two consecutive tests.

- If >150 mg/dl (8.3 mmol/L) on two consecutive tests—seek consultation
- If (very) low obtain a serum blood sugar but do not delay treatment

Treatment

Initial blood sugar < 50 mg/dl (2.8 mmol/L)

↓

Begin IV of D10 at 80 ml/kg/day and repeat blood sugar within 30 minutes of first test.

↓

Check blood sugar every 30 minutes until > 50 mg/dl (2.8 mmol/L) on two consecutive tests

Initial IV fluid and rate—D10 without electrolytes at 80 ml/kg/day pump

$$\frac{\text{Weight in kg} \times 80}{24 \text{ (hours)}} = \frac{\text{ml per hour to run}}{\text{the IV via an infusion}}$$

↓

Repeat blood sugar < 50 mg/dl (2.8 mmol/L) after 1 hour of IV therapy increase the IV rate to 100 ml/kg/day

↓

Once > 50 mg/dl (2.8 mmol/L) screen every 1 to 2 hours until transported or as needed based on patient's condition.

HYPOTHERMIA MANAGEMENT

- Extremely vulnerable infants include low birth weight, those requiring prolonged resuscitation.
- Preventing cold stress is a challenge. Heat loss occurs by conduction, convection and radiation (Fig. 36.4).

Prevention: Use pre-warm scales and X-ray plates, cautiously use radiant heat cover scale with warm blanket.

Move away from drafts, raise sides on warmer, close port holes and raise room temperature.

Dry thoroughly, replace wet linens, cautiously use radiant heat, do not bathe infant if showing signs of compromise.

Guidelines for Re-warming

Severely hypothermic infants: Temperature < 35°C (95°F)

↓

Incubator or radiant warmer: Core temperature goal—37°C (98.6°F)

↓

While re-warming-monitor vital signs constantly as infant is at risk for apnea, hypotension, RDS, metabolic acidosis, shock, death.

MAS is respiratory distress occurring soon after birth in an infant born from a meconium stained milieu with compatible radiological findings which cannot be otherwise explained (Fig. 36.5).

Incidence

Despite changing strategies, meconium staining of the amniotic fluid (MSAF) happens in approximately 10–15% of childbirths with

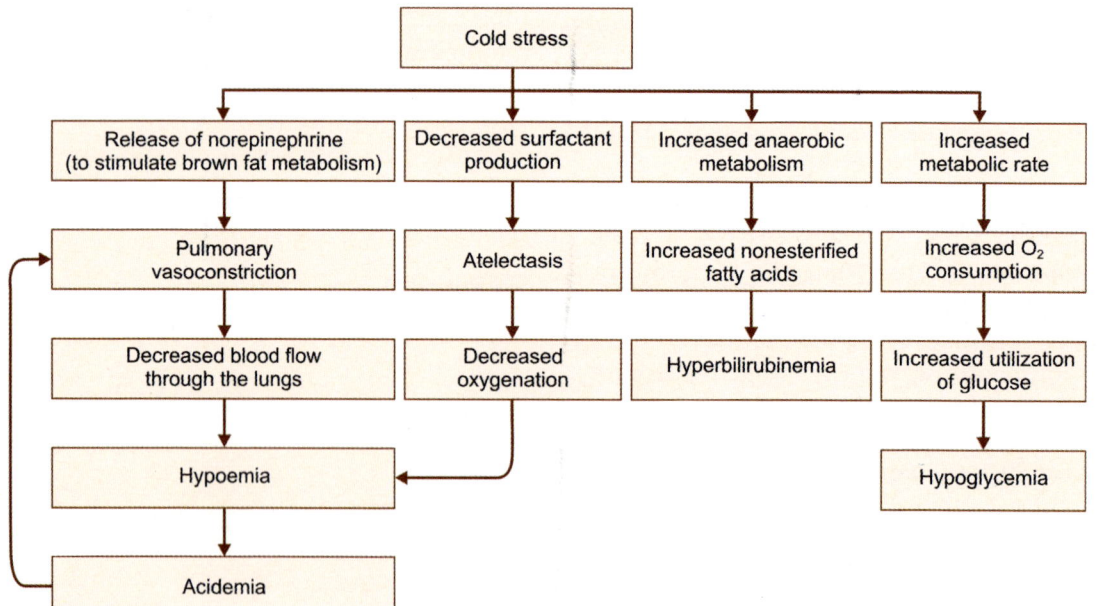

Fig. 36.4: Hypothermia management

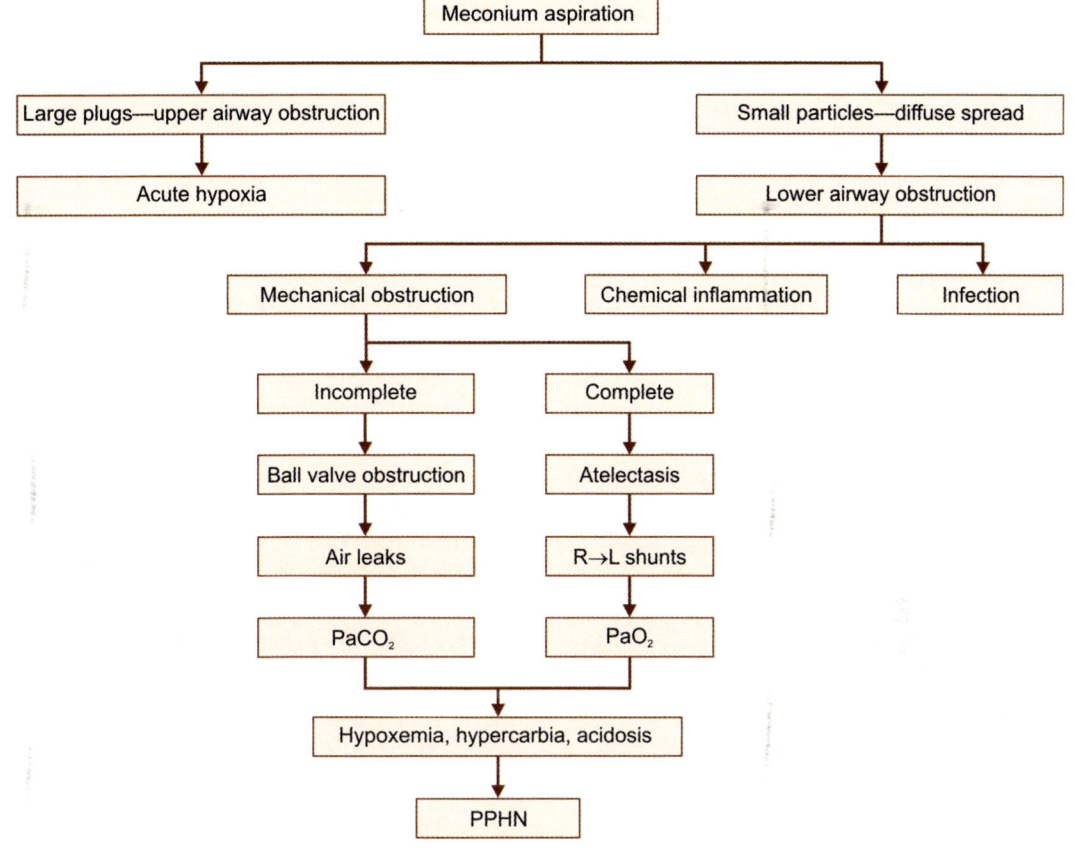

Fig. 36.5: Meconium aspiration syndrome

incidences ranging from 5 to 25%.[2] MAS develops in approximately 4–10% of the infants born from a MSAF milieu.

Neonatal Management MAS

Although meconium staining and the MAS are common neonatal problems, the appropriate management in the delivery room and subsequently remains controversial.

- *Airway clearing:* It is likely that MAS will develop in a small minority of apparently health meconium stained infants, but there is no way of identifying these neonates at risk during childbirth. The 2005 Joint Committee of the American Academy of Pediatrics and American Heart Association delineated neonatal resuscitation guidelines recommend that tracheal toileting be performed on meconium stained newborns soon after delivery if the neonate is depressed.

- The features of fetal depression delineated are absent/depressed respiration, heart rate < 100/minute and hypotonia.

- It is further recommended that the initial suctioning should not exceed five seconds. If no meconium is retrieved, repetitive suctioning is not required. However, if meconium is retrieved and no bradycardia is present, it is recommended to reintubate and perform suction under oxygen cover.

In case of bradycardia, positive pressure ventilation is to be administered and airway toileting considered later.

Because moderate amounts of meconium may remain in the stomach and be aspirated later, it is advisable to perform a gastric lavage after the baby has stabilized. Saline lavage and chest physiotherapy performed with due caution in the stabilized baby may assist the removal of tenacious secretions.

- *Ventilatory support:* One-third of the infants with MAS require ventilatory support. Because air leaks are a major problem in this condition, high concentrations of oxygen are necessary initially. Continuous positive airway pressure (CPAP)/Bubble CPAP could be beneficial if air trapping is not a major problem. If CPAP does not suffice, mechanical ventilation using low inspiratory pressures, short inspiratory and long expiratory times and rapid rates have been advocated to maintain blood gases within normal limits.

- *High frequency ventilation* (HFV) by providing effective gas exchange at low tidal volumes has been found advantageous in treating MAS. Its benefits include less barotraumas, increased mobilization of airway secretions, quicker attainment of respiratory alkalosis and fewer histopatho-logical changes.

- *Surfactant therapy:* Surfactant deficiency in MAS is a consequence of altered function rather than a deficiency state. Meconium displaces surfactant from the alveolar surface and inhibits its surface tension lowering function. In high concentrations, it has a direct cytotoxic effect on the type 2 pneumocytes used in this condition. It has also been observed that surfactant therapy in MAS restored the distended terminal airspaces of the lungs and kept the spaces from irregular overdistension. Surfactant replacement by bolus or slow infusion in infants with severe meconium aspiration syndrome improved oxygenation and reduced the severity of respiratory failure, air leaks and need for extracorporeal membrane oxygenation.

Doses from 100–200 mg/kg of phospholipid have been used in various studies with repeat dosages being provided 6–8 hourly till oxygenation improves. Although there was no increase in acute morbidity in these infants, transient oxygen desaturation and endotracheal tube obstruction occurred during bolus administration in nearly one-third of the surfactant treated infants. A review of randomised control trials (RCTs) evaluating its effects in infants with MAS suggested that surfactant administration may reduce the severity of respiratory illness and decrease the number of infants with progressive respiratory failure requiring support with extra corporeal membrane oxygenation (ECMO). The relative efficacy of surfactant therapy including KL-4 surfactant compared to, or in conjunction with, other approaches to treatment including nitric oxide, liquid ventilation, surfactant lavage and high frequency ventilation remains to be tested. There are reports which suggest that surfactant when combined with an adjuvant-PEG/dextran is more efficacious.

- *Bronchoalveolar lavage (BAL):* The efficacy of lung lavage by bronchoscopy in removing large quantities of meconium and improving lung functions is increasingly being documented. Surfactant lavage for meconium aspiration was evaluated in a small, randomized trial.

Trends toward lower duration of ventilation and severity of illness were reported. The recent reports suggest that surfactant is more effective than saline as a lavage fluid. The use of surfactant/dextran mixture has been reported to aid the meconium clearance ability of surfactant. There are reports of perfluorocarbon lavage followed by partial liquid ventilation being a more efficacious method as compared to surfactant lavage alone.

- *Inhaled nitric oxide* (INO) is currently considered the most effective therapy in the management of PPHN which often accompanies MAS. The recommended dosage of INO is 20 parts per million (ppm). Effective use of INO requires adequate lung expansion to optimize its delivery within the lungs. Hence effective ventilation is required to achieve the full benefits of INO when there is significant parenchymal disease of the lungs as occurs in MAS.

- *Steroid therapy:* Meconium in the airway evokes an inflammatory response characterized by the presence of elevated cell counts and pro-inflammatory cytokines, viz. interleukin (IL-1B), IL-6, tumor necrosis factor-α (TNF-α). Reduction in the levels of these cytokines has been found to correlate with improved lung function. Steroids provided by both the intravenous as well as inhaled routes have been observed to suppress this inflammatory response and thus improve pulmonary functions in babies with MAS. Given its easy availability and inexpensive nature, this form of therapy holds promise in its application in the neonatal intensive care unit (NICUs) of the developing nations.

- *Extra corporeal membrane oxygenation (ECMO):* In the 1990s, substantial work has been done assessing the usefulness of ECMO in neonates with MAS wherein it has been proved effective in reducing both death and severe disability in neonates. Further studies have indicated that MAS patients had a significantly lower number of complications vs no MAS patients on ECMO. These data support the consideration of relaxed ECMO entry criteria for MAS.

- *Antibiotics:* Meconium is almost always sterile. Yet several workers routinely administer antibiotics to the babies with MAS, the rationale being:
  a. Meconium produces a chemical pneumonitis with segmental atelectasis mimicking bacterial pneumonitis.
  b. There is the possibility that infection may be the stimulation for *in utero* meconium passage.
  c. *In vitro* enhancement of bacterial growth by meconium suggests the increased risk

of superimposed bacterial infection in MAS. However, the consensus opinion does not favor the routine use of antibiotics in babies with MAS.

- *Supportive care:* It is necessary to maintain an optimal thermal environment and minimal handling because these infants are agitated easily and become hypoxemic and acidotic quickly. Careful attention should be paid to systemic blood pressure and blood volume. Volume expansion, transfusion therapy and systemic vasopressors are critical in maintaining systemic blood pressure greater than pulmonary blood pressure, thereby decreasing the right to left shunt through the patent ductus arteriosus.

Newer therapies: INO as a pulmonary vascular relaxing agent has been used to treat PPHN, a common accompaniment of MAS. Studies suggest that though mortality statistics did not alter significantly, sustained improvement in oxygenation with nitric oxide and better oxygenation at initiation with ECMO may have important clinical benefits. It has been speculated that adopting specific lung expansion strategies with nitric oxide may lead to reduced use of the more invasive ECMO. Novel pharmacologic interventions like pentoxiphylline by anti-inflamatory property of preventing meconium induced polymorph degranulation, CC10 and tezosentan are awaiting trials with sufficient power before they come to be used regularly in this scenario.

Management of the afflicted neonate is a daunting task requiring critical care support. Several modalities of monitoring and treatment are available, but these are yet to be substantiated with quality scientific investigation. Our understanding of this rather complex though common entity is as yet incomplete, making it a fertile ground for research.

REFERENCES

1. Williams Obstetrics, 24th edition.
2. MJAFI Medical Journal Armed Forces India 2010; 66: 152–157.

Neonatal Jaundice and Hyperbilirubinemia

Baraturam Bhaisara

Jaundice is the visible manifestation of bilirubinemia, caused by the accumulation of bilirubin in the skin and mucous membranes. It is a common problem in neonates with an incidence of 70–80%.[1] Most of these neonates develops pathological hyperbilirubinemia during the first week of life. Premature babies have much higher incidence of neonatal jaundice requiring therapeutic intervention than term babies. Although most jaundice in neonates is physiologically normal, it is a cause of concern for the physician and a source of anxiety for the parents. It is important to detect pathological causes of jaundice and those babies at risk of significant hyper-bilirubinemia with the aim of preventing bilirubin encephalopathy.

Types of Jaundice

Physiological

This is attributed to physiological immaturity of neonates to handle increased bilirubin production. Other attributing factors could be higher concentration of red blood cells, shorter life span of newborn red blood cells, slower metabolism, circulation and excretion of bilirubin. Usually appears 2 to 4 days after birth, resolving after 1 to 2 weeks. No treatment is required but baby should be observed closely for signs of worsening jaundice.

Breastfeeding Failure Jaundice

This common type of jaundice is not related to characteristics of breast milk but rather to the pattern of breastfeeding. Decreased frequency of breastfeeding is associated with exaggeration of physiological jaundice. Encouraging a mother to breastfeed her baby at least 10–12 times a day would be helpful.

Breast Milk Jaundice

It develops 5 to 7 days after birth and peaks at day 14. A suggested cause is an increased concentration of β-glucuronidase and other unidentified factors in breast milk which inhibits conjugation or enhances intestinal absorption. These babies with TSB (total serum bilirubin) beyond 10 mg/dl after third week of life should be investigated for prolonged jaundice. Mothers should be advised to continue breastfeeding at frequent intervals.

TSB levels usually decline over a period of time.

Pathological (Non-physiological)

TSB concentrations have been defined as non-physiologic if concentration exceeds 5 mg/dl on first day of life in term neonates,10 mg/dl on second day, or 12–13 thereafter. Treatment is required in the form of phototherapy or exchange blood transfusion. One should investigate to find the cause of pathological jaundice.

Presence of any of the following signs denotes that the jaundice is pathological:

- Clinical jaundice detected before 24 hours of age
- Serum bilirubin more than 15 mg/dl
- Clinical jaundice persisting beyond 14 days of life
- Clay/white color stool and/or dark urine staining the clothes yellow
- Direct bilirubin > 2 mg/dl at any time

Risk Factors for Jaundice

A simple pneumonic for risk factors is JAUNDICE

J   Jaundice within the first 24 hr of life
A   A sibling who was jaundiced as neonate
U   Unrecognized hemolysis
N   Non-optimal sucking/nursing
D   Deficiency of G6PD
I   Infection
C   Cephalhematoma/bruising
E   East-Asian/North-Indian

Assessment

Jaundice may be a sign of serious illness.

Review the History

Family history of significant hemolysis

Clinical examination for following physical findings:

- Weight (small for gestational age)
- Prematurity
- Hydration including elimination (number of wet nappies and stools)
- Dark urine and light stools
- Microcephaly
- Pallor/petechae
- Cephalohematoma
- Hepatosplenomegaly
- Evidence of hypothyroidism

Look for risk factors or precipitating factors:

- Preterm birth
- Asphyxia
- Acidosis

Visual Inspection of Skin

Kramer's dermal staining may be used as a clinical guide to the level of jaundice (Table 37.1). Newborn should be examined in good light. Dermal staining in newborn babies progresses in a cephalo-caudal direction. The skin should be blanched with digital pressure and the underlying color of skin and subcutaneous tissue should be noted. Newborns detected to have yellow discoloration of the skin beyond the legs should have an urgent laboratory confirmation for level of TSB.

Table 37.1	Guide to dermal staining with level of bilirubin (modified from Kramer's original article)[2]	
Area of body		*Level of Bilirubin*
Face		4–6 mg/dl
Chest, upper abdomen		8–10 mg/dl
Lower abdomen, thighs		12–14 mg/dl
Arms, lower legs		15–18 mg/dl
Palms, soles		15–20 mg/dl

Bilirubin Encephalopathy

Bilirubin enters the brain as free (unbound) bilirubin or as bound to albumin in the presence of disrupted blood–brain barrier.

Acute Bilirubin Encephalopathy

Refers to the clinical manifestations of bilirubin toxicity seen in the first few weeks after birth. Clinically it can be divided into three phases:

a. *Early phase:* Lethargy, hypotonia, poor feeding, high-pitched cry.
b. *Intermediate phase:* Irritability, hypertonia of extensor muscles, seizures, fever. All infants who survive this phase develop chronic bilirubin encephalopathy (clinical diagnosis of kernicterus).

c. *Advanced phase:* Pronounced opisthotonos, high-frequency hearing loss, shrill cry.

Investigations

Bilirubin Measurement

a. *Total serum bilirubin (TSB):* This should be measured in all infants with jaundice in order to decide treatment options.
b. *Transcutaneous bilirubin (TcB):* This method is non-invasive and based on reflectance data of multiple wavelengths from bilirubin stained skin. TcB has a linear correction to TSB and may be useful as a screening device to detect significant jaundice and decrease the need for frequent TSB levels.
c. *Other relevant investigations*
   Routine:
   • CBC with peripheral smear examination
   • Blood group (maternal and baby)
   • DCT (direct Coombs' test)

   Consider in special cases:
   • Microbiological cultures
   • TSH
   • G6PD deficiency
   • Hb electrophoresis
   • Metabolic work up

Differential Diagnosis

Jaundice visible at <24 hours is a medical emergency
Measure serum bilirubin within 2 hours of identifying obvious or suspected jaundice
• Commence phototherapy while awaiting serum bilirubin results.
• Urgent neonatology/pediatric reference
• Exclude pathological causes of jaundice. Organise transfer to nearest referral service.

Jaundice first visible after 24 hours to 10 days
Most common cause is benign physiological jaundice. Other causes include:
• Dehydration
• Sepsis
• Hemolysis

• Polycythemia
• Breakdown of extravagated blood (e.g. bruising)
• Increased entero-hepatic circulation
• Metabolic diseases

Onset of jaundice after 10 days of age and prolonged jaundice
• Hypothyroidism
• Hemolytic anemia
• Hereditary spherocytosis
• Pyloric stenosis or gastrointestinal obstruction

Differential Diagnosis for Conjugated Hyperbilirubinemia

• Congenital obstruction and malformations of the biliary system
• Idiopathic neonatal hepatitis
• Infections
• Metabolic disorders
• Prolonged parenteral nutrition
• Requires urgent referral to a neonatologist/pediatrician.

Management

Management of jaundice is directed towards reducing the level of bilirubin and preventing CNS toxicity.

A. Prevention

1. *Maternal and labor history* (Table 37.2)
2. *Pregnancy, labor and birth:* Test all pregnant women for ABO, Rh (D) blood types during pregnancy:
   a. *If maternal red blood cell antibodies are noted antenatally, test cord blood for:*
      • Blood group including Rh type
      • Direct antiglobulin test (Coombs' test)
      • CBC for hemoglobin and hematocrit
      • Discuss with neonatologist
   b. *If the mother has not had antenatal blood tests send:*
      • Maternal blood for blood group (ABO/Rh), and

Table 37.2	Significance of maternal and labor information[5]
Information	*Significance*
Unexplained illness during pregnancy	Consider congenital infection (TORCH)
Diabetes mellitus	Increased incidence of jaundice
Drug ingestion during pregnancy	Sulfonamides, nitrofurantoins, antimalarials may increase hemolysis in G6PD-deficient infants
Vacuum extraction	Increased incidence of jaundice and cephalohematoma
Oxytocin-induced labor	Increased incidence of jaundice
Delayed cord clamping	Increased incidence of jaundice among polycythemic infants
Apgar score	Increased incidence of jaundice among asphyxiated infants

- Baby's cord blood for blood group, Rh type and Coombs' test
3. *Breastfeeding:*
  - Support all women for breastfeed.
  - Encourage demand feeding or at least 3–4 hourly or as age appropriate.
  - Consider referral to lactation consultant/lactation support group to provide the mother with feeding support.
  - Promote the ingestion of colostrum to increase stooling to prevent reabsorption of bilirubin.
  - EBM (expressed breast milk) is the feed of choice even in sick infant too.

B. Treatment

Hyperbilirubinemia can be treated (Table 37.3) with phototherapy, exchange transfusion and pharmacological agents. Adequate hydration is also an important consideration. It is important to also treat the underlying illnesses that may be causing jaundice.

Phototherapy

Special blue lamps with a peak output at 425 to 475 nm are the most efficient for photo-

Table 37.3	Treatment modalities to reduce serum bilirubin concentration[5]

- Hydration
- Phototherapy
- Exchange transfusion
- Others (immunoglobulin, phenobarbitone, metal protoporphyrins)

therapy. Effective phototherapy depends on:

- Light spectrum. Blue green spectrum is most effective.
- Irradiance (energy output) special blue tubes 10–15 cm above the infant will produce an irradiance of at least 35 Uw/cm^2 per nm.
- Distance from the infant.
- Extent of skin area exposure.
- TSB level at start of PT. Higher the TSB, the more rapid decline in TSB with PT.
- *Cause of jaundice:* Jaundice due to hemolysis or obstructive cause.

Side Effects of Phototherapy

- Insensible water loss
- Watery diarrhea
- Low calcium
- Retinal damage
- Mutations, DNA breaks have been described in cell culture.

Phototherapy for neonates born at <35 weeks of gestation: It is generally recommended to treat hyperbilirubinemia at lower levels in low birth weight (LBW) infants in comparison to term infants (Fg. 37.2).

Exchange Transfusion

Based on the chart (Fig. 37.1) when phototherapy fails to bring down bilirubin level below cut off range, excess transfusion is recommended.

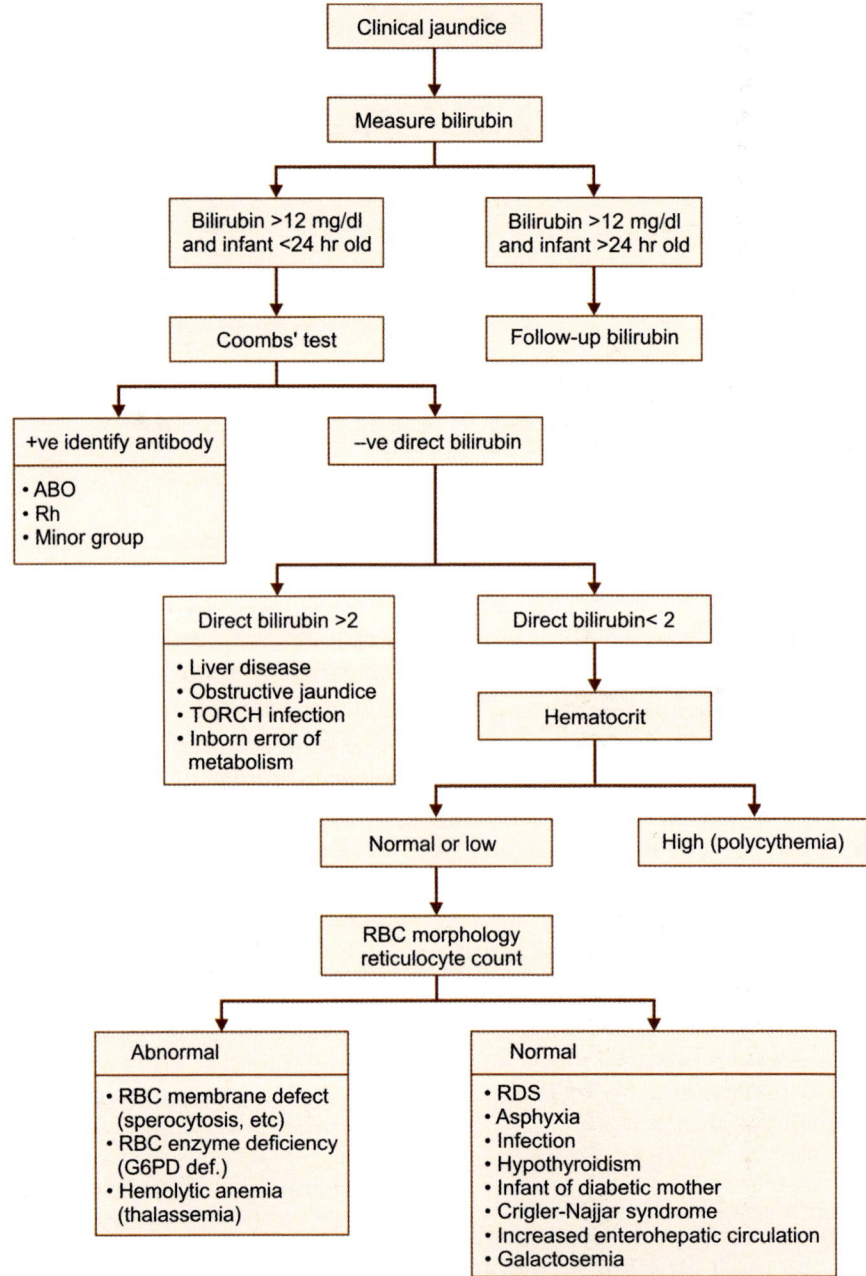

Fig. 37.1: Approach to neonatal jaundice[6]

Discharge Planning

- Reassure that neonatal jaundice is common and usually transient
- Parents and carers should also seek advice from a healthcare professional if their baby:
  1. becomes jaundiced
  2. has worsening jaundice
  3. has jaundice persisting beyond 14 days
  4. is passing pale chalky stools or dark urine
  5. is not feeding well
  6. shows signs of dehydration

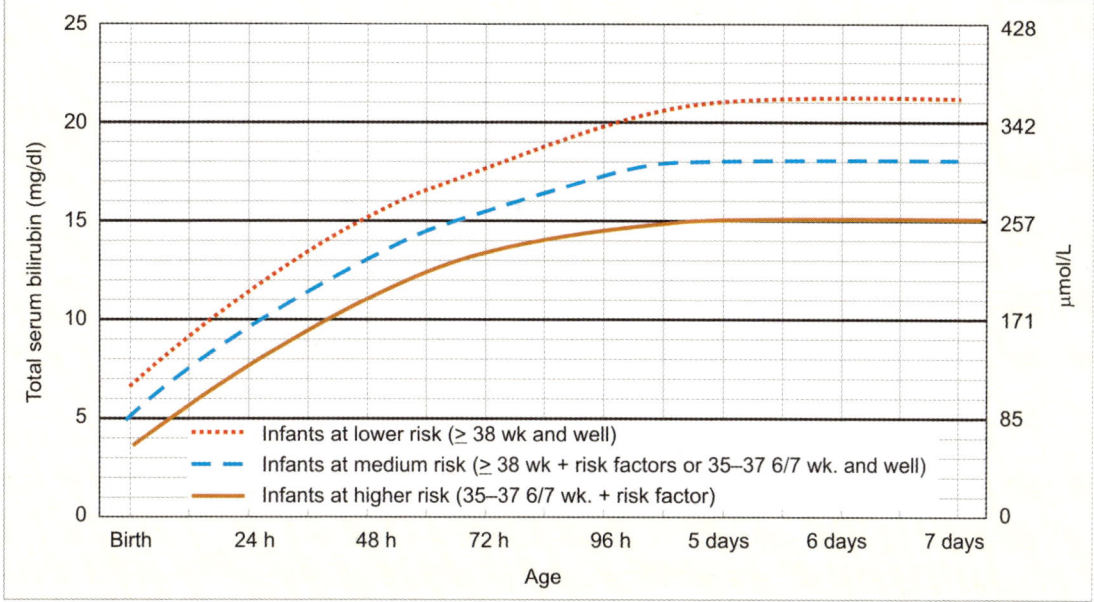

- Use total bilirubin. Do not subtract direct (conjugated) bilirubin.
- Risk factors: Isoimmune hemolytic disease, G6PD deficiency, asphyxia, significant lethargy, temperature instability, sepsis, acidosis, or albumin <3 g/dl
- For well infants 35–37 6/7 wk., can adjust TSB levels for intervention around the medium risk line. It is an option to intervene at lower TSB levels for infants closer to 35 weeks and at higher TSB levels for those closer to 37 weeks.
- It is an option to provide conventional phototherapy in hospital at TSB levels 2–3 mg/dl below those shown.

Fig. 37.2: Guideline for phototherapy in hospitalized infants > 35 weeks[6]

Table 37.4	Guidelines for management of hyperbilirubinemia in the healthy term newborn (TSB mg/dl)[5]			
	Serum total bilirubin level (mg/dl)			
Age (h)	Consider phototherapy	Phototherapy	Exchange transfusion if intensive phototherapy*fail	Exchange transfusion and phototherapy
25–48	≥12	≥15	≥20	≥25
49–72	≥15	≥18	≥25	≥30
>72	≥17	≥20	≥25	≥30

*Intensive PT should produce a decline in STB of 1–2 mg/dl within 4–6 hr and should continue to decline below the threshold for exchange transfusion

Recommended follow-up: Within the first 24 hours of life—do not discharge a baby with visible jaundice.

Follow-up can be planned as per table form is given in Table 37.5.

Follow-up assessment

Let neonatologist/pediatrician evaluate the baby. Follow-up assessment must include:

- Baby's weight and percentage change from birth weight.
- Review of feeding history to determine adequacy of intake.
- Voiding and stooling pattern.
- Presence or absence of jaundice.
- Clinical judgment to determine the need for total serum bilirubin level measurement.

Table 37.5	Suggested follow-up policy[7]	
Scenario	Age at discharge	Follow up
None of risk factors *present	24–72 h	48 h after discharge
Any risk factor* present	>72 h	F/U Optional
	24–48 h	24 h after discharge**
	49–72 h	48 h after discharge**
	73–120 h	48 h after discharge**

*Risk factors: History of jaundice needing treatment in previous sibling, setting of blood group incompatibility, visible jaundice at discharge, gestation <38 completed weeks, high prevalence of G6PD deficiency, primipara mother, weight loss at discharge >3% per day or >7% cumulative weight loss.
**may need a repeat visit depending on physician's assessment.

- Overall look of the infant (sick look, or alert and awake, or drowsy) to rule out sepsis and overt jaundice.

KEY POINTS

- Promote and support successful breast-feeding.
- All neonates should be monitored clinically for appearance of jaundice during first postnatal week.
- In cases of discharge before 72–96 hr from the hospital, a thorough assessment of risk factors for severe jaundice should be done in all babies.
- In neonates with significant jaundice, investigations should include blood groups of mother and baby, a Coombs' test, evidence for hemolysis and G6PD assay in areas known to have high prevalence of G6PD deficiency.
- We need to assess jaundice risks the way we assess other risks, refer to pediatrician/ neonatologist.

REFERENCES

1. IPS, Chung M, Kulig J, et al. An evidence-based review of important issues concerning neonatal hyperbilirubinemia. Pediatrics 2004; 114:e130–53.
2. Kramer LI. Advancement of dermal icterus in the jaundiced newborn. Am J Dis Child 1969; 118:454–8.
3. American Academy of Pediatrics, Provisional Committee for Quality Improvement and Subcommittee on Hyperbilirubinemia. Practice parameter: Management of hyperbilirubinemia in the healthy term newborn. Pediatrics. 1994; 94:558–562.
4. Bhutani V, Gourley GR, Adler S, Kreamer B, Dalman C, Johnson LH. Noninvasive measurement of total serum bilirubin in a multiracial predischarge newborn population to assess the risk of severe hyperbilirubinemia. Pediatrics. 2000; 106(2). Available at: www.pediatrics.org/cgi/content/full/106/2/e17.
5. Avery's Diseases of the Newborn, Part XV-Hematologic system & Disorder of Bilirubin metabolism 1246 - Elsevier.
6. Gregory MLP, Martin CR, Cloherty JP. Neonatal hyperbilirubinemia. In: Cloherty JP, Eichenwald EC, Hansen AR, Stark AR, editors. Manual of neonatal care. 7th ed. Philadelphia: Lippincott Williams & Wilkins 2012.
7. NNF Clinical Practice Guidelines 2010 Management of Neonatal Hyperbilirubinemia G. Guruprasad, Deepak Chawla, Sunil Agarwal, Anil Narang, Ashok K Deorari, pp 141–150.
8. Management of hyperbilirubinemia in the newborn infant 35 or more weeks of gestation. Pediatrics 2004; 114:297–316.
9. Bhutani VK, Johnson L, Sivieri EM. Predictive ability of a pre-discharge hour-specific serum bilirubin for subsequent significant hyperbilirubinemia in a healthy term and near-term newborn. Pediatrics 1999; 103:6–14.
10. American Academy of Pediatrics Subcommittee on Hyperbilirubinemia. Clinical Practice Guideline: Management of hyperbilirubinemia in the newborn infant, 35 or more weeks of gestation. Pediatrics. 2004; 114:297–316. DOI: 10.1542/peds.114.1.297.
11. Robert M, Kliegman, Bonita MD, Stanton MD, Joseph St. Geme Nelson. Textbook of Pediatrics, 19th edition, Elsevier Saunder, Neonatal Hyperbilirubinemia, pp 603–608.

Child with Neurodisability and Subsequent Pregnancy

Neelu Desai, Monika Chhajed

INTRODUCTION

Neurological disorders are common in children and often have significant morbidity, mortality, disability and poor quality of life. This has implications not only on the child but also on the family and the entire community.

In a community survey in rural north India amongst infants and children, the prevalence of neurological disorders was as high as 30/1000 with developmental/motor disabilities predominating. Cerebral palsy, autism, neuro-muscular diseases, epilepsy and various genetic/metabolic disorders comprise a few of the varied spectrum of neurological disorders encountered in children. Some of these disorders are treatable; others preventable to a certain extent but many which have do not have any definite cure and impose a tremendous burden on the family and society. Mother is often the epicentre of the entire struggle and to plan another child becomes a nightmare for the family. The fear of raising a second child with similar problems and the tremendous resources spent on the previous child deters them to procure more children.

In the current chapter, we would review a few common neurological disorders in children where appropriate counselling and/or management could lead to a better outcome in the subsequent pregnancies.

CEREBRAL PALSY (CP)

Cerebral palsy is a non-progressive disorder of posture and movement control due to lesions or anomalies of the developing brain. With an approximate worldwide incidence of 1.5–2.5 per 1000 which has remained constant over the years, cerebral palsy unfortunately remains one of the commonest neurological disabilities in children. Though antenatal factors are implicated as the commonest aetiology in the developed world, perinatal and postnatal factors still predominate in developing countries. Some of these causes are preventable if appropriate, timely measures are taken. With recent advances in neonatal and perinatal medicine, there may have been a change in subtype of CP with patients with choreoathetoid CP declining due to better care of Rh isoimmunisation while spastic diplegic CP may have risen due to enhanced survival of the preterm neonate.

Some common causes of CP are mentioned in Table 38.1.

A few measures which could decrease the incidence of neurodisability in these cases are as follows:

Mother

- Delivering high-risk mothers with medical illnesses/threatened preterm labor/other high-risk pregnancies in tertiary centers with advanced neonatal care facilities.

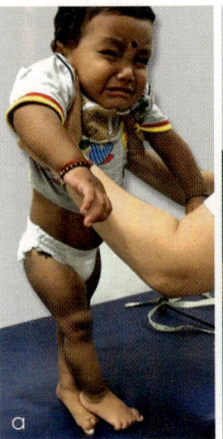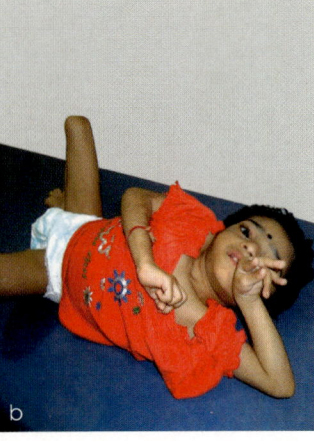

Fig. 38.1: a. Spastic diplegic CP due to prematurity with lower limb spasticity more than upper limb, b. Dystonic CP due to kernicterus

Table 38.1	Etiology of CP	
Antenatal	*Perinatal*	*Postnatal*
IUGR	HIE/NE	Infections
Prematurity	Intracranial hemorrhage	Stroke
Intrauterine infections	Kernicterus	Trauma/Non-accidental injury
CNS Malformations	Pregnancy complications	
Placental/maternal problems		

IUGR: Intrauterine growth retardation, HIE: Hypoxic ischemic encephalopathy, NE: Neonatal encephalopathy

- The neuroprotective role for antenatal magnesium sulfate therapy given to women at risk of preterm birth before 30 weeks gestational age is well established. This therapy could be tried in expectant mothers with threatened preterm delivery.
- Appropriate care of mothers with risk of Rh incompatibility in children by administering Rh immunoglobulin. Early monitoring and intervention of these children to prevent damaging effects of hyperbilirubinemia.
- Intrapartum close monitoring of high-risk pregnancies and timely intervention.

Neonate

- Therapeutic hypothermia for neonates with HIE: It is well established that moderate hypothermia improves survival without cerebral palsy or other disability by 40% and reduces death or neurological disability by nearly 30% at 18 months. It blunts secondary adverse effects of reperfusion injury and modifies secondary neuronal damage by reducing cerebral metabolic rate. Total body cooling and selective head cooling are both effective methods. 72 hours of moderate hypothermia started within 6 hours of birth is recommended.
- Avoiding postnatal steroids
- Various experimental therapies that have been tried are phenobarbital, topiramate, erythropoietin, glutamate antagonists, melatonin, N-acetylcysteine, vitamin C, vitamin E, 50% xenon inhalation and autologous cord blood transplantation. Definitive evidence is lacking for their effectiveness due to lack of multicenter, well controlled trials.

Though most causes of cerebral palsy are not recurrent, certain genetic polymorphisms have recently evoked interest. Many cases of CP are associated with genetic alterations (mutations) that may contribute to susceptibility to CP. Polymorphisms for cytokine genes may exacerbate or attenuate inflammation-associated neural damage. Thus a similar insult can have such different consequences in different fetuses and low Apgar scores and umbilical artery pH values are often poor predictors of eventual outcome.

Various metabolic or genetic disorders which are slowly progressive can mimic CP. Prognosis varies in each and some of them are amenable to treatment. Genetic counseling of families is indispensable in these metabolic disorders as often they are autosomal recessive in inheritance with 25% recurrence risk in each pregnancy. With the advent of next generation sequencing techniques, many of these disorders can be proven genetically and antenatal diagnosis is possible in the subsequent pregnancies. Certain red flags in history and examination in CP can lead to further evaluation for a genetic/metabolic disorder.

PERINATAL STROKE

One cause of hemiplegic CP which has intrigued health professionals for many decades is perinatal ischemic stroke. This can be defined as a group of heterogeneous conditions with impairment of cerebral blood flow due to venous or arterial thromboses or embolization, between 20 weeks of fetal life through 28th postnatal day, which can be confirmed by neuroimaging or neuropathological studies. The incidence of perinatal stroke has been estimated around 1 in 1600 to 5000 births. The commonly seen types of ischemic stroke in the perinatal period are arterial ischemic stroke and cerebral sinovenous thrombosis. Perinatal stroke can lead to cerebral palsy, epilepsy, cognitive impairment and sensory deficits.

The etiological factors include a large variety of maternal and neonatal risk factors as well as prothrombotic coagulation factors.

An extensive literature search and study of a retrospective cohort of 134 newborn infants with stroke suggests that embolism is the most common identifiable cause for stroke in general (25%) followed by preceding trauma (10%) and infection (8%). Other causes, such as asphyxia, acute blood loss, extracorporeal

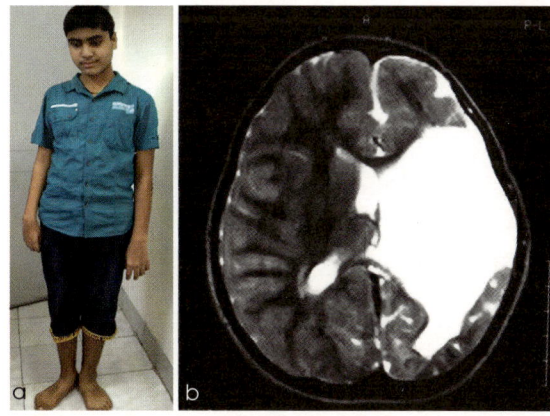

Fig. 38.2: a. Child with right hemiplegic CP, b. MRI brain shows a left middle cerebral artery infarct

membrane oxygenation, genetic disorders or prothrombotic conditions, are seen in less than 5% of cases.

MTHFR gene mutation, factor V Leiden and prothrombin gene mutations causing fetal thrombophilia have been found as important risk factors causing cerebral palsy.

Measures to Prevent Neonatal Stroke

i. As embolism seems to be an important cause for stroke, prevention is possible in some instances, e.g. by avoiding routine use of deep venous lines in preterm infants, etc.
ii. Avoidance of vaginal breech delivery at term.
iii. Avoidance of birth trauma following instrumental traction.

Table 38.2		
Maternal	*Perinatal*	*Neonatal*
Pre-eclampsia	Emergency cesarean section	Infection
IUGR	Assisted delivery—forceps, etc.	Polycythemia
Maternal diabetes	Perinatal asphyxia	Prothrombotic factors
Hypertension	Trauma	Cardiac disorders
Maternal drug abuse		Extracorporeal membrane oxygenation (ECMO)
Maternal infection		Pulmonary hypertension
Antenatal trauma		Indwelling catheters
Prothrombotic disorders		

iv. Appropriate management of maternal diabetes, infection, hypertension, etc.

v. A prothrombotic screen can be performed in infants and mothers with history of early thrombosis or with bad obstetric history. The main goal for prothrombotic screening is to prevent recurrence in the child as well as subsequent pregnancy.

Regarding primary prevention, the potential association between maternal thrombophilia, placental vasculopathy and perinatal stroke has led some authors to suggest that the use of maternal thrombo-prophylaxis may reduce the risk of perinatal stroke. This, however, remains speculative at the present time and awaits further insights into the relationship between these factors, as well the results of large prospective studies designed to address pregnancy outcomes in women receiving thromboprophylaxis during pregnancy.

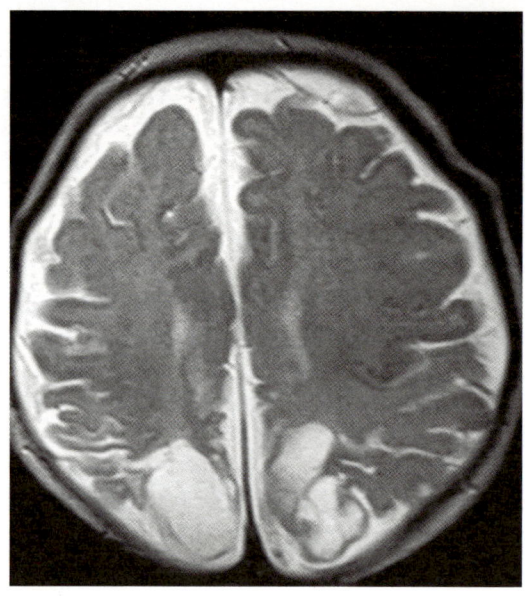

Fig. 38.3: MRI brain showing occipital encephalomalacia due to severe neonatal hypoglycemia. Child presented with developmental delay, microcephaly, visual impairment and infantile spasms at 7 months age

NEONATAL HYPOGLYCEMIC BRAIN INJURY

Hypoglycemia in the first few days of life can have deleterious consequences on the growing brain. Often doctors and paramedical staff insist on exclusive breastfeeding even in LSCS deliveries or low birth weight babies. Beyond doubt, this is ideal for optimum nutrition of baby. However in some babies, where feeding has not been established due to either Cesarean delivery, maternal illness or nipple problems, etc, delayed institution of feeding can lead to hypoglycemia. Hypoglycemia in this period causes irreversible damage to the vulnerable brain. This in later life causes a myriad of problems including epilepsy often refractory, cognitive impairment, autism, microcephaly, apraxia of hand use and visual impairment.

An obstetrician's or pediatrician's role is crucial in these cases where top feeding could be instituted if feeding is not established and risk of hypoglycemia is high (low birth weight, infant of diabetic mother, etc.).

AUTISM

Autism is a developmental disorder characterized by impaired social interaction, verbal and non-verbal communication, and by restricted and repetitive behaviors. These symptoms become evident before a child turns three years old.

Incidence of autism has been rising over the years. A recent report by Centers for Disease Control and Prevention says one in 68 US children has an autism spectrum disorder (ASD), a 30% increase from 1 in 88 two years ago. However, symptoms and their severity vary widely across the three core areas.

The exact cause of autism is not well understood but various genetic and environmental factors have been implicated. Autism affects information processing in the brain by altered connections and organization.

Red Flags for Autism

• No social smile or other joyful expressions by six months or thereafter

- No back-and-forth sharing of sounds, smiles or other facial expressions by nine months
- No babbling by 12 months
- No pointing, showing, reaching or waving by 12 months
- No words by 16 months
- No meaningful, two-word phrases (not including imitating or repeating) by 24 months
- Any loss of speech, babbling or social skills at any age

Early detection and treatment in autism improves outcome, often dramatically. One of the most important things one can do as a parent or caregiver is to identify the red flags of autism and consult a pediatrician/ developmental pediatrician or pediatric neurologist early.

Early intensive behavioral intervention improves learning, communication and social skills in young children with autism spectrum disorders (ASD). Many of the sensory processing problems can be addressed with occupational therapy and/or sensory integration therapy.

Both genetic and environmental factors have been implicated in autism and more than 200 autism susceptibility genes have been identified. Recurrence risk of autism is 5–15% after one affected child (7.7 % if male, 20% if female) and 25% after 2 affected children. The concordance rate among monozygotic (MZ) twins is 60 to 90% and 0 to 20% in dizygotic (DZ) twins. A few chromosomal disorders and several single gene disorders are also associated with an increased risk for autism, e.g. Fragile X, Rett syndrome, tuberous sclerosis, Prader-Willi syndrome and Angelman syndrome. Family members of an autistic child may show minor difficulties in social interactions and communication.

Parents should be counselled about the recurrence risk in subsequent pregnancy though no genetic or other tests can confirm whether the next child would be affected.

EPILEPSY

Epilepsy affects 0.5–1% of population and around 5–7% of children have epileptic seizures. It is often associated with many disabilities (learning, behaviour, physical, social, emotional, etc.), injuries and death. Diagnosis of epilepsy is essentially clinical. There is mild increased incidence of epilepsy in family members (2–4 times higher). However, there is no increased risk of pregnancy related complications (pregnancy induced hypertension, preterm delivery, antepartum hemorrhage, LSCS). Seizure freedom for 9 months prior to pregnancy is suggestive of high likelihood (84–92%) of remaining seizure free during pregnancy (Level B). All women who are planning pregnancy should take folic acid.

Discontinuing/changing antiepileptic drugs (AEDs) is not a reasonable option during pregnancy (physical injury, serious drug adverse reactions, and polytherapy exposure). AEDs produce a pattern of malformations which overlap amongst the individual AEDs. Common malformations are:

- *Sodium valproate (VPA):* Major birth defects (neural tube defects, facial clefts and hypospadias) and poor cognitive outcomes with *in utero* exposure. Changing from VPA to another AED should be done well before pregnancy. It is advisable to avoid VPA as part of polytherapy during the first trimester. Limiting the dosage of VPA or Lamotrigine (LTG) during the first trimester could be done to lessen the risk of malformations (Level B).
- Monitoring levels of lamotrigine, carbamazepine and phenytoin during pregnancy should be considered.
- Newborns exposed to enzyme-inducing AEDs *in utero* routinely receive vitamin K at delivery to prevent hemorrhagic disease of newborn which in itself can cause major neurologic disability.

- VPA, phenobarbitone, phenytoin, and carbamazepine are not transferred into breast milk to great extent.

NEURAL TUBE DEFECTS

Neural tube defects (NTD) are a heterogenous group of congenital malformations resulting from incomplete or improper fusion of the neural tube during embryonic development. They are classified by the presence or absence of a layer of skin covering the anomaly as closed or open type respectively. Cranial presentations include anencephaly, encephalocele (meningocele or meningomyelocele), craniorachischisis totalis etc while spinal presentations include spina bifida aperta, myelomeningocele, congenital dermal sinus, lipomyelomeningoceles, diastematomyelia, diplomyelia and caudal agenesis. Open neural tube defects can result in severe neurological disabilities, including paraplegia, hydrocephalus, sexual dysfunction, skeletal deformities and sometimes cognitive impairment.

The exact etiology of neural tube defects is unknown. Environmental factors thought to be involved in increasing the risk include maternal age, maternal diabetes and obesity,

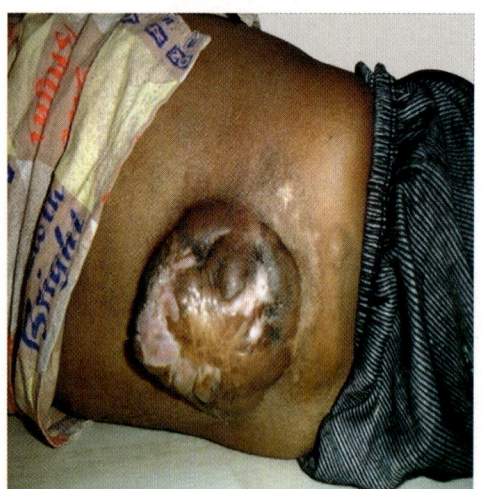

Fig. 38.4: Child with paraparesis and bowel/bladder incontinence due to meningomyelocele at lumbosacral region

low socioeconomic class, pollutants, and exposure to antiepileptic drugs like sodium valproate and carbamezapine.

There is a strong evidence for protective role of folate in this disorder. It has been observed that multiple members of the same family are affected with a broad spectrum of NTDs, suggesting the possibility of a common genetic etiology. The recurrence risk for myelomeningocele ranging from 2% to 5% in siblings was noted in a study. Methylenetetra-hydrofolate reductase (MTHFR) gene mutations have been implicated as risk factors for neural tube defects (NTDs).

Neural tube defects are usually diagnosed prenatally through laboratory or imaging tests and should be utilized where previous child is affected with the disorder. Prenatal laboratory tests include "Triple screen" blood test which includes alpha-fetoprotein (AFP), human chorionic gonadotropin (hCG) and estriol. This test generally is done during the second trimester. Testing of amniotic fluid may also show a high level of AFP, as well as high levels of acetylcholinesterase. This test may be done to confirm high levels of AFP seen in the triple screen blood test. The amniotic fluid also can be tested for chromosomal abnormalities, which might be the cause of the abnormal AFP level. Prenatal ultrasound imaging is usually able to detect almost all types of neural tube defects.

The evidence supports that folic acid-containing supplements reduces the risk for NTD-affected pregnancies. It is advised that every women in their reproductive years planning a pregnancy should undergo periconceptional supplementation with 0.4 mg (400 µg) folic acid daily. Those women who have had a previous history of NTDs are advised to undergo at least 4.0 mg folic acid supplementation daily. The supplementation should start at least one month but preferably two to three months prior to conception. In a recent study it was found that NTD risk estimates were lowest for women whose diets were rich in choline, betaine, and methionine. Recent clinical trials have shown that prenatal correction of open spina bifida (OSB) via open fetal surgery was associated with improved infant neurological outcomes relative to postnatal repair, though at the expense of increased maternal morbidity.

NEUROMUSCULAR DISORDERS

Though many neuromuscular disorders can affect children, the two common ones are Duchenne muscular dystrophy (DMD) and spinal muscular atrophy (SMA). Both these disorders are inherited and have a dismal prognosis. While the former is X-linked in inheritance, the latter is transmitted by autosomal recessive inheritance. Children affected present with delayed motor milestones, gait difficulty and progressive muscle weakness. Death can occur due to respiratory insufficiency in both while cardiac involvement can also cause death in DMD.

Parents with affected children should be counselled about the recurrence risk in subsequent pregnancy and a genetic confirmation should be done in the index case especially if more children are desired. Once the causative mutation is identified, prenatal diagnosis can be accomplished in the future pregnancies by amniocentesis or chorionic villus sampling of the fetus. Medical termination can be ensued if the fetus is affected.

HEARING LOSS

Hearing impairment is seen in around 1 in 500 infants either at birth or during early childhood. Hearing loss can be classified as congenital or acquired, prelingual or postlingual, progressive or nonprogressive, conductive or sensorineural, syndromic or nonsyndromic, and familial or sporadic. Around 400 syndromes have been described of which hearing loss is a component. Branchio-oto-renal syndrome, Usher syndrome, Pendred syndrome and Waardenburg are some of the important syndromes with hearing loss as a part of it. About 70–80% of genetic deafness is nonsyndromic.

In the case of hereditary deafness, establishment of a specific etiologic diagnosis is important as some cases of syndromic deafness have specific treatments or need of specific diagnostic tests. The Jervelle, Lange-Neilsen syndrome, Usher syndrome, Branchio-oto-renal syndrome, and Alport syndrome are examples of genetic forms of deafness in which serious complications involving other

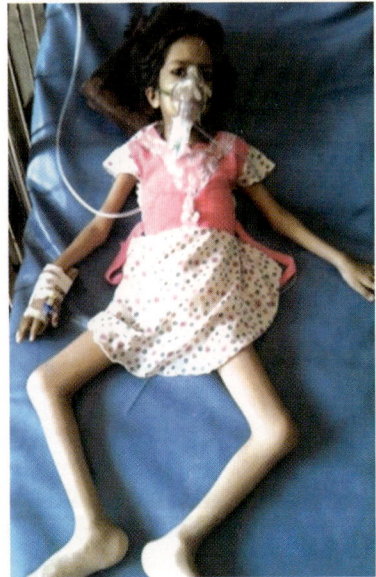

Fig. 38.5: Child with spinomuscular atrophy type 2, both parents were carriers of the gene mutation

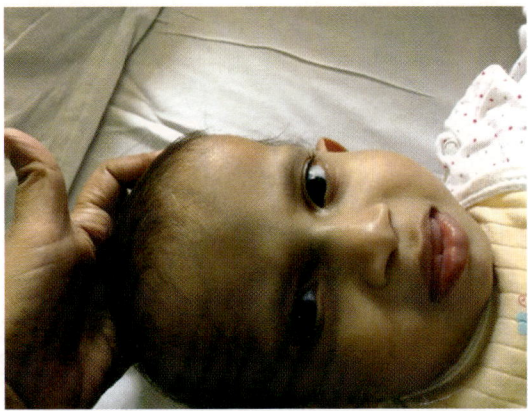

Fig. 38.6: Child with Waardenburg syndrome with white forelock of hair and profound sensorineural deafness

organ systems may arise. In nonsyndromic deafness also genetic tests can at times diagnose a specific form of NSD. These tests are especially important in confirming a genetic etiology in families where there is only one affected child and no family history of deafness. Connexin testing has rapidly become the standard of care for the management of such cases. Reliable information regarding whether deafness will recur in the immediate family or in the children of the proband can only be provided when the genetic form of deafness and its mode of inheritance are known.

SUGGESTED READING

1. Kumar R, Bhave A, Bhargava R, Agarwal GG . Survey of Neurological Disorders in Children Aged 9–15 Years in Northern India. J of Child Neurol, 2015;1–11.

2. Govaert P, Ramenghi L, Taal R, Dudink J, Lequin M. Diagnosis of perinatal stroke II: mechanisms and clinical phenotypes. Acta Paediatr. 2009 Nov;98(11):1720–6.

3. Elizabeth A. Chalmers. Perinatal stroke—risk factors and management. British Journal of Haematology 2005; 130; 333–343.

4. Shaw GM1, Carmichael SL, Yang W, Selvin S, Schaffer DM. Periconceptional dietary intake of choline and betaine and neural tube defects in offspring. Am J Epidemiol. 2004 Jul 15;160(2):102–9.

5. Pedreira DA1, Zanon N2, Nishikuni K3, Moreira de Sá RA4, Acacio GL5, Chmait RH6, Kontopoulos EV7, Quintero RA7. Endoscopic surgery for the antenatal treatment of myelomeningocele: the CECAM trial. Am J Obstet Gynecol. 2016 Jan;214(1):111.e1–111.e11.

6. Walter E. Nance.The genetics of deafness. Mental retardation and developmental disabilities research reviews 2003;9:109–119.

7. Stephen W. Hone, Richard J. H. Smith. Genetics of hearing impairment. Semin Neonatol 2001; 6:531–541.

8. Kumar R, Bhave A, Bhargava R, Agarwal GG. Prevalence and risk factors for neurological disorders in children aged 6 months to 2 years in Northern India. Dev Med Child Neurol 2013; 55;348–356.

9. Doyle LW, Crowther CA, Middleton P, Marret S, Rouse D. Magnesium sulphate for women at risk of preterm birth for neuroprotection of the fetus. Cochrane Database Syst Rev 2009 Jan 21;(1).

10. MacLennan AH, Thompson SC, Gecz J. Cerebral palsy: causes, pathways, and the role of genetic variants. Am J Obstet Gynecol. 2015 May 21.

11. Appendix B: AAN Summary of Evidence-Based Guideline for Clinicians: Management Issues for Women With Epilepsy—Focus on Pregnancy: Obstetrical Complications and Change in Seizure Frequency. Continuum (Minneap Minn). 2016 Feb;22(1 Epilepsy):283–4.

Emerging Infections in Pregnancy—Swine Flu, Zika Virus

Reena J Wani, Rashmi Jalvee

Infection control is an essential part of any labor room, NICU and hospital facility and has also been an important factor in outcomes in health care. However, emergence of newer diseases has been a cause for concern, hence we are focussing on 2 infections which are a cause for global concern and have impacted maternal and child health.

SWINE FLU

Swine flu is a zoonotic disease which was originally transmitted from pigs to humans and now spreads from one human to another. It is a highly contagious respiratory disease in pigs caused by one of several swine influenza A viruses.The outbreak of swine flu among humans first occurred in Mexico in 2009 after which there were rising number of cases throughout the world and was declared as a global pandemic by the WHO.

The virus is an orthomyxovirus, an enveloped virus with spike like glycoproteins called hemagglutinin and neuraminidase, hence the description H1N1. The virus is known for its ability to mutate. The pandemic of swine flu was caused by the SIV subtype H1N1. But other subtypes H1N2, H1N3, H3N1, H3N2 and H2N3 can also cause the illness.

Mode of Transmission

The infection is highly contagious and is thought to spread by airborne exposure to droplets from an infected person while coughing or sneezing or by touching the nose or the mouth after contact with a contaminated surface. Infected people start shedding virus one day before the onset of symptoms and shed it at least until symptom resolves.

Clinical Features

Incubation period varies from one and seven days.

- High grade fever above 38°C
- Headaches, myalgias
- Cough, sore throat
- Watery diarrhea

Most patients present with a mild disease characterised by fever (at least 100.4°F), cough and myalgia and recover with symptomatic treatment.

Some cases may develop severe rapidly progressive course with more severe symptoms like:

- Dyspnea
- Hemoptysis
- Chest pain
- Confusion with altered consciousness
- Pneumonia

Ventilatory support is often required in these patients.

The alterations in the maternal immune response during pregnancy puts pregnant women at higher risk of developing pulmonary complications like ARDS and pneumonia especially during second and third trimester. However, there is no evidence to suggest pregnant women are more susceptible to the virus.[1] They are more likely to be hospitalised, and the risk of having a preterm delivery, stillbirths and maternal mortality increases.

Diagnosis

Specimen for diagnosis should be collected within the first 4 to 5 days of illness.

Preferred specimens include nasopharyngeal swab or aspirate and oropharyngeal swab. In intubated patients, endotracheal aspirate or bronchioalveolar lavage should be obtained. Available laboratory test includes:

- *Rapid antigen test:* Not very sensitive; does not differentiate between various types of influenza A virus. Detects influenza viral nucleoprotein antigen and can provide result within 30 minutes.
- *Real-time reverse transcription:* PCR for virus isolation
- Chest X-ray
- Fourfold rise in H1N1 influenza virus specific neutralising antibodies.

Complications

- Tracheitis, bronchitis
- Bronchiolitis, bronchopneumonia (primary influence pneumonia)
- Secondary bacterial pneumonia.
- Exacerbation of underlying asthma and COPD.
- Toxic cardiomyopathy, worsening of underlying CCF and coronary artery disease.
- Encephalitis

- Post-influenzal demyelinating encephalopathy, peripheral neuropathy.
- Post-influential asthenia and depression.

Treatment[2,3]

Drug therapy: Once the patient is tested positive for swine flu, treatment should be initiated immediately. Drugs should be administered within 48 hours of the first symptom.

Anti-viral drugs effective against swine flu include:

- *Oseltamivir:* Administered orally: 75 mg twice daily for five days (for treatment) or 75 mg daily for 7 to 10 days (for post-exposure prophylaxis).
- *Zanamivir:* Administered as inhaler—two inhalation (2×5 mg = 10 mg) twice a day for 5 days

These are Category C drugs. Both these agents are neuraminidase inhibitors which inhibit the ability of virus to release progeny virus particles. Zanamivir acts directly on the respiratory tract with no absorption in the blood stream thus being drug of choice in pregnant women.

Oseltamivir can cause nausea and vomiting, and rarely confusion, hallucinations and self injury.

Peramivir, another neuraminidase inhibitor, is under clinical trials.

Supportive therapy should include:
- Plenty or oral fluids/intravenous fluids
- Paracetamol for fever, myalgia and headache
- Oxygen therapy/ventilatory support
- Antibiotics for secondary infection
- Vasopressors for shock
- For sore throat, short course of topical decongestants, saline nasal drops, throat lozenges and steam inhalation
- Aspirin is strictly contraindicated in any influenza patient due to its potential to cause Reye's syndrome

Most pregnant women with swine flu tolerate labor and delivery with adequate hydration and pain relief.

Decision to delivery will depend on obstetric indication. However, in critically ill patients women can be delivered by cesarean section to help mechanical ventilation. Multi-disciplinary approach involving obstetrician, intensivists and neonatal team is recommended.

Drug Prophylaxis

The best way to reduce risk of transmission of the virus is to observe good respiratory hygiene, i.e. covering mouth and nose while sneezing or coughing. Hands should be washed frequently with soap and water. Pre- and post-exposure prophylaxis with anti-virals are effective in preventing infection but may inhibit develop of immunity and prolonged repeated prophylaxis may cause drug resistance.

Public Guidelines by Indian Medical Association[4]

- Avoid touching face and mouth with your hands
- Avoid touching or having close contact with person who is sneezing or coughing
- Patients who have symptoms should stay indoors and avoid crowded locations till their symptoms subside
- Chemoprophylaxis is indicated in:
  - Close household contacts of confirmed or suspected case who are at high risk for complications (chronic medical conditions, persons >65 y or <5 y, pregnant women)
  - All health care personnel coming in contact with suspected, probable or confirmed cases

Oseltamivir is the drug of choice given at dose of 75 mg for 10 days.

Vaccine

Vaccine contains inactivated H1N1 virus that has been developed from the bird flu virus H1N5. The inactivated virus does not cause any harm to the fetus or mother and helps develop active immunity.

Live attenuated vaccine is for people in the age group 3 to 50 years and is not meant for pregnant woman and those with immuno-compromised condition.

Groups considered to be high-priority to receive prophylactic vaccination with inactivated vaccines are:[5]

- Pregnant woman
- Household contacts and caregivers for children less than six months of age
- Health care and emergency medical service personnel
- People associated with higher risk or medical complication from influenza.

KEY POINTS

- H1N1 has emerged as a predominant virus causing seasonal flu.
- Children, pregnant women and immuno-compromised individuals are particularly at high risk and should be treated aggressively
- Early medical treatment can prevent morbidity and mortality associated with the infection.
- Vaccination is available and should be offered.

ZIKA VIRUS

Zika is an infectious disease caused by Zika virus. There is increasing incidence of transmission of Zika virus in South and Central America over the past few months.[6] In India, 3 cases of Zika virus disease are reported so far from Ahmedabad, Gujarat.

Zika, a flavivirus is transmitted by the bite of Aedes mosquito. Anyone who is living in or travelling to an area where Zika virus is found is at risk for infection, including pregnant women.

Mode of Transmission

- Spread to people primarily through the bite of an infected Aedes species mosquito, most commonly *Aedes aegyptii*.
- It can be transmitted from a pregnant mother to her baby during pregnancy or around the time of birth.

Signs and Symptoms

The incubation period of Zika virus varies between 3 and 12 days for most people.

- More than 75% of those infected are asymptomatic[6]
- Fever, rash, joint pains, and conjunctivitis (red eyes), retro-orbital pain are the common symptoms
- Full recovery occurs in most patients within a week.

There is no evidence that pregnant women are more vulnerable to either acquiring the infection or that infection is more severe in pregnant women.

Once a person has been infected with Zika, they are likely to be protected from future infections. Rarely, severe complications like Guillain-Barré syndrome and other neuro-logical and autoimmune syndromes have been reported.[6,7]

Diagnosis

Diagnosis of Zika is based on a person's recent travel history, symptoms, and test results.

The mainstay of testing for the virus in maternal serum is RT-PCR for symptomatic patients with onset of symptoms within the previous week.[8]

This test is available at National Centre for Disease Control, New Delhi and National Institute of Virology, Pune.

In MCGM, Kasturba hospital PCR lab has been up scaled for diagnosis of Zika.

Antibody testing is less reliable due to potential cross-reaction with antibodies against other similar viruses (e.g. dengue or yellow fever, which are often co-located)

Guidelines for Testing for Zika

- Patients having symptoms fever, rash, joint pain, red eyes for 7 days and tested negative for malaria, dengue.
- Anyone who is living in or travelling to countries where Zika virus is found.
- Pregnant women having fever and other symptoms suggestive of Zika.

Table 39.1	Recommendations for pregnant women who have travelled to an area of Zika transmission[8]	
H/O travel and symptoms consistent with Zika virus disease during or within two weeks of travel	Test for Zika infection and other travel associated infections (including malaria)	• *Zika virus positive:* Referral to a fetal medicine service for further assessment. • *Zika virus negative:* Consider serial (4-weekly) fetal ultrasound scans to monitor fetal growth and anatomy
H/O travel and whose symptoms consistent with Zika virus disease have resolved during presentation	Testing for Zika not recommended	Should be offered serial (4-weekly) fetal ultrasound scans
H/O travel and asymptomatic while travelling and for two weeks after their return	Testing for Zika not recommended	Serial fetal ultrasound should be considered as infection is often associated with minimal symptoms in the majority

Congenital Zika virus syndrome includes microcephaly, intracranial calcifications, ventriculomegaly, arthrogryposis, and abnormalities of the corpus callosum, cerebrum, cerebellum and eyes.

Treatment

There is no specific antiviral treatment for Zika virus. Treatment is symptomatic.

- Women are advised plenty of rest.
- Drink fluids to prevent dehydration
- Acetaminophen or paracetamol to reduce fever and pain
- Aspirin or other non-steroidal anti-inflammatory drugs should be avoided.

Prevention and Control

There is currently no vaccine or drug available to prevent Zika infection.

Several countries like US, UK now advises that pregnant women should consider avoiding travel to countries where Zika outbreaks are ongoing, to reduce the risk to their babies.[9]

As the infection spreads by mosquito bite, vector control activities have to be undertaken. Reducing the contact between man and mosquitoes

- Full sleeve light colored clothing
- Mosquito repellants
- Use of physical barriers—screens, closed doors and windows
- Mosquito nets

- Travellers to take basic precautions to protect themselves from mosquito bites

Surveillance activity is ongoing in all MCGM maternity homes for detection of microcephaly in newborn babies. No baby has been found with such birth defect.

REFERENCES

1. Stirrat GM. Pregnancy and Immunity. BMJ 1994; 308:1385–1386.
2. Clinical management of human infection with pandemic (H1N1) 2009: revised guidelines: WHO, November 2009. Available at http://www.who.int/csr/resources/publications/swine flu/ clinical management/2n/index.html.
3. Pharmacological management of pandemic Influenza (H1N1). WHO 2009 Part 1: Recommendations. Rev Feb 2010.
4. Indian medical Association. "Treat Swine flu like ordinary flu" Prakash et al, 1st January 2015.
5. Meeting of the strategic advisory Group of Experts on immunisation, Apr 2010: Conclusions and Recommendations. Weekly Epidemiological Report, WHO 2010:85; 197–212.
6. CDC. Zika virus. Atlanta, GA: US Department of Health and Human Services, CDC; 2016. http://www.cdc.gov/zika/index.html.
7. Oehler E, Watrin L, Larre P et al. Zika virus infection complicated by Guillain-Barre syndrome: Case report, French Polynesia, December 2013. Euro Surveil 2014;19:4–6.
8. Interim RCOG/RCM/PHE/HPS clinical guidelines. Zika Virus Infection and Pregnancy. Information for Healthcare Professionals.
9. National Travel Health Network and Centre. Zika virus: update and advice for pregnant women. http://travelhealthpro.org.uk/zika-virus-update-and-advice-forpregnant-women.

Metabolic Disorders of Newborn

Madhuri Mehendale, Riddhi Shah

Metabolic disorders of the neonate comprise a myriad of problems encountered in treatment of newborn due to altered body mechanisms. These are important as many depend on the maternal management and are predictable and preventable. Metabolic disorders pose a bigger risk as if not detected or treated in the newborn timely, can cause challenges in the

Table 40.1	Etiology and risk factors
a. Hyperinsulinemic hypoglycemia	
i. Infant of diabetic mother (IDM)—most common	
ii. Congenital genetic—mutations in genes encoding ATP sensitive potassium channel and sulfonylurea receptor	
iii. Secondary	Birth asphyxia
	Disorders of glycosylation
	Erythroblastosis, Rh incompatibility, exchange transfusion
	Developmental syndromes
	Maternal tocolytic therapy with beta-sympathomimetic drugs
	Abrupt cessation of high glucose infusion
b. Large for gestational age infants—born to a non-diabetic mother	
c. Reduced stores	
i. Prematurity	
ii. Intrauterine growth restriction	
iii. Inadequate caloric intake	
iv. Delayed onset of feeding	
d. Increased utilization	
i. Perinatal stress	Sepsis
	Shock
	Asphyxia
	Hypothermia
	Respiratory distress
	Postresuscitation
ii. Carbohydrate metabolism defects	
iii. Endocrine deficiencies (adrenal/hypothalamic/hypopituitarism)	
iv. Amino acid metabolism defects	
v. Polycythemia	

management of high-risk infants and more dangerously, might lead to long-term neurological changes.

HYPOGLYCEMIA

Hypoglycemia is the most common metabolic problem in newborns with incidence estimated to be 16% for large for gestational age (LGA) infants and 15% of small for gestational age (SGA) infants.

Current evidence does not define a specific level of glucose that can separate normal from abnormal. Operational thresholds defined by Cornblath, et al. "Healthy full term neonate <24 hours of age, 30–35 mg/dl and after 24 hours, 45–50 mg/dl, and for infants with any abnormal signs and symptoms, 45 mg/dl".[1]

Pathophysiology

Pederson's maternal hyperglycemia fetal hyperinsulinism hypothesis

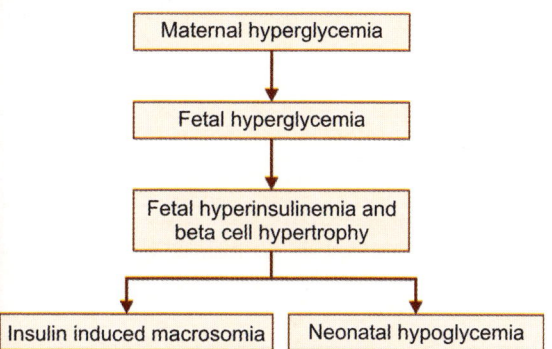

Other factors are decreased catecholamine and glucagon secretion, decreased substrate mobilization due to reduced hepatic glucose production and decreased fatty acid oxidation. Physiologically, normal fetal glucose levels are 2/3rd of the maternal values, which fall for first two hours and normalize by 3–4 hours.[3]

Anticipation and prevention are indispensable for the management of hypoglycemia. Prevention of neonatal hypoglycemia is by routine screening in high risk newborns, early breastfeeding and treatment of precipitating factors like hypothermia and sepsis. Hypoglycemia is found to correlate with fetal macrosomia, elevated maternal and cord blood HbA1c as well as with cord blood C-peptide and immunoreactive insulin levels. This suggests a tight maternal blood glucose control in the last trimester may prevent or reduce neonatal hypoglycemia. Mothers should not receive high doses of glucose peripartum as it may trigger an insulin response in the hyperinsulinemic fetus. Maternal glucose levels should be maintained at an average 120 mg/dl near delivery.[4]

Diagnosis is not clinical due to the non-specific nature of the symptoms. Hence, screening is important for infants with high risk factors like infant of diabetic mothers, preterm and small for gestational age. Most cases present on day 2 or day 3. Screening should be done between 30 and 60 minutes of birth and should be continued till postnatal day 3 or 4. Reagent strips even with reflectance meter are not quite reliable, especially at very low glucose levels, as they measure whole blood glucose which is 15% lower than plasma levels. Valid confirmatory laboratory glucose determination is a must for diagnosis.[2]

Screening in infants of diabetic mothers is all the more important as risk of significant hypoglycemia is 47%, risk of hypocalcemia is 22%, of hyperbilirubinemia is 19% and of polycythemia is 34%. Blood glucose should be screened at 1, 2, 3, 6, 12, 24, 26, and 48 hours by reagent strips and values <40 mg/dl are

Table 40.2	Symptoms of hypoglycemia
a. Abnormal movements	Tremors, jitteriness, irritability, seizures
b. Abnormal behavior	Apathy, refusal to feed, lethargy, limpness, weak or high pitched cry
c. Autonomic disturbances	Recurrent apnea, tachycardia, cyanosis, pallor

confirmed by a clinical laboratory test. Hematocrit levels are checked at 1 and 24 hours of age. Calcium levels are checked if infant is jittery or appears sick and bilirubin levels are measured if infant appears clinically jaundiced. Infants with normal glucose level are fed well starting first hour of life with milk or orally 10% dextrose 5 ml/kg, hourly for three, four hours.[5]

Asymptomatic hypoglycemia can be treated by conservative management. Infants with levels in the early thirties will respond to feeding. Levels should be followed up 1 hour later. Early milk feeding is preferable to glucose solution. Early breastfeeding enhances gluconeogenesis.[3]

Many hospitals have computerized provider order entry systems for calculating glucose infusion rate. Recheck after 20–30 minutes of loading dose and then hourly. Additional boluses may be needed. If stable levels, the infusion may be tapered with increasing feeds. Central venous catheter may be needed to give adequate glucose in an acceptable fluid volume.

Persistent hypoglycemia is that which does not resolve in 2 to 3 days or requires more than 8–10 mg/kg/min. Endocrine evaluation for insulin, cortisol, betahydroxy butrate and free fatty acid levels should be done of fetal blood collected in the hypoglycemic state. It is treated by adding hydrocortisone 5 mg/kg/day intravenous. Diazoxide, octreotide and glucagon are also used. Persistent hypoglycemia may result in neurodevelopmental problems in 30–70% cases. MRI scan may be done for babies with symptomatic hypogly-cemia, although its role is yet to be determined, as stated in a report by National Institute of Child Health and Human Development. However, these babies should have a close follow-up for neurodevelopmental status.

Untreated symptomatic hypoglycemia may be fatal. Hence, timely intervention is important and ensures dramatic recovery. However, it certainly cannot replace watchful screening and prevention.

HYPERGLYCEMIA

Defined as whole blood glucose levels >125 mg/dl or plasma glucose values >145 mg/dl. It is usually associated with premature infants receiving intravenous glucose and is associated with hyperos-molarity, osmotic diuresis and subsequent dehydration. It causes water to move from intracellular to extracellular compartment and could result in intracranial hemorrhage. It might be due to drugs like steroids, phenytoin, theophylline and diazoxide. Neonatal diabetes mellitus is rare, so are congenital pancreatic lesions. Decreasing glucose infusion rates appropriately and monitoring are the mainstay of treatment. Parenteral nutrition with amino acids should begin and feeding may be continued. Insulin infusion is used only if levels exceed 250 mg/dl.[2]

HYPOCALCEMIA

It is the second most common metabolic disorder in neonates. Neonatal hypocalcemia is defined when the total serum calcium is below 7 mg/dl or the ionized calcium is less than 4 mg/dl irrespective of age and gestation.

Table 40.3	Intravenous therapy for hypoglycemia
Indications	Inability to tolerate oral feeding
	Symptomatic
	Oral feedings do not maintain normal glucose levels
	Glucose levels <25 mg/dl
Urgent dose	2 ml/kg of 10% dextrose (200 mg/kg glucose over 1 minute)
Continuing dose	3.6 ml/kg/hr 10% dextrose (6 mg/kg/min)

Cord calcium levels are higher than maternal levels which suggest that calcium is actively transported through the placenta in the third trimester. Calcitonin and parathormone do not cross the placenta and the calcium transport is regulated by parathormone related peptide (PTHrP). Hence, after delivery, levels start decreasing and reach up to 7.5 to 8.5 mg/dl in healthy term babies by day 2 of life. PTH levels increase gradually and normalize the calcium by 3rd day of life.

Body Ca exists as skeleton (99%) and in extracellular fluid (1%). Ca in the extracellular fluid may be bound to albumin (40%), bound to anions (10%) or free ionized form (50%). Ionized serum calcium plays an important role in blood coagulation, neuromuscular excitability, cell membrane integrity and function, and cellular enzymatic and secretory activity. Hence total serum calcium estimation is unreliable as free ionized levels are affected by hypoalbuminemia and acid–base imbalances, both of which are common in sick babies and also with prematurity. Total calcium levels may be falsely low by 0.8 mg/dl with hypoalbuminemia of every 1 g/dl. However, these correction nomograms are not reliable and ionized calcium levels must be asked for in suspected and at risk cases.

Screening in premature and infants of diabetic mothers is at 12, 24 and 48 hours after birth. If hypocalcemia is detected, total serum phosphorus and magnesium levels must be evaluated.

In VLBW infants symptoms are less common, although levels of ionized calcium less than 4 mg/dl are common. Whereas, in term infants symptoms occur readily at that concentration. Early onset is usually asymptomatic or incidental finding, while late onset show symptoms and signs.

Clinical signs are non-specific and include apnea, jitteriness, increased extensor tone, clonus, hyperreflexia and stridor. These being more common with the early onset type, the late onset type may present with seizures. Usually seizures on day 3 of life in an otherwise alert and active newborn are seen with late hypocalcemia. They must be differentiated from other causes of neonatal seizures, e.g. "fifth-day" fits. Late onset also may present as partial breastfeeding, abnormal movements and lethargy.

Treatment of symptomatic babies is by intravenous calcium gluconate 10%, 2 ml/kg at less than 1 ml/minute to a maximum dose of 10 ml in term neonates and 5 ml if preterm. Asymptomatic hypocalcemia or bolus corrected

Table 40.4	Etiology of hypocalcemia
Early onset (within 24–72 hours)	1. Prematurity
	2. Pre-eclampsia
	3. Infant of diabetic mother
	4. Perinatal stress/asphyxia, prolonged labor
	5. Maternal: Anticonvulsants (phenobarbitone, phenytoin sodium), maternal hyperparathyroidism, maternal nifedipine use
	6. Iatrogenic (alkalosis, blood transfusions, diuretics, phototherapy)
Late onset (classic tetany)	1. Increased phosphate load: Cow milk, renal insufficiency, formula feeds (wrong Ca:P ratio)
	2. Hypomagnesemia
	3. Maternal vitamin D deficiency
	4. Malabsorption/hepatobiliary disease
	5. Hypoparathyroidism: DiGeorge's syndrome, CATCH 22 syndrome, maternal hyperparathyroidism
	6. Phototherapy, iatrogenic

symptomatic cases are maintained on intravenous calcium gluconate 2 ml/kg 6 hourly for 48 hours followed by oral supplements.[6]

Precautions to be taken as follows:

- Slow infusion with careful cardiac monitoring otherwise it may cause bradycardia and cardiac arrest in asystole
- Extravasation in subcutaneous tissue could result in severe necrosis and subcutaneous calcifications
- Umbilical vein infusion to be avoided fearing hepatic necrosis if the catheter is in a portal vein branch
- Should never be mixed with bicarbonate solutions or calcium carbonate will precipitate

Refractory hypocalcemia deserves parathyroid studies and defects of vitamin D metabolism are treated with vitamin D analogs (rare).

Neonate with hypocalcemia associated with hyperphosphatemia (late onset) can be fed human milk or low phosphate formula and oral calcium supplements.

Outcome is good in hypocalcemic cases, even those presenting with seizures as there are no neurodevelopmental sequelae.

HYPERCALCEMIA

Neonatal hypercalcemia, defined as >11 mg/dl total serum calcium, may be asymptomatic. Severe hypercalcemia (total serum calcium >16 mg/dl or ionized calcium >1.8 mmol/L) requires immediate medical intervention. It is caused by iatrogenic factors, extreme prematurity, hyperparathyroidism, hyperthyroidism, hypervitaminosis D of the mother, William's syndrome and during the recovery phase of acute renal failure.

Clinically, it presents with hypotonia, encephalopathy, poor feeding, vomiting, constipation, polyuria, probably in a setting of family history of hypercalcemia. Work-up includes serum and urine phosphate levels, urinary calcium: creatinine ratio and specific hormone levels (PTH/calcitriol). Treatment involves emergency medical management with isotonic saline solution (volume expansion) and promoting urinary excretion by intravenous furosemide.

HYPOMAGNESEMIA

Serum magnesium levels of <1.6 mg/dl is known as hypomagnesemia. Although it is uncommon, it may be seen with late onset hypocalcemia in babies who are small for gestational age or infants of diabetic mothers or after diarrhea.

Symptoms include apnea and poor motor tone. It does not present as seizures per se, but in its presence hypocalcemic seizures become unresponsive to calcium therapy. Such babies should be given magnesium sulfate 50%, 50 to 100 mg/kg intravenously over 1–2 hours with heart rate monitoring. The dose may be repeated after 12 hours after repeating calcium levels.

HYPERMAGNESEMIA

Magnesium sulfate therapy to mother or antacids given to neonate could cause hypermagnesemia (>3 mg/dl). It causes curariform effects like apnea, respiratory depression, hypotonia, lethargy, hyporeflexia, poor suck reflex and delayed meconium passage due to decreased intestinal motility. Treatment is by removing the source of excessive management and care to be taken that feeding reestablished only after suck and intestinal mobility have started. Respiratory support should be provided.

METABOLIC ACIDOSIS

Late metabolic acidosis is common in preterm babies fed high-casein containing formulas due to the high solute load in the feeds and the immature renal acidification. Classical presentation is in 2nd–3rd week of life with failure to thrive, tachypnea, circumoral pallor and sluggishness.

Table 40.5	Etiology of metabolic acidosis
With increased anion gap (>15 mEq/L)	Renal failure, inborn errors of metabolism, lactic acidosis (due to asphyxia or cardiopulmonary disease), toxin exposure
With normal anion gap (<15 mEq/L)	Premature infants manifesting renal tubular acidosis, diarrhea, hyperalimentation

It is managed with oral bicarbonate therapy for 2 weeks or till 35 weeks conceptional maturity and with suitable formula feed modifications. Treating the underlying cause is most important.

NEONATAL HYPERBILIRUBINEMIA

Hyperbilirubinemia affects almost 60% of term babies and 80% of pre-term infants within the first week. Subjective assessment of bilirubin level is unreliable. Kernicterus is a rare complication of unconjugated hyperbilirubinemia that can lead to major long-term neurological sequelae.

Maternal high risks should prompt suspicion in cases like birth trauma such as cephalhematoma, significant bruising (breakdown of heme), Rh Negative blood group, viral serology positive, family history of hemolytic disease (ABO/G6PD, spherocytosis).

Jaundice before 24 hours or after 3 days of birth is more likely to be pathological. Physiological jaundice which is an exaggerated normal response usually clears in 2 weeks in a term neonate or up to 3 weeks in a preterm.[7]

Management includes phototherapy, treatment of underlying sepsis, phenobarbitone and in rare cases, exchange transfusion.

INBORN ERRORS OF METABOLISM

Inborn errors of metabolism (IEM) are caused due to a partial or complete inhibition to an essential pathway in the body's metabolism. There are a large number of conditions included in this group of disorders. Most of these disorders are inherited as autosomal recessive.

Parental history may suggest consanguineous parents, previous unexplained neonatal deaths or sudden infant death syndrome (SIDS), family history (e.g. relatives with undiagnosed 'syndrome'), at-risk ethnicity. Diagnosis is tricky as symptoms are non-specific. Neonates may have poor feeding/suck, vomiting, hypotonia, respiratory compromise/apnea, progressive encephalopathy and seizures.[8]

Management requires involvement of a metabolic pediatrician and can be very complicated at times.

SUGGESTED READING

1. Cornblath M, Howdan JM, Williams AF, et al. Controversies regarding definition of neonatal hypoglycaemia: suggested operational thresholds. *Paediatrics* 2008; 122:65–74.
2. John P. Clogher tym. *Manual of neonatal care.* 7th ed. New Delhi: Wolter Kluwers Pvt Ltd;2014.
3. David H. Adamkin, et al. Clinical Report—Postnatal Glucose Homeostasis in Late Preterm and Term Infants. *Pediatrics* Volume 127, Number 3, March 2011
4. American College of Obstetricians and Gynaecologists (ACOG). ACOG Practice Bulletin. Clinical management guidelines for obstetrician–gynecologists. Number 137, August 2013. Gestational Diabetes Mellitus. *Obstet Gynecol* 2013;122:406–16.
5. American College of Obstetricians and Gynaecologists (ACOG). ACOG Practice Bulletin. Clinical management guidelines for obstetrician-gynecologists. Number 173, September 2016. Fetal Macrosomia. *Obstet Gynecol* 2016;128:e43–53.
6. Teena C. Thomas, et al. Transient Neonatal Hypocalcemia: Presentation and Outcomes. *Pediatrics* Volume 129, Number 6, June 2012.
7. Karen Muchowski. Evaluation and Treatment of Neonatal Hyperbilirubinemia. *Am Fam Physician* 2014 Jun 1;89(11):873–78.
8. Morteza Pourfarzam and Fouzieh Zadhoush. Newborn Screening for inherited metabolic disorders; news and views. *J Res Med Sci* 2013 Sep; 18(9): 801–08.

Down's Baby— Obstetrician to Blame?

Uday Thanawala

In 2014 an obstetrician from Mumbai was sued by the husband of a 40-year-old patient who delivered a Down's baby. He sought ₹ 25 lakh compensation, "the cost to bring up my daughter is too high" he said.

"I was assured by the doctor that everything would be fine. When I confronted her with the problem later, she told me that I would have never been able to afford those tests," he added.

Screening for Down's syndrome is not the standard of care as yet. So is the obstetrician totally absolved from the responsibility when a Down's baby is born?

Yes—if the doctor had offered screening for Down's and the patient after understanding what it was all about—declined—which was recorded; then the obstetrician has done his bit.

1. Let us Establish it is Important to Screen for this Disorder

Trisomy 21 is the most common trisomy at the time of birth. Also called Down syndrome, it is associated with moderate to severe intellectual disabilities and may also lead to digestive disease, congenital heart defects and other malformations. Incidence is about 1 per 1000 babies born each year. It is named after John Langdon Down, the British doctor who fully described the syndrome in 1866. The genetic cause of Down syndrome, an extra copy of chromosome 21.

How do They Fare in Later Life?

The average IQ of a young adult with Down syndrome is 50, similar to the mental age of an 8 or 9-year-old child, but this varies widely. Some children with Down syndrome are educated in typical school classes while others require more specialized education. In adulthood about 20% in the United States do paid work in some capacity. Life expectancy is around 50 to 60 years in the developed world with proper health care. In fact in Australia there is a movement not to test for Down's because there is no need to abort them. In today's developed world a Down's baby can have a reasonable life.

In a developing country like ours the situation is different. Having a Down's is an emotional and financial drain on the family. One needs to speak to the mother of any Down's child and she will tell you how she has devoted all her life looking after the child.

One must also consider the fact that no expectant mother wants a defect in her baby. So when expectant mothers are visiting us the foremost question for which they are seeking an answer is: Is my baby normal? A Down's syndrome baby possibly is not and is a special child with special needs—and the couple may not be able to cope with this, thus it is imperative for them to know that methods are available which can help them to diagnose a Down's fetus—and they should be made fully aware of these.

2. Whom to Screen? When to Screen? How to Screen?

Whom to Screen?

Though incidence is higher as age advances but since younger women generally have more children, about 75–80% of children with Down syndrome are born to younger women. Keeping this in mind it is advisable to offer universal screening. Screening only above 35 years, or only the once who you feel can afford the test is not the right approach—one could land up in a legal issue as the doctor above.

Counseling every woman could be a challenge in a busy OPD, so one needs to have posters/flyers for patient information. Make the patient aware—let them decide—whatever their decision—take it in writing and respect it. The author has had a 40-year-old IVF pregnancy—infertility conception—high NT denied screening as she had decided to continue. The baby was a Down's but was welcomed by the family.

Counseling will also involve telling them that the blood test and NT are screening methods, but confirmatory tests, CVS or amniocentesis are invasive. These tests being invasive have a small (1%) risk of abortion. Thus if the patient is sure she does not want to undergo any invasive tests (which many patients do) there is no point in offering them the biochemical screening blood test.

When to screen?

Generally in most cities patients are presenting in early gestation. Offering them the first trimester combined screening makes sense because of higher sensitivity. Moreover, if detected as high risk—doing a confirmatory CVS, followed by an earlier termination if positive for Down's is possible. Quadruple and Triple markers have low pick up rate and are only offered to women presenting after the first trimester till 20 weeks.

How to Screen?

Look at Table 41.1.

Maternal age has only a 30% detection rate. Adding NT value to maternal age the detection rate goes up to 80%.

Maternal age + NT + double marker test gives us up to 95% detection rate for a false positive of 5%. A small increase is there if we add nasal bone and Doppler studies.

Thus today, screening is offered at the time of 11–13.4 weeks scan. This scan is important for evaluating the full structural anatomy as well measuring the NT.

Nuchal translucency: Measured between 11 and 13.5 weeks on USG (CRL 45–85 mm). There are strict criteria to be observed while doing a NT, some of which are:
- Proper mid-sagittal plane.
- 75 percent magnification.
- Neutral position of head
- Proper placement of calipers

Only a high NT does not call for a termination. Although increased fetal NT thickness is associated with abnormalities and fetal death, the majority of babies survive and develop normally. After the diagnosis of increased NT the aim must be to distinguish as accurately and quickly as possible between those that are likely to have problems from those where the baby is likely to be normal.

Biochemical markers: The double marker test is offered for the biochemical markers of PAPP-A and serum beta-hCG.

Combined screening: NT along with the result of the double marker test is a good screening method.

Table 41.1	Performance of different methods of screening for trisomy 21		
Method of screening		Detection rate (%)	False-positive rate (%)
MA		30	5
First trimester			
MA + fetal NT		75–80	5
MA + serum free β-hCG and PAPP-A		60–70	5
MA + NT + free β-hCG and PAPP-A (combined test)		85–95	5
Combined test + nasal bone or tricuspid flow or ducts venosus flow		93–96	2.5
Second trimester			
MA + serum AFP, hCG (double test)		55–60	5
MA + serum AFP, free β-hCG (double test)		60–65	5
MA + serum AFP, hCG, uE3 (triple test)		60–65	5
MA + serum AFP, free β-hCG, uE3 (triple test)		65–70	5
MA + serum AFP, hCG, hCG, uE3 inhibin A (quadruple test)		65–70	5
MA + serum AFP, free β-hCG, uE3, inhibin A (quadruple test)		70–75	5
MA + NT + PAPP-A (11–13 weeks) + quadruple test		90–94	5

MA: Maternal age; NT: Nuchal translucency, β-hCG: β-human chorionic gonadotrophin, PAPP-A: Pregnancy-associated plasma protein-A

The patient's age, gestational age, weight along with other details like smoking, diabetes or bleeding during last 15 days are all entered in the requisition form.

Including the above variables along with the NT reading are fed in the computer with the result of the biochemical screen. The software arrives at a risk.

Each laboratory gives a value—generally 1:250 is the cut off. If the risk is 1:90 means out of 90 similar women screened, only 1 will have Down's. But this is above the cut off—and an invasive test either CVS or amniocentesis is indicated.

What if the risk comes to 1:260? Just a bit above the laboratory cut off?

One can do one of the following:
1. Let cut offs be cut offs and not do anything further.
2. FMF has a defined an intermediate risk 1:51–1:1000 as per FMF guidelines and look for additional markers:
   - HR
   - Nasal bone
   - DV
   - Facial angle

3. *Do integrated screen:* First trimester biochemistry + NT + triple test all are integrated and given a risk. For this the laboratory must have the software to integrate all the values.
4. Offer non-invasive prenatal testing

NIPT (Non-invasive prenatal testing): DNA from the fetus circulates in the mother's blood. This cell-free DNA (cfDNA) results from the natural breakdown of fetal cells (presumed to be mostly placental) and clears from the maternal system within hours of giving birth.

Testing maternal blood for this DNA is a non-invasive prenatal testing (NIPT). Though it is not a diagnostic test, because it is 99.5% specificity; 99.9% sensitivity; and not 100%—it is still a expensive screening modality.

Understanding NIPT: Katie Stoll is a genetic counselor in Washington State. On the blog, The DNA Exchange, she examines the role of the incidence rate for Down syndrome and a test's Positive Predictive Value (PPV).[2]
- In population of 100,000, 35-year-old women who have an incidence rate of 1-in-250–400 will be pregnant with a child with Down syndrome (100,000 × 1/250 = 400).

- NIPS labs report a sensitivity rate of 99.5%, meaning 99.5% of those actually carrying a child with Down syndrome will be detected by NIPS.
- Therefore, of the 400, 35-year old moms, 398 will receive a "positive" NIPS result (400 × 99.5% = 398). Note as well that 2 will receive a "negative" NIPS report—a false negative, since they are carrying a child with Down syndrome.
- NIPS labs also report a 99.9% specificity rate (the percentage of those pregnancies not carrying a child with Down syndrome that will receive a negative NIPS report).
- In Stoll's example, there are 99,600 moms not carrying a child with Down syndrome (100,000 moms—the 400 carrying a child with Down syndrome = 99,600).
- Of those 99,600 moms, 99,500 will receive a negative report (99,600 × 99.9% = 99,500).
- This then means 100 will receive a "positive" NIPS result (99,600–99,500 = 100), making these 100 false positives.
  - So, in this example, there were 400 pregnancies actually carrying a child with Down syndrome. Of these, 398 would receive a positive NIPS result, but 100 false positives would also be reported, making for a total of 498 positive NIPS reports when only 400 pregnancies were actually carrying a child with Down syndrome.
  - This means that a positive NIPS report means the mother has a one-in-five chance of a having a false positive (100 false positives/498 = 20%, or 1-in-5). And, this false positive rate goes up; the lower the incidence rate. Current testing, however, can only provide a highly accurate recalculation of the probability that the fetus has Down syndrome.

As a result, what had been hoped to be noninvasive prenatal diagnosis (NIPD) can only be called noninvasive prenatal testing (NIPT).

So—when to use NIPT?

Indications for Considering the Use of Cell-free Fetal DNA

- Maternal age 35 years or older at delivery
- Fetal ultrasonographic findings indicating an increased risk of aneuploidy
- History of a prior pregnancy with a trisomy
- Positive test result for aneuploidy, including first trimester, sequential, or integrated screen, or a quadruple screen.
- Parental balanced robertsonian translocation with increased risk of fetal trisomy 13 or trisomy 21.

ACOG Recommendations for Use of NIPT[3]

- Cell-free fetal DNA testing should be an informed patient choice after pretest counseling and should not be part of routine prenatal laboratory assessment.
- Cell-free fetal DNA testing should not be offered to low-risk women or women with multiple gestations because it has not been sufficiently evaluated in these groups.
- A negative cell-free fetal DNA test result does not ensure an unaffected pregnancy, though the likelihood of having a Down's after a negative NIPT is extremely remote .
- A patient with a positive test result should be referred for genetic counseling and should be offered invasive prenatal diagnosis for confirmation of test results.

Thus on screening for Down's the ISPD Position Statement (4 April 2013) states[4]:

1. Ultrasound nuchal translucency at 11–13 completed weeks combined with serum markers at 10–13 weeks.

2. Extending option (1) to include other first trimester sonographic markers, provided ultrasound performance has been prospectively validated by the center where the screening is to be performed.

3. A 'contingent' test whereby women with borderline risks from option (1) have option, (2) at a specialist center and risk is subsequently modified.

4. Four maternal serum markers (quadruple test) at 15–19 weeks, for women who first attend after 13 weeks 6 days. (Integrated screening can be offered when CVS is not available. A serum integrated test when NT measurement is unavailable.)

5. Contingent second trimester ultrasound to modify risks for aneuploidy for women having options (1), (4) or (5).

6. cfDNA screening for women classified as high risk by any of the above options (1–6). Or those with maternal age; presence of an ultrasound abnormality suggestive of trisomy 21, 18 or 13; family history of a chromosome abnormality that could result in full trisomy 21, 18 or 13; and history of a previous pregnancy/livebirth with trisomy 21, 18 or 13.

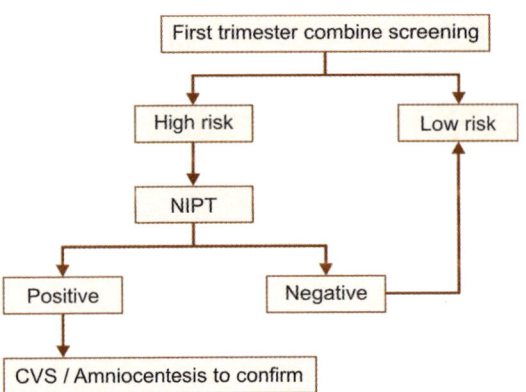

Patients above 40 years or those with history of previous baby being Downs should be offered NIPT/or definite testing by CVS or amniocentesis.

CONCLUSION

If a Down's baby is born—once does tend to think that the obstetrician is to blame. Not so if the doctor has:

1. Counseled the patient for Down's screening in the first/early second trimester.

2. Patient had understood what Down's syndrome is and had refused screening/ or did not come in time for screening

3. If screening was done and was low risk— we must remember that the combine screening with a pick up rate of 95% or even NIPT with a pick up rate of 99.5% are ultimately screening tests. Like one cannot abort only based on results of a positive screen—a negative screen does not guarantee an unaffected child. The author has had a 25-year-old primigravida delivering a Down's—even when NT was normal and combined screening had a risk of 1:21000! Rare but can happen. That is the importance of explaining to the patient want risk assessment is that a low risk does not eliminate the risk—it simply says that may be an invasive testing is not warranted.

REFERENCES

1. Prenatal diagnosis 2011;31:7–15. Published online in Wiley Online Library (wileyonlinelibrary.com) DOI: 10.1002/pd.2637?

2. How accurate is the new blood test for Down syndrome mark leachjuly 30, 2013; *downssyndromeparenting*.com?

3. The American College of Obstetricians and Gynecologists Committee on Genetics. The Society for Maternal-Fetal Medicine Publications Committee Number 545, December 2012.

Breastfeeding—
Hiccups in Practice

Prashant Gangal, Mangala Wani

In rich and poor countries alike, breastfeeding saves lives and gives children the healthiest start. While breastmilk is nature's perfect food for babies, the act of breastfeeding does not always come naturally. Hence, for breastfeeding success, women need access to skilled support and guidance by health care workers trained in lactation counselling.[1]

Obviously the staff should be well aware of hiccups in practice. They need to spend a lot of time and effort to help all mothers postnatally and more so if it is a cesarean delivery. The well-known 'ten steps to successful breastfeeding' provide excellent guidelines (Fig. 42.1). A maternity service which fulfils all 'ten steps' is called a 'baby-friendly hospital' as per baby-friendly hospital initiative (BFHI: Documents updated in 2009). A maternity home policy based on the 'ten steps' is enclosed (Fig. 42.1).

By and large maternity home staff works hard to help mothers exclusively breastfeed their babies because they are consciously or subconsciously aware about benefits of breastfeeding. It is crucial that they do not get frustrated or tired at motivating and helping the mothers to breastfeed successfully.

Out of recommended 2 years of total duration of breastfeeding, a mother spends only about 3 days in a maternity home and the remaining approximately 725 days at home. Hence it is logical that motivation and breastfeeding help should be more focused post-discharge by her family and health care providers. However, on deeper thought one would realize that the initial days are very important. What good things she will practice in these initial days, she will continue to practice at home for rest of the 700 days. Colostrum which is the prime defence for the baby is secreted in the first few days. Research suggests that early, frequent removal of the colostrum determines future milk production potential. What starts well ... ends well.

The practically encountered breastfeeding 'Hiccups' in the:

a. Antenatal period

b. Labor room

c. Postnatal period

Participation of Lactation Consultant (IBCLC: International Board Certified Lactation Consultant), Lactation Counsellor (certified by a credible institute or organization after formal/informal training) or Mother Support Group Leader greatly augments the efforts of maternity service to help mothers breastfeed successfully. Mother Support Group Leader is an experienced mother who has undergone training to help new mothers to breastfeed successfully.

Guidelines for Mothers on Breastfeeding

Based on the 'TEN STEPS TO SUCCESSFUL BREASTFEEDING' by WHO and UNICEF [Family members are requested to help mothers to follow the guidelines]

1. This maternity service is bound to help mothers for successful breastfeeding. This will be achieved by policy consisting of following strategies.

2. Our staff is trained to help you to breastfeed successfully.

3. Guidance for appropriate baby care and successful breastfeeding is offered before and after delivery (pre-delivery and post-delivery counseling).

4. Soon after delivery hold your baby close to give skin-to-skin body contact (in about 5 minutes) in the delivery room to facilitate the baby to start breastfeeding within an hour of delivery (early contact and breastfeeding).

5. a. Learn to attach and position the baby properly and comfortably while breastfeeding. This will help to prevent pain, soreness and cracked nipples. Learn to express breastmilk and store it prior to discharge.

   b. Do not apply any cream on sore or cracked nipples.

6. a. Do not give glucose/jaggery/sugar/plain water or honey before the first breastfeed (no prelacteal feeds).

   b. The breastmilk during the first 3–5 days (colostrum) after delivery, though scanty, is highly nutritious, protective and also sufficient for the baby.

   c. Give only breastmilk (exclusive breastfeeding) till the baby has completed 6 months of age. Water, top milk, honey, vitamins and almonds should not be given during this period and continue breastfeeding at least till second birthday

   d. Do not give your baby gripe water, balkadu, glucose water or tonics for teething

7. Always keep your baby with you in the same bed (bedding in).

8. a. Breastfeed whenever your baby is hungry and without any restriction of time (demand feeding). However, breastfeed at least 10 times/24 hr till breastfeeding is established (first 1–2 weeks).

   b. Establishment of breastfeeding is indicated by baby starting to urinate frequently (at least 6–7 times/24 hr) and gaining weight to cross birth weight latest by fifteenth day. After this period adequacy of feeding in an exclusive breastfeed baby is indicated by baby urinating at least 6–7 times in 24 hours and gaining at least 500 gm in a month.

   c. Mother should feed on one side as long as possible because the milk which comes initially is rich in water and sugar (foremilk), while the milk in the later part is rich in fats (hind milk).

9. Do not use bottles, pacifiers, animal milk (formula powder/ready) or baby food free or subsidised samples of these items are not accepted in this maternity home. Their advertisement in any form is also prohibited here.

10. For any breastfeeding problem we recommend an expert. kindly consult the doctor or our Mother Support Counsellor even after discharge. You are requested to note their phone numbers from the receptionist / nurses

11. Do not put restrictions on mother's diet and fluid intake

12. Introduce appropriate, hygienically prepared, homemade, mashed complementary foods in adequate quantities at the end of six months and continue to breastfeed till the second birthday and beyond.

Fig. 42.1: Tips to successful breastfeeding

A. Antenatal Period

A-1. Lactation Counselling

- Group session for mothers.
- In early 3rd trimester because this gives her an opportunity to attend the session at a later date if she misses the first one.
- Mother and/or mother-in-law also attends.
- Separate session for fathers (Father Support Group) and family.
- Frequency depends on the delivery rate and the space available for the session.
- >70–80% of those delivering should have attended the session.
- *Content:* Frequently asked questions by the mothers (Refer ShishuPoshan App)
- Display and discuss the 'Breastfeeding Policy'.

A-2. Breast Examination

In view of possible reduced confidence arising from antenatal examination and lack of effective pre-delivery remedies (for inverted or non-protractile nipples) routine antenatal breast examination is not recommended. It should be done only if mother expresses some doubt about the nipples or the breast.[2]

A-3. Short/Flat/Inverted Nipples

Nipple protractility improves by the time of delivery. No basis exists for the use of breast shells or Hoffman's exercises as antenatal treatment for inverted or non-protractile nipples.[2] Helping mother to attach the baby with good technique in the early postpartum period is likely to be more effective than antenatal interventions.

A-4. Clothes Facilitating Breastfeeding

Clothes that permit good skin-to-skin contact: Sari-Blouse, shirt or gown (full front opening), two-piece night suit. Feeding gowns with half opening central zip/buttons or zips on both the sides are discouraged.

A-5. Good Websites and Apps

bpnimaharashtra.org; bpni.org; llli.org; waba.org.my; hetv.org

A 20153 study reviewed various interventions which can improve breastfeeding rates. Greatest improvements were seen when counseling or education was provided concurrently in home and community, health systems and community, health systems and home settings, respectively. Baby-friendly hospital support at health system was the most effective intervention to improve rates of any breastfeeding.

B. Labor Room

B-1 Recommendation for Initiation of Breastfeeding

UNICEF and WHO recommend that the baby should be placed in skin-to-skin contact with the mother in about 5 minutes after birth in order that baby can initiate breastfeeding within an hour of delivery. This first skin-to-skin contact should be continued till baby finishes its first breastfeed.

B-2 Breast Crawl (Fig. 42.2)

Every newborn, when placed on its mother's abdomen, soon after birth, has ability to find her mother's breast all on its own and to

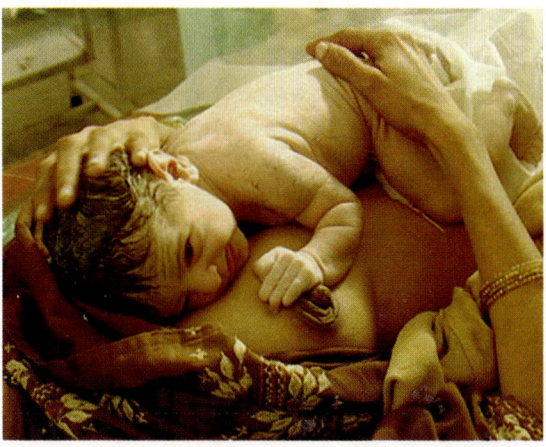

Fig. 42.2: Breast crawl

decide when to take the first breastfeed. This is known as the 'Breast Crawl'. (www.breastcrawl.org)

(www.breastcrawl.org)

B-3 Do's and Don'ts for Success of the Breast Crawl

For the mother

- Use drugs for labor analgesia judiciously. Epidural Anesthesia/Programmed Labour Protocol is not known to affect/hamper breastfeeding.
- Do not wash/wipe breast before feeding.
- Raise mother's head on a pillow to facilitate mother–baby visual contact.

For the baby

- A baby who has cried well does not need oro-nasal suction.
- Dry the baby thoroughly except for the hands.
- Do not pass orogastric/nasogastric tube or do gastric suction as a routine.
- If the weather is cold then baby and mother should be covered together with a cloth, so that they keep warm while continuing with skin-to-skin contact.
- Unnecessary routines like the vitamin K injection, wrapping, weighing and routine anthropometric measurements should be delayed till after the first breastfeed.
- Baby bath is best delayed to beyond 24 hours.

The infant's rooting and sucking reflexes are particularly strong immediately after birth, and a mother is usually keen to see and touch her child. Encouraging skin-to-skin contact immediately after birth increases bonding, stimulates breast milk secretion, and releases oxytocin facilitating expulsion of the placenta.

B-4 Initiation in a Cesarean Delivery

Breastfeeding can be initiated in the operation theatre even while the cesarean is in progress or soon after the operation is over. If not, it can be initiated as soon as the mother is shifted in the ward. Occasionally when a mother needs general anesthesia, feeding is initiated as soon as she is able to respond and definitely not later than 4 hr after surgery.

Note: Review by Moore[4] supports current practices as recommended by the BFHI, in which SSC is encouraged for the first hour after birth.

C. Postnatal Period

1. Less milk
2. Cracked nipples
3. Positioning
4. Engorgement, mastitis, breast abscess
5. NICU transfer
6. Myths

C-1. Less Milk

Less milk volumes in the first few days match the smaller stomach capacities of the newborn. As the stomach capacity increases so does the milk volume.

Volume in ml/feed: 3 kg baby[5]

Day	1	2	3	4	5	6	7
Volume	2–10	5–15	15–30	30–60	45–60	50–60	55–65

- The colostrum has lesser calorie value (65 calories/100 ml) as compared to mature milk (67 calories/100 ml). Smaller volumes of colostrum cannot fulfil the energy requirements of the newborn (100 calories/kg). Hence the newborn depends on the glycogen stored in the liver and muscles for meeting energy requirements and also on the brown fat, especially for temperature control.
- Blood sugar levels are directly proportional to frequency of feeding and bilirubin levels are inversely proportional to frequency of feeding.
- The recommended frequency of feeding <15 days of life (till breastfeeding is established) is at least 10 times in 24 hours.[6]
- *To maintain this frequency:* Teach the mothers various techniques to wake up a sleepy baby and to identify the feeding cues.[6]

- Frequent feeding is the key to prevent hypoglycemia as well as to prevent babies from crying. Fear of hypoglycemia and crying drives a maternity home to staff to give supplements. Hence, frequent feeding is the key to promote exclusive breastfeeding in a maternity home.
- Monitor the babies for adequacy of exclusive breastfeeding.
- *Weight:* Conventionally babies lose 5–7% of birth weight (over 2–4 days) before they begin to gain weight. Reported mean infant weight loss ranges widely from 3.79 to 8.6%.[7] A recent study[8] on a large sample shows:

Hours post-delivery

Delivery	24	48	72	96
Vaginal	4.2%	7.1%	6.4%	--
Cesarean	4.9%	8.0%	8.6%	5.8%

Lower Threshold for Intervention[9]

- Loses 10% of the birth weight

- Does not start to gain weight by 9 days
- Fails to regain the birth weight by 2 weeks of age.

Additional intensive personal breastfeeding support by dedicated breastfeeding support group: Positioning, feeding techniques and use of hand expression with additional cup feeding (Fig. 42.3).

Higher Threshold for Intervention[9]

- Loses 12.5% of the birth weight—biochemical testing and rehydration therapy
- Fails to regain the birth weight by 2 weeks of age—other nutritional supplements

Babies are best monitored by regular weighing (0, 4, 7, 14, 28 days) on a digital scale with minimal sensitivity of 5 grams.

- *Stool:* In first 1–2 weeks of life stooling is a better indicator of adequacy of feeding. If a child passes 1 stool on day 1, 2 stools on day 2, 3 stools on day 3 and 4 stools on day 4 with color changing from black green stools

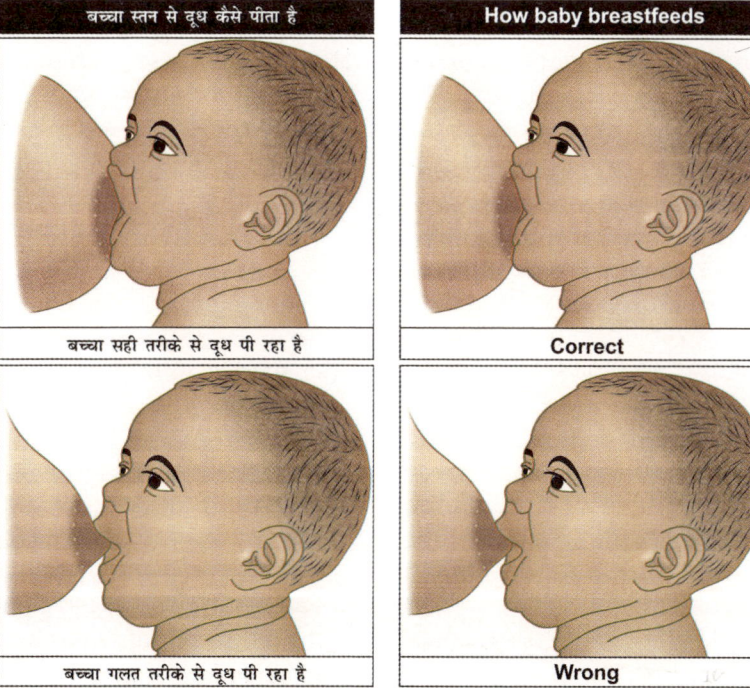

बच्चा स्तन से दूध कैसे पीता है   How baby breastfeeds

बच्चा सही तरीके से दूध पी रहा है   Correct

बच्चा गलत तरीके से दूध पी रहा है   Wrong

Fig. 42.3: Breastfeeding method

on day 1 to yellow frothy stools on day 5; it is indicative of adequate feeding. If a child continues to pass green stools by day 7, it indicates inadequate feeding.

- *Urine:* First urine is usually passed by 24 hours (maximum 24–48 hours). Subsequently baby should urinate at least once in any 24-hour period for first few days of life. Breastmilk 'comes in' (increases in quantity and begins the change from colostrum to mature milk) about 3–5 days after delivery, after which baby starts urinating more frequently (at least 6–7 times in 24 hours)

Normal urine and stool output[5]

Day	1	2	3	4	5	6	7
Number of urine output	1	2	3	4–6	4–7	5–8	>6
Number of stools	1	2	3	4–5	4–6	4–8	>5

C-2. Cracked Nipples

- Most common reason for the cessation of breastfeeding.
- Normally breastfeeding should not hurt. If the baby's attachment to the breast is not proper, then it causes pain/soreness/cracks.
- The prevention/remedy is learning correct attachment from the first breastfeed. The mother should apply hind-milk to the cracked/sore nipple and leave it open to the air for some time.
- Frequent washing of nipple and areola with soap and water can cause drying and cracks by removing the natural oily substance which normally covers this area. Routine once a day cleaning of the breasts during bath is sufficient.
- Nipple may get cracked at the base (semi-circular cracks) if the child is taken away abruptly from the breast while feeding. To unlatch, the mother should insert her little finger in the corner of baby's mouth and detach the baby slowly.
- *Nipple shield:* Old funnel types of shield made of rubber or latex should not be used. The newer ultrathin silicone nipple shields

are helpful for flat nipples after all attempts by the mother have been unsuccessful in spite of help at latching by an expert. Of course, it will be useful for truly inverted nipples. Shields are also useful for preterm babies who have a latching problem or for babies who have developed preference for a bottle nipple and refuse the breasts. Short-term use of nipple shields will preserve the breastfeeding relationship while baby is learning to breastfeed.

A study[10] compared various treatment modalities for nipple pain and concluded that there was insufficient evidence to recommend any intervention. Regardless of the treatment used, for most women nipple pain reduced to mild levels after approximately 7 to 10 days postpartum. It is important to prevent nipple trauma and pain in the early postpartum period.

C-3. Positioning

This is the most important component of the postnatal support.

Note: If child is >1 month old and it is certain that the baby is getting enough milk, it may not be wise to instruct the mother regarding attachment and positioning.

The mother also needs to be shown different ways in which she can hold the baby (Figs 42.4a and 42.4b).

C-4 Engorgement, Mastitis, Abscess and Blocked Duct

- The mother starts producing copious milk (i.e. mature milk) from 3 to 5 days after the delivery.
- Breasts become full and little heavy just prior to a breastfeed (full breasts, Fig. 42.5).
- Excessive production of milk or incomplete removal (infrequent suckling or poor attachment)/delayed emptying (child overslept) will cause heaviness, hardening and pain (engorgement, Fig. 42.6). Engorgement can be prevented by timely expression of milk. Hence, it is essential that

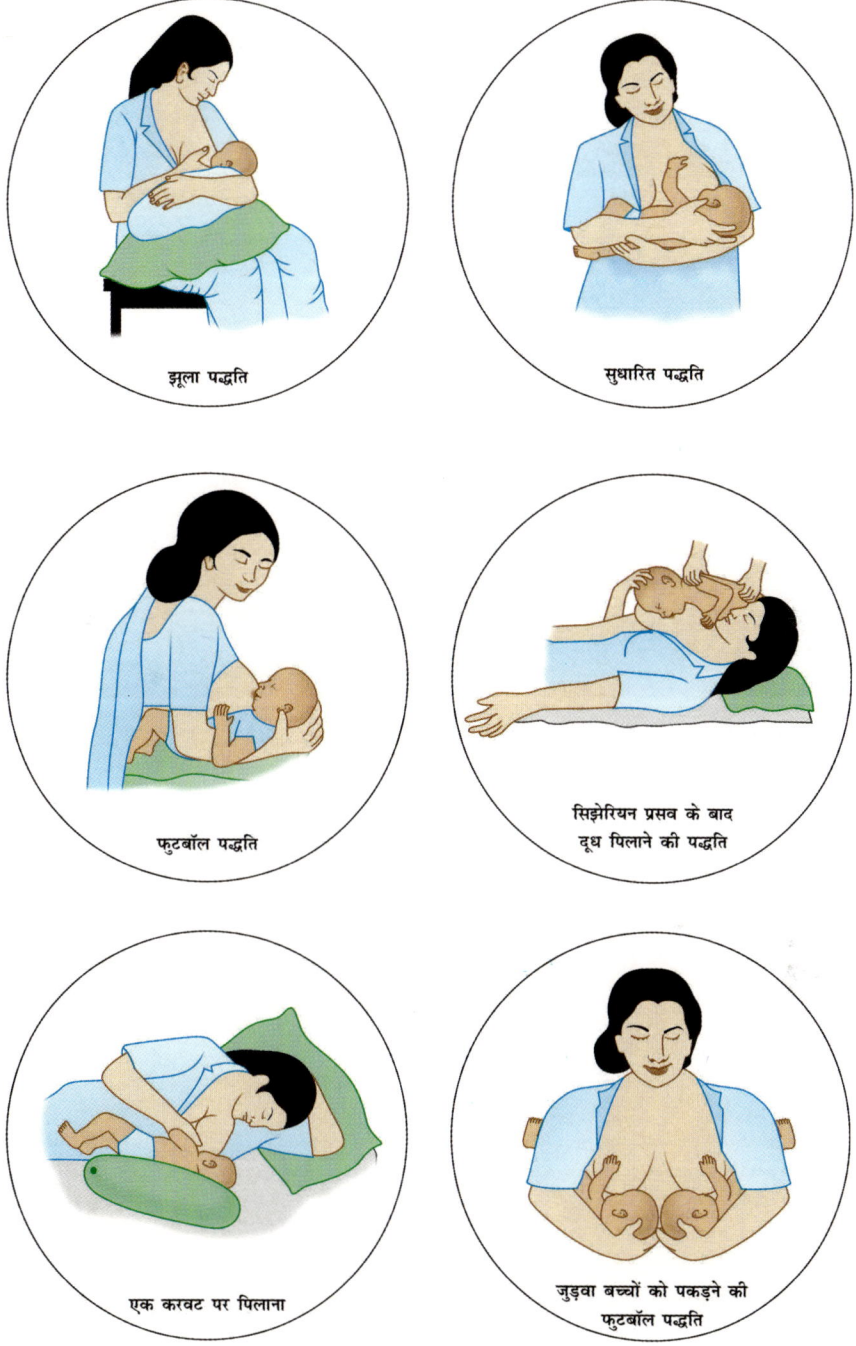

झूला पद्धति

सुधारित पद्धति

फुटबॉल पद्धति

सिझेरियन प्रसव के बाद
दूध पिलाने की पद्धति

एक करवट पर पिलाना

जुड़वा बच्चों को पकड़ने की
फुटबॉल पद्धति

Fig. 42.4a: बच्चे को पकड़ने की अलग–अलग पद्धति

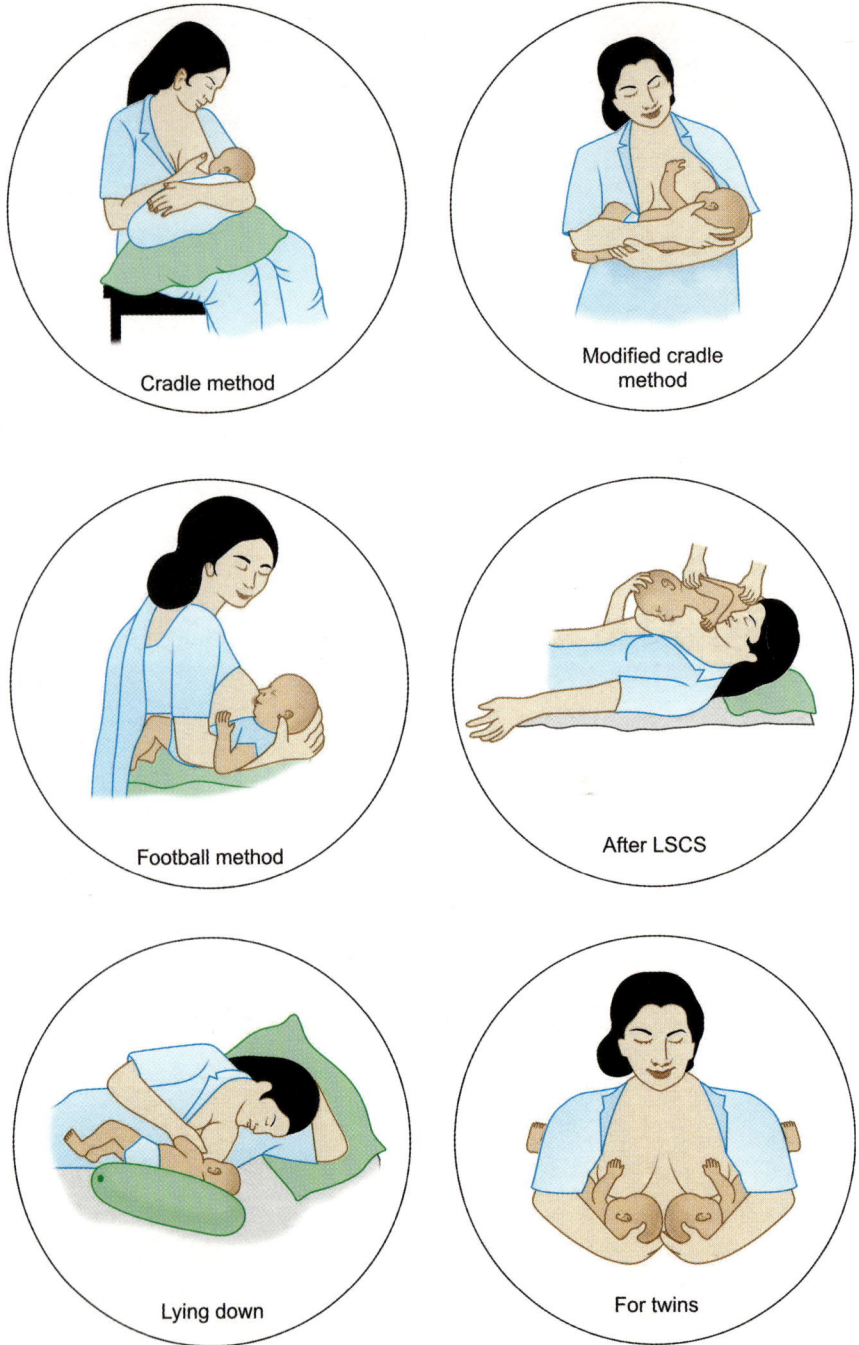

Cradle method

Modified cradle method

Football method

After LSCS

Lying down

For twins

Fig. 42.4b: Different ways to hold the baby

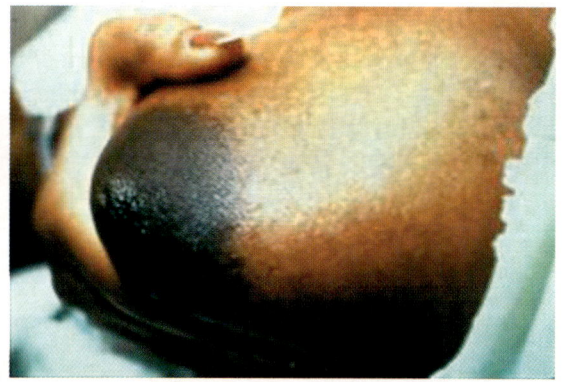

Fig. 42.5: Full breast (*Source:* UNICEF–WHO BFHI, 2009 Documents)

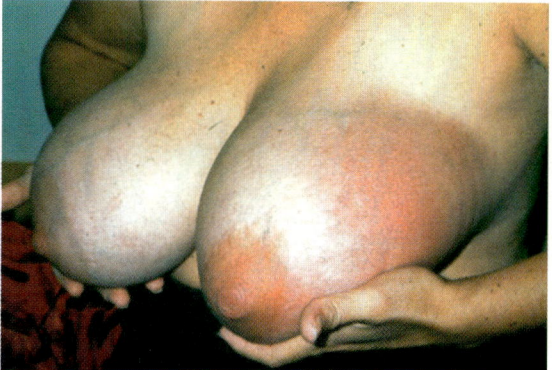

Fig. 42.7: Mastitis

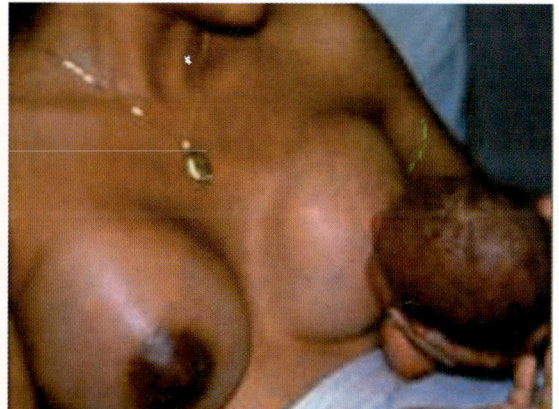

Fig. 42.6: Engorgement (*Source:* UNICEF–WHO BFHI, 2009 Documents)

Clinically

	Full breast	Engorgement	Mastitis	Abscess
Redness	-	+	++ Particular Part	++/+
Fever	-	+/– (if persist for 24 hours)	++	++/+/-
Pain	-	+	++	++/+
Ease of expression	Easy	Difficult	Difficult	Easy from other parts

Management

Interventions	Full breast	Engorge-ment	Mastitis	Abscess
Rest	+	++	++	++/+
Analgesics	–	+/++	++	++/+
Antibiotics	–	–	++	++/+
Breast emptying breast	+	++	++	+
Use of cabbage leaves	–	++	+	+
Surgical options	–	–	–	+/++

Irusen and Rohwer[11] have compared the surgical management of breast abscess by 2 methods:

Criteria	Needle aspiration	Incision and drainage
Days to complete resolution	Decrease healing time	
Breastfeeding continuation	More likely	Less likely
Failure	More common	Less common

Note: Insufficient evidence to determine whether needle aspiration is a more effective option to I & D.

every mother knows this technique of expression (Flowchart 42.1).

- Engorgement restricted to a part of the breast gives a lumpy feel. Engorgement of the breast tissue in the armpit will produce a lump there.
- Unattended engorgement can lead to increasing pain, redness over a part of the breast and fever (mastitis, Fig. 42.7).
- If neglected, this may progress to pus formation (breast abscess).
- *Blocked ducts:* Localised tenderness and/or a firm red area in the breast usually due to inadequate milk removal from one duct. It could be due to external pressure like a tight fitting bra or ineffective drainage of the breast.

Flowchart 42.1: Techniques of expression

Early feeding cues

- Restlessness
- Rapid eye movements
- Soft cooing or sighing sounds
- Sucking movements
- Sucking sounds
- Hand-to-mouth movements
  Crying is a late feeding cue and may interfere with effective breastfeeding.

Flowchart 42.2: Correct breastfeeding technique

Attachment

- Maximum possible areola in baby's mouth (lower part more)
- Mouth wide open
- Lower lip turned outward
- Chin touches the breast

Baby's position

- Turned towards the mother
- Good skin-to-skin contact
- Head and body in one line
- Neck, back and buttocks well supported

Mother's position

- Sitting comfortably
- Holding breast in big 'C' grip
- Touches nipple to upper lip by bringing nipple in front of nose and gives mouthful of breast as soon as the baby opens the mouth—widely
- Interacting with baby while feeding

Flowchart 42.3

How frequently child should feed

First 7–15 days
- At least 10 times in 24 hours
- Pro-active feeding: Waking baby for feeds
- Recognizing early feeding cues

Strategies to wake the infant

- Remove blankets
- Remove clothes
- Change the infant's diaper if soiled
- Place the infant skin-to-skin
- Massage the infant's back, abdomen, arms and legs

C-5 NICU Transfer

- Prior to transfer give skin-to-skin contact if condition permits
- Allow suckling if no distress
- Feed expressed breastmilk during transfer
- Motivate the mother to express milk 2–3 hourly (with proper technique) and transport it to NICU
- It is difficult to collect milk in a bowl in the initial few days. Hence, a syringe can be used to aspirate and collect milk droplets expressed at the nipple.
- Transfer mother to a facility near NICU at the earliest
- Mother should be allowed to visit, touch, hold the baby
- *Direct breastfeeding:* As soon as baby's condition and maturity permits (under supervision)
- Breast milk can be fed by infant feeding tube to sick babies in small recommended quantities (trophic feeding)
- At discharge babies should be exclusively or predominantly breastfed.

C-6 Myths

- *Mother should not feed lying down:* Mothers can feed in any comfortable position.
- *Mother should avoid certain foodstuffs post-delivery:* Mothers should be permitted to eat homemade foods of their choice.
- *Mothers should feed equally from both sides during a breastfeed:* Mother should feed completely from one side (20–25 minutes) before feeding from the other.
- *Excessively crying baby means that the mother has less milk:* A baby may cry excessively for various reasons apart from hunger. After establishment of lactation (> 1–2 weeks) if an exclusively breastfed baby passes urine > 6 times/24 hours and gains >500 gm/month, then it can be safely concluded that baby is well fed.

REFERENCES

1. UNICEF. From the First Hour of Life: Making the sense for improved infant and young child

feeding everywhere, Part 1: Focus on Breastfeeding. New York, United Nations Children's Fund (UNICEF) Aug 2016.

2. Alexander JM, Grant AM, Campbell MJ (1992) Randomised controlled trial of breast shells and Hoffman's exercises for inverted and non-protractile nipples. British medical journal, 304:1030-1032.

3. Sinha B, Chowdhury R, Sankar MJ, Martines J, Taneja S, Mazumder S, et al. Interventions to improve breastfeeding outcomes: a systematic review and meta-analysis. Acta Paediatrica. 2015; 104:114–134.

4. Moore ER, Anderson GC, Bergman N, Dowswell T. Early skin-to-skin contact for mothers and their healthy newborn infants. Cochrane Database of Systematic Reviews 2012, Issue 5. Art. No.: CD003519. DOI: 10.1002/14651858.CD003519. pub3.

5. Homes AV. Establishing successful breastfeeding in the newborn period. Pediatric Clinics of North America 2013;60:147–168.

6. 6 ILCA, Overfield M, Carol A. Ryan C, Amy Spangler, Mary Rose Tully. IN Mary Rose Tully. IN Teach mothers to recognize and respond to early infant feeding cues and confirm that the baby is being fed at least 8 times in each 24 hours in Clinical Guidelines for the establishment of exclusive breastfeeding. International Lactation Consultant Association, NC, USA, 2005, Page 10.

7. Thulier D. Weighing the Facts: A Systematic Review of Expected Patterns of Weight Loss in Full-Term, Breastfed Infants Journal of Human Lactation 2016;32(1):28–34.

8. Flaherman VJ, Schaefer EW, Kuzniewicz MW, Li SX, Walsh EM, Paul IM. Early weight loss nomograms for exclusively breastfed newborns. Pediatrics. 2015;135(1):e16-e23. doi:10.1542/peds.2014–1532.

9. Macdonald P D, Ross SRM, Grant L and Young D. Neonatal weight loss in breast and formula fed infants. Archives of Disease in Childhood Fetal and Neonatal Edition 2003;88:F472.

10. Dennis CL, Jackson K, Watson J. Interventions for treating painful nipples among breastfeeding women. Cochrane Database of Systematic Reviews 2014, Issue 12. Art. No.: CD007366. DOI: 10.1002/14651858.CD007366.pub2.

11. Irusen H, Rohwer AC, Steyn DW, Young T. Treatments for breast abscesses in breastfeeding women. Cochrane Database of Systematic Reviews 2015, Issue 8. Art. No.: CD010490. DOI: 10.1002/14651858.CD010490.pub2.

Kangaroo Mother Care— Practical Aspects

Arpita Thakker Adhikari, Krishna Ashok Shetye

Kangaroo mother care (KMC) is a simple method of care for low birth weight infants that includes early and prolonged skin-to-skin contact with the mother (or a substitute caregiver) and exclusive and frequent breastfeeding. This natural form of human care introduction stabilizes body temperature, promotes breastfeeding, prevents infection and other morbidities. This also leads to early discharge, better neurodevelopment and encourages bonding between mother and infant.

History of KMC

A team of pediatricians started KMC in Instituto Materno Infantil (IMI) in Bogota, Colombia in 1978 where it was developed as an alternative to inadequate and insufficient incubator care for those preterm newborn infant who had overcome initial problems and required only to feed and grow. In 2003, WHO formally endorsed KMC and published KMC practice guidelines. KMC was introduced in India, in 1994 in BJ Medical College and Hospital, Ahmedabad. This was followed by experiences in KEM Mumbai and at the All India Institute of Medical Sciences (AIIMS). A KMC India Network (www. kmcindia.org) comprising six institutions was started in 2003.

Key Features of KMC

- Early, continuous and prolonged skin-to-skin contact between the mother and the baby
- Exclusive breastfeeding (ideally)
- It is initiated in hospital and can be continued at home
- Small babies can be discharged early
- Mothers at home require adequate support and follow-up
- It is a gentle, effective method that avoids the agitation routinely experienced in a busy ward with preterm infants.

KMC must not be confused with routine early skin-to-skin care at birth, which WHO recommends for every newborn in the first hour of life to keep it warm and to initiate breastfeeding.

Components and Pre-requisites for KMC

The two main components of KMC are:
1. Skin-to-skin contact.
2. Exclusive breastfeeding.

The two prerequisites for KMC are:
1. Support to the mother in hospital and at home.
2. Post-discharge follow-up.

KMC satisfies all five senses of the baby. The baby feels warmth through skin-to-skin contact (touch), listens to mothers voice and heartbeats (hearing), sucks breast milk (taste), has eye contact with the mother (vision) and smells mother's odor (olfaction).

Requirements for KMC Implementation

- Training of nurses, physician and other staff involved in the care of the mother and baby.
- Educational materials such as information sheets, posters and video films on KMC in local language.
- Reclining chairs in nursery and postnatal wards, and beds with adjustable back rest should be arranged. Mothers can provide KMC sitting on any comfortable chair or sofa or in semi-reclining posture on a bed with the help of pillows (Fig. 43.1).

Eligibility Criteria

Baby

All stable LBW babies are eligible for KMC. However, very sick babies should be under radiant warmer or incubator and later when hemodynamicaly stable can be shifted to KMC. It can even be initiated in babies who are otherwise stable but may still be on intravenous fluids, tube feeds, oxygen and even on CPAP.

Mother

All mothers can provide KMC, irrespective of age, parity, education, culture and religion.

Fig. 43.1: Sleeping and resting in KMC

Counseling

The following points must be taken into consideration when counseling on KMC:

- *Willingness:* The mother must be willing to provide KMC;
- *Full-time availability to provide care:* Other family members can offer intermittent skin-to-skin contact but they cannot breastfeed;
- *General health:* If the mother suffered complications during pregnancy or delivery or is otherwise ill, she should recover before initiating KMC;
- Being close to the baby: She should either be able to stay in hospital until discharge or return when her baby is ready for KMC;
- Supportive family: She will need support to deal with other responsibilities at home; and
- Supportive community: This is particularly important when there are social, economic, family constraints.

Initiation of KMC

The timing of initiation of KMC depends on the birth weight and stability of the infant.

1. *Birth weight more than 1,800 g and less than 2500 g:* These infants are generally stable at birth. Therefore, in most such cases KMC can be initiated soon after birth in the postnatal ward. The neonate weighing less than 2000 g should be given priority for initiation of KMC.

2. *Birth weight more than 1,200 g and less than 1800 g:* Many infants of this group have significant problems in the neonatal period. It might take a few days before KMC can be initiated. Such infants may need care in a special newborn care unit (SNCU) or a newborn intensive care unit (NICU). Intermittent KMC can be given to a hemodynamically stable infant receiving IV fluids, antibiotics and oxygen. KMC should be practiced under medical supervision. The duration may be gradually increased and thereafter the infant may be transferred to a dedicated KMC ward.

3. *Birth weight less than 1,200 g:* These infants frequently experience serious prematurity related morbidity often starting soon after birth. It may take days to weeks before the infant's condition allows initiation of KMC. Duration of KMC should be gradually increased based on the tolerance of infant.

Mothers clothing: KMC can be provided using any front-open, light dress as per local culture. For example saree and blouse, gown, front open kurta, shirt or shawl.

Baby's clothing (Fig. 43.2): Baby is dressed with cap, socks, nappy and front open sleeveless shirt.

THE KMC PROCEDURE

Kangaroo Positioning (Figs 43.3 and 43.4)

- Baby should be placed between the mothers breast in an upright position.
- Head should be turned to one side and in a slightly extended position.
- Hips should flexed and abducted in a frog position, the arms should also be flexed.
- Baby's abdomen should be at the mother's epigastrium. Mother's breathing stimulate the baby thus reducing the occurrence of apnea.
- Support the baby's bottom with sling/ binder. Tie the cloth firmly enough so that

Fig. 43.3: Baby in KMC

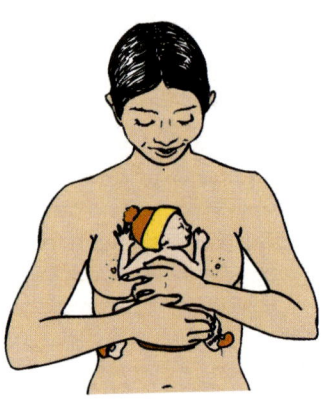

Fig. 43.4: Carrying pouches for KMC baby

when the mother stands up the baby does not slide out. Make sure that the tight part of the cloth is over the baby's chest.

Monitoring: Babies receiving KMC should be monitored carefully especially in the initial days. Nursing staff should ensure that baby's neck is neither too flexed nor too extended, the airway is clear, breathing is regular, color is pink and baby is maintaining temperature. Mother should be involved in observing the baby during KMC so that she can continue monitoring at home.

Feeding: Mother should be explained how to breastfeed during the procedure. Holding the

Fig. 43.2: Dressing baby in KMC

baby near breast stimulates the milk production. She may express milk while the baby is in KMC position. The baby could be fed with paladai, spoon or tube depending on the clinical condition.

Privacy: KMC unavoidably leads to some exposure on the part of mother. This can make the mother nervous and could be demotivating. The staff must respect mothers sensitivities in this regard and accept culturally acceptable privacy standards.

Moving the baby in and out (Fig. 43.5): Show the mother how to move the baby in and out of the binder. As the mother gets familiar with this technique, her fear of hurting the baby will disappear.

- Hold the baby with one hand placed behind the neck and on the back;
- Lightly support the lower part of the jaw with her thumb and fingers to prevent the baby's head from slipping down and blocking the airway when the baby is in an upright position;
- Place the other hand under the baby's buttocks.

Caring for the baby in KMC: Babies can receive most of the necessary care, including feeding, while in kangaroo position. They need to be moved away from skin-to-skin contact only for:

Fig. 43.5: Moving baby in and out

- Changing diapers, hygiene and cord care; and
- Clinical assessment, according to hospital schedules or when needed.

Daily bathing is not needed and is not recommended. If local customs require a daily bath and it cannot be avoided, it should be short and warm (about 37°C). The baby should be thoroughly dried immediately afterwards, wrapped in warm clothes, and put back into the KMC position as soon as possible.

During the day the mother carrying a baby in the KMC position can do whatever she likes: She can walk, stand, sit, or engage in different recreational, educational or income-generating activities. Such activities can make her long stay in hospital less boring and more bearable.

She has to meet, however, a few basic requirements such as cleanliness and personal hygiene (stress frequent handwashing). She should also ensure a quiet environment for her baby and feed him regularly.

Length and Duration of KMC

Short: 4 hours daily
Extended: 5–8 hours daily
Long: 9–12 hours daily
Continuous: More than 12 hours daily.

Duration to be counted as cumulative completed hours during a 24-hour period.

Skin-to-skin contact should start gradually, with a smooth transition from conventional care to continuous KMC. Sessions that last less than 60 minutes should, however, be avoided because frequent changes are too stressful for the baby. The length of skin-to-skin contacts gradually increases to become as continuous as possible, day and night, interrupted only for changing diapers, especially where no other means of thermal control are available.

When the mother needs to be away from her baby, he can be well wrapped up and placed in a warm cot, away from draughts, covered by a warm blanket, or placed under an appropriate warming device, if available.

Fig. 43.6: Father in KMC

During those breaks family members (father or partner, grandmother, etc.), or a close friend, can also help caring for the baby in skin-to-skin kangaroo position (Fig. 43.6).

When the mother and baby are comfortable, skin-to-skin contact continues for as long as possible, first at the institution, then at home. It tends to be used until the baby reaches term (gestational age around 40 weeks) or 2500 g. Around that time the baby also outgrows the need for KMC. She starts wriggling to show that she is uncomfortable, pulls her limbs out, cries and fusses every time the mother tries to put her back skin-to-skin. This is when it is safe to advise the mother to wean the baby gradually from KMC. Breastfeeding, of course, continues. Mother can return to skin-to-skin contact occasionally, after giving the baby a bath, during cold nights, or when the baby needs comfort. KMC at home is particularly important in cold climates or during the cold season and could go on for longer.

Observing the Well-being of Baby in KMC

Once the baby has recovered from the initial complications due to preterm birth, is stable and is receiving KMC, the risk of serious illness is small but significant. The onset of a serious illness in small babies is usually subtle and is overlooked until the disease is advanced and difficult to treat. Therefore, it is important to recognize those subtle signs and give

prompt treatment. Teach the mother to recognize danger signs and ask her to seek care when concerned. Treat the condition according to the institutional guidelines.

Danger Signs

- Difficulty breathing, chest in-drawing, grunting
- Breathing very fast or very slowly
- Frequent and long spells of apnea
- The baby feels cold: Body temperature is below normal despite rewarming
- Difficulty feeding: The baby does not wake up for feeds anymore, stops feeding or vomits
- Convulsions
- Diarrhea
- Yellow skin

Reassure the mother that there is no danger if the baby:

- Sneezes or has hiccups;
- Passes soft stools after each feed;
- Does not pass stools for 2–3 days.

Special Situations

There may be special situations where despite the newborn being sick KMC can be given with some precautions.

Sick LBW infants: In such cases kangaroo mother care may be given only under close and constant supervision in centers that are well-versed with the practice of KMC. Hemodynamically stable preterm infants on prolonged ventilation or on CPAP can also be given KMC.

Transport: Ideally, transport incubators with appropriate monitoring equipment are the best method to transport sick infants. However, in case they are not available, the best method to keep a preterm/LBW infant warm during transport after initial stabilisation is by continuous skin-to-skin contact with the mother/family member.

Post-discharge follow-up: Close follow-up is fundamental pre-requisite of KMC practice. In general a baby should be followed up once weekly till 40 weeks of post-conceptional age or the baby reaches 2.5 kg. Thereafter follow-up once in two weeks till 3 months of age and later once in 1–2 months during first year of life.

The baby should gain adequate weight (15–20 gm/kg/day up to 40 weeks and 10 gm/kg/day subsequently).

Several studies have been performed in the world supporting the merits of KMC, including a few in India as well. The following are benefits of KMC with evidence-based studies supporting it.

1. *Breastfeeding:* Four randomized controlled trial a cohort study carried out in low-income countries looked at the effect of KMC on breastfeeding. Three studies found that the method increased the prevalence and duration of breastfeeding.

2. *Thermal control:* Four studies carried out in low-income countries showed that prolonged skin to skin contact between mother and her preterm/LBW infant provided effective thermal control and was associated with reduced risk of hypothermia.

3. *Early discharge:* Studies show that KMC cared babies gain weight faster than others and speeds up the discharge process.

4. *Less morbidity and mortality:* Four published randomized controlled trials (RCT) comparing KMC with conventional care have shown results with no difference in survival between the two groups. Although the evidence shows that KMC does not necessarily improve survival, it does not increase mortality. Of these, two studies have shown a lower rate of serious illness and hospitalization with use of KMC in first year of life.

5. *Pain relief:* KMC helps in reducing stress during painful procedures like heel pricks and adhesive tape removal.

6. *Beneficial to parents:* KMC helps both infants and parents. Mothers have reported significantly less stress during kangaroo care than when baby is receiving conventional care. They have described a sense of empowerment, confidence and a feeling that they can do something positive for their preterm infants in different settings and cultures. Fathers too said that they feel relaxed, comfortable and contended while providing kangaroo care.

CONCLUSION

KMC is protective against a wide variety of adverse neonatal outcomes. This safe, low-cost intervention has the potential to prevent many complications associated with preterm birth and may also provide benefits to full-term newborns. The consistency of these findings across study settings and infant populations provides support for widespread implementation of KMC as standard of care for newborns.

SUGGESTED READING

1. World Health Organization, Department of Reproductive Health and Research. Kangaroo Mother Care: A Practical Guide. Geneva, Switzerland: World Health Organization; 2003.

2. Government of India. Guidelines for operationalisation of Kangaroo mother care, 2014.

3. Website of KMC India network. Guidelines for parents and health providers are present online at www.kmcindia.org

4. Charpak N, Ruiz-Pelaez JG, Figueroa de CZ, Charpak Y. A randomized controlled trail of kangaroo mother care: results of follow-up at 1 year of corrected age. Pediatrics 2001 Nov; 108(5): 1072–9.

5. Nanavati RN, Balan R, Kabra NS. Effect of kangaroo mother care versus expressed breast milk administration on pain associated with removal of adhesive tape in very low birth weight neonates: a randomized controlled trial. Indian Pediatr 2013 Nov 8;50(11):1011–5.

6. Nimbalkar SM, Chaudhary NS, Gadhavi KV, Phatak A. Kangaroo Mother Care in reducing pain in preterm neonates on heel prick. Indian J Pediatr 2013Jan;80(1):6–10.

7. Chidambaram AG, Manjula S, Adhisivam B, Bhat BV. Effect of Kangaroo mothercare in reducing pain due to heel prick among preterm neonates: a crossovertrial. J Matern Fetal Neonatal Med. 2014 Mar;27(5):488–90.

8. Ramanathan K, Paul VK, Deorari AK, Taneja U, George G. Kangaroo Mother Care in very low birthweight infants. Indian J Pediatr 2001 Nov; 68 (11):1019–23.

9. Chwo MJ, Anderson GC, Good M, Dowling DA, Shiau SH, Chu DM. A randomized controlled trial of early kangaroo care for preterm infants: effects on temperature, weight, behaviour and acuity. J Nurs Res 2002 Jun?10 (2):129–42.

10. Parmar VR, Kumar A, Kaur R, Parmar S, Kaur D, Basu S, Jain S, Narula S. Experience with Kangaroo mother care in a neonatal intensive care unit (NICU) in Chandigarh, India. Indian J Pediatr 2009 Jan;76(1):25–8.

Index